Rubber and the Making of Vietnam

FLOWS, MIGRATIONS, AND EXCHANGES

Mart A. Stewart and Harriet Ritvo, *editors*

The Flows, Migrations, and Exchanges series publishes new works of environmental history that explore the cross-border movements of organisms and materials that have shaped the modern world, as well as the varied human attempts to understand, regulate, and manage these movements.

MICHITAKE ASO

Rubber and the Making of Vietnam
An Ecological History, 1897–1975

The University of North Carolina Press *Chapel Hill*

This book was published with the assistance of a grant from the Association for Asian Studies First Book Subvention Program.

The University of North Carolina Press has been a member of the
Green Press Initiative since 2003.

Library of Congress Cataloging-in-Publication Data
Names: Aso, Michitake, author.
Title: Rubber and the making of Vietnam : an ecological history, 1897–1975 /
    Michitake Aso.
Other titles: Flows, migrations, and exchanges.
Description: Chapel Hill : University of North Carolina Press, [2018] |
    Series: Flows, migrations, and exchanges | Includes bibliographical references and index.
Identifiers: LCCN 2017049079 | ISBN 9781469637143 (cloth : alk. paper) |
    ISBN 9781469637150 (pbk : alk. paper) | ISBN 9781469637167 (ebook)
Subjects: LCSH: Rubber plantations—Vietnam—History—20th century. |
    Rubber Plantations—Environmental aspects—Vietnam. | Rubber plantations—
    Health Aspects—Vietnam. | Vietnam—Politics and government—20th century.
Classification: LCC HD9161.V52 A86 2018 | DDC 338.1/738952095970904—dc23
    LC record available at https://lccn.loc.gov/2017049079

Cover illustration: Bleeding young rubber trees on the plantation of Messieurs Écail and Tullié (photograph from *Annuaire du syndicate de caoutchouc de l'Indochine 1926* used by permission of Archives nationales d'outre-mer/ANOM).

To my father's memory, and our son's future.

Contents

Illustrations, Maps, and Tables

Acknowledgments

I love to read the acknowledgments section, as it tells me a history of the book that I am about to read. But this section is like the visible tip of an iceberg: it represents just a small number of the people who have contributed to, and provided necessary distraction from, the writing of this book.

My book would not have been possible without the generous financial support of centers, foundations, and universities. Grants from the Center for Southeast Asian Studies and the Robert F. & Jean E. Holtz Center at the University of Wisconsin–Madison enabled an initial summer of research in France. Funding from the Fulbright-Hays Doctoral Dissertation Research Abroad Program and a Vilas Travel Grant allowed for extended research in the archives and in the field, while a Blakemore-Freeman Language Fellowship helped me get a better grip on Vietnamese language and literature. A Mellon/American Council of Learned Societies Dissertation Completion Fellowship granted me critical time to write, and a dissertation prize from the International Union of the History and Philosophy of Science gave me writing momentum. I am forever indebted to Greg Clancey for agreeing to host me as a postdoc at the Asia Research Institute and Tembusu College, National University of Singapore in 2011–12. There I met some amazing colleagues, and friends, and I learned about the joys of laksa noodles on the island where rubber was introduced to Southeast Asia. In 2014–15, I spent a year as a fellow at the Institute for Historical Studies, University of Texas at Austin. Seth Garfield, Courtney Meador, the other fellows, and friends in Austin demonstrated the hospitality of the Lone Star State. Over the years, essential research and conference trips were funded by the D. Kim Foundation as well as the University at Albany's start-up funds and College of Arts and Sciences Conference and Travel Support Awards. Colleagues at several colleges and universities kindly invited me to give talks that sharpened the quality of my ideas, including Ben Kiernan and Erik Harms at Yale; Fae Dremock at Ithaca College; the graduate students at Cornell; Christian Lentz at the University of North Carolina, Chapel Hill; Andrew Stuhl at Bucknell; and Jean-François Klein at Institut national des langues et civilisations orientales, Paris.

My journey as an environmental historian began when my parents recommended that I take a course taught by Carolyn Merchant at the University of

California, Berkeley. Little did they suspect the results of their suggestion. The roots of this particular book started to grow in 1999 when I first arrived in Việt Nam. Friends in Biên Hòa, Huế, and Hà Nội welcomed me to their country, and homes, and I am grateful for all that they have shared with me for nearly two decades. It is a testament to the abilities of my Thầy and Cô (teachers) that after several years my nonsense Vietnamese phrases began to mean something to those outside the classroom. I must especially thank Cô Thuận and Cô Hương (and their families) for help with research, including the transcription of my interviews. Without archivists and librarians, historians could not do the work they do. I want to thank those in Cambodia, France, Singapore, Switzerland, the United States, and Việt Nam (especially at the National Archives) who showed me kindness, and courage, in their often thankless jobs. They were extremely helpful on a professional level and many have become good friends. Olivia Pelletier at the Archives nationales d'outre-mer in Aix-en-Provence expedited permissions to reproduce images, while Aline Pueyo at IMTSSA and Serge Volper at the CIRAD library pointed me in the right direction. I would also like to thank the University of Social Sciences and Humanities in Hồ Chí Minh City, in particular Ngô Thanh Loan, chair of the Geography Department, for agreeing to host my project and for sponsoring my visa. People in the rubber industry were incredibly generous with their time, and I appreciate those at the Vietnam Rubber Group, especially Phạm Thanh Hòa, and the Rubber Research Institute of Vietnam who showed me the current state of the industry and helped me find people who had worked on plantations from the time of the French. I admire these interviewees for their willingness to share memories that were not always pleasant to recall. Southern scholars proved their hospitality, and Hồ Sơn Đài, Huỳnh Lứa, Nguyễn Văn Lịch, Trần Quang Toại, Phạm Văn Hy, and Mạc Đường kindly shared their profound knowledge of the south. In Hà Nội, the National University and the Institute of Vietnamese Studies and Development Sciences hosted my research, and dedicated teachers from IVIDES continued my language instruction. Conversations with Tạ Thị Thúy about plantations and agriculture proved very enlightening, as did chats with folks at the History Research Institute (Viện Nghiên Cứu Lịch Sử). Nguyễn Mạnh Dũng, Trương Hoàng Trương, Lê Thu Hằng, and Nguyễn Thị Thanh Lưu furnished timely help with diacritics, while young scholars Hào and Nguyệt carried out crucial research.

Mentors and colleagues in the United States and France nurtured my book along the way. Gregg Mitman, Richard Keller, and Warwick Anderson were there from the beginning and wisely knew when to encourage and when to crack the whip. Scholars at the University of Wisconsin–Madison, National

University of Singapore, and University of Texas at Austin read portions of my work, gave comments on talks, and tossed around ideas over food and drink. Interdisciplinary projects draw on the expertise of many individuals. In the fields of history of science, technology, environment, and medicine, David Arnold, Christophe Bonneuil, Peter Coclanis, Joan Fujimura, Peter Lavelle, Stuart McCook, Suzanne Moon, Nancy Langston, Ron Numbers, Michael Osborne, Hans Pols, Megan Raby, Sigrid Schmalzer, Chris Shepherd, Frank Uekötter, and Jeannie Whayne were all generous with their time. In Southeast Asian studies, David Biggs, Mark Bradley, Fabien Chébaut, Haydon Cherry, David del Testa, Aline Demay, Erich DeWald, Claire Edington, Frédéric Fortunel, Jamie Gillen, Annick Guénel, Hazel Hahn, David Hunt, Eric Jennings, Charles Keith, Al McCoy, Ed Miller, Laurence Monnais, Michael Montesano, Martina Nguyen, Mytoan Nguyen, Philippe Peycam, John Phan, Phùng Ngọc Kiên, Jason Picard, Stéphanie Ponsavady, Gerard Sasges, Markus Taussig, Allen Tran, Nu Anh Tran, Michael Vann, Meredith Weiss, Béatrice Wisniewski, Alan Ziegler, and the Ecole française d'extrême-orient scholars in Hà Nội and Sài Gòn shared sources, ideas, and the joys and frustrations of research and writing. One of the greatest joys of researching a book is the discovery of the generosity of fellow travelers. Chris Schweidler shared months' worth of research on colonial medicine, Steffen Rimner supplied key documents, while François Denis Fievez, Pierre Morère, Sylvie Malye, Eric Panthou, and Nam Pham passed on contacts, material, and memories.

My book came to fruition through thoughtful and detailed feedback, and I have been incredibly lucky to find a welcoming intellectual community in Albany. In the University at Albany History Department, fellow environmental historians Kendra Smith-Howard and Chris Pastore, current and former chairs Nadia Kizenko and Richard Hamm, and colleagues Carl Bon Tempo, Rick Fogarty, Kori Graves, Ryan Irwin, and Maeve Kane (and quite frankly the whole department) were models of how to balance research, teaching, and service. Writing groups in Madison, Austin, Connecticut, and Albany provided a stream of good-natured critique. Pierre Brocheux, Pam McElwee, Michael Needham, Michele Thompson, and Ed Wehrle read and commented on significant portions, if not all, of the manuscript. At UNC Press, the Flows, Migrations, and Exchanges series editors Mart Stewart and Harriet Ritvo have been true supporters of my project, and Brandon Proia kept me on track. Comments from two anonymous readers made my book infinitely better. All errors are my own unless I can prove otherwise.

None of this research and writing would be as enjoyable if not for the companionship and good cheer of friends and colleagues from Berkeley,

Madison, Sài Gòn, Hà Nội, Singapore, Austin, and Albany. You know who you are. I would be remiss, however, if I did not mention Ivan Small and Kellen Backer, who have been there at key moments in my life. My grandparents passed away long before this book was finished, but they welcomed me into their lives in Japan and the United States. I am sad that my dad, who passed away in 2015, neither got to see the completion of this book nor meet his newest grandson. I hope he knows that both are here. For my mom, I can only say thank you since a child can never hope to repay a parent; and for my sister, you are my hero (but you know that already). Lastly, I thank Caroline Herbelin. We had a romantic start, meeting in Seattle, then Paris (in April), and Provence (for the summer), and for the past ten years you have been an intellectual and emotional partner in the writing of this book. This is a long time to wait for anything, but good training in the patience that we'll need to raise our son.

Abbreviations in the Text

AFC	Agriculture, forêt et commerce
Agefom	Agence française d'outre-mer
AMI	Assistance médicale indigène
ANOM/CAOM	Archives nationales d'outre-mer/Centre des archives d'outre-mer
APCC/I	Association/Annales des planteurs de caoutchouc en Cochinchine/Indochine
ASV	Associated States of Vietnam
BCN	Bộ Canh Nông
BEI	Bulletin économique de l'Indochine
BIF	Bienhoa industrielle et forestière
BLĐ	Bộ Lao Động
BNF	Bibliothèque nationale de France
BNL	Bộ Nông Lâm
BNN	Bộ Nông Nghiệp
BSPCI	Bulletin du Syndicat des planteurs de caoutchouc de l'Indochine
BTC	Bộ Tai Chính
BVT	Bộ Vật Tư
CC	Conseil colonial
CIRC	Comité International de Réglementation du Caoutchouc
CORDS	Civil Operations and Revolutionary Development Support
COSVN	Central Office of South Vietnam
CVT/TLĐLĐVN	Confédération vietnamienne du travail/Tổng Liên Đoàn Lao Động Việt Nam

CVTC/TLĐLCVN	Confédération vietnamienne du travail chrétien/ Tổng Liên Đoàn Lao Công Việt Nam
ĐNB	Đông Nam Bộ
DRV	Democratic Republic of Vietnam
EFEO	Ecole française d'Extrême-Orient
EPA	Ecole pratique d'agriculture
ESASI	Ecole supérieure d'agriculture et de sylviculture de l'Indochine
FEATM	Far Eastern Association of Tropical Medicine
FTEO	Forces Terrestres en Extrême-Orient
GGI	Gouvernement/Gouverneur général de l'Indochine
Goucoch	Gouverneur de la Cochinchine
Ha	hectare
HCM	Hồ Chí Minh
HS	hồ sơ
HVHBD	Hội Văn Hóa Bình Dân
ICP	Indochinese Communist Party
IDEO	Imprimerie d'Extrême-Orient
IFC	Institut français du caoutchouc
IMTSSA	Institut de médecine tropicale du Service de santé des armées
INDONF	Indochine, Nouveau fonds
IRCC/I/V	Institut de recherches sur le caoutchouc au Cambodge/en Indochine/au Vietnam
Kg	kilogram
LDP	Land Development Program
MDC	Ministère/ministre des colonies
NAC	National Archives of Cambodia
NARA	National Archives and Records Administration (United States)
NAVN	National Archives of Vietnam (Trung Tâm Lưu Trữ Quốc Gia)

NLF	National Liberation Front
NR	natural rubber
NXB	Nhà Xuất Bản
PAVN	People's Army of Vietnam
PLAF	People's Liberation Armed Forces
POW	prisoner of war
PRC	People's Republic of China
PRG	Provisional Revolutionary Government
PTT	Phủ Thủ Tướng
RC	Route coloniale
RRIV	Rubber Research Institute of Vietnam
RSA	Résident supérieur en Annam
RSC	Résident supérieur au Cambodge
RST	Résident supérieur au Tonkin
RVN	Republic of Vietnam
SE	Service économique
SIPH	Société indochinoise des plantations d'hévéas
SPCI	Syndicat des planteurs de caoutchouc de l'Indochine
SPTR	Société des plantations des terres rouges
SR	synthetic rubber
SRV	Socialist Republic of Vietnam
TARA	Mission Interagency Committee on Plantations
TP	Travaux public or Thành phố
TUDD	Tổng Ủy Dinh Điền
UPCI	Union des planteurs de caoutchouc en Indochine
VCP	Vietnamese Communist Party
VRPA	Vietnamese Rubber Planters' Association
WHO	World Health Organization

Rubber and the Making of Vietnam

.

Introduction

It was like being in a wilderness, but yet not. Dolly had visited Huay Zedi several times and had come to love the electric stillness of the jungle. But this was like neither city nor farm nor forest: there was something eerie about its uniformity; about the fact that such sameness could be imposed upon a landscape of such natural exuberance. She remembered how startled she'd been when the automobile crossed from the heady profusion of the jungle into the ordered geometry of the plantation. "It's like stepping into a labyrinth," she said to Elsa.

—Amitav Ghosh, *The Glass Palace*

The contribution of Dr. A. Yersin to Vietnamese agriculture at the beginning of this century is really great. Through this contribution, one can see a Yersin who opens the way, a Yersin who always goes in the lead. We admire Yersin not only for his contribution but moreover for his resolute audacity to be oriented always toward new horizons.

— Đặng Văn Vinh

The story of Alexandre Yersin (1863–1943) is an unlikely place to start a story about rubber. Born in Switzerland, Yersin trained as a medical researcher in Paris before heading to the French colonies in Asia in 1890, where he gained fame for his role in the discovery of the plague bacillus in Hong Kong. In 1895, he established a laboratory in the coastal town of Nha Trang, which became one of the three branches of the Pasteur Institute in French Indochina (Cambodia, Laos, Tonkin, Annam, and Cochinchina). He also explored the hinterlands around Nha Trang and played an important role in establishing the colonial hill station of Đà Lạt.[1]

Although Yersin is more celebrated for his medical and veterinary discoveries, he also had a keen interest in agriculture. Yersin established a farm outside Nha Trang at a place called Suối Dầu, meaning "the spring of the Dipterocarps" (a genus of tropical trees). Yersin wanted this agricultural plot to serve as a testing ground to demonstrate the types of plants that could be grown in Indochina and to show how they could be acclimatized to the local soil and climate. He had many crops planted, including *Eloeis guineensis* (palm oil), Liberian coffee, cacao, cinnamon, manioc, and medicinal plants, but the most agriculturally successful turned out to be *hevea brasiliensis*, a latex-secreting plant species originally found in the Amazonian forest. "I am today convinced," Yersin wrote to a fellow Pasteurian in Paris, "that hévéa brasiliensis grows best in our plot."[2]

Yersin studied rubber production from *hevea*, but he was not interested solely in the scientific aspects of tropical agriculture; he also viewed rubber as a financial resource for the Pasteur Institute's other activities, and gradually expanded the area that was under rubber cultivation on the Suối Dầu concession. In 1897, Yersin received 200 saplings from Saigon's botanical garden. By 1900 Suối Dầu had about 30 hectares (ha) in cultivation, which increased to 307 ha by 1914. He was not alone in using rubber to supplement income; by the 1920s a range of institutions, from the Paris Foreign Missions Society to the mental asylum in Biên Hoà, grew rubber. Yersin also recognized the connections between production and consumption, and began to collaborate with the Michelin Company. In 1905, Michelin paid Yersin 28 francs 50 centimes for 1,316 kilograms (kg) of rubber from Suối Dầu's first tapping. By the 1940s, the site was producing one hundred metric tons of dry rubber content annually, a tiny amount compared with the total production of Indochina, but enough to finance the activities of the institute.[3]

Several decades later, the Vietnamese Communist Party (VCP) initiated *đổi mới*, or a process of renovation aimed at addressing the many hardships that Vietnamese society had experienced since the end of the Vietnam War in 1975. Under the weight of U.S. economic sanctions and the reduction of support from communist allies such as the Soviet Union, Việt Nam struggled to recover from the disruption caused by over thirty years of warfare. While both the Democratic Republic of Vietnam (DRV) and the Republic of Vietnam (RVN) had agreed on denouncing the period of French colonial rule, rubber production remained a key component of Vietnam's economy, and soon after 1975 the VCP restarted rubber production on plantations.[4]

After *đổi mới*, the Vietnamese government signaled its readiness to reconsider certain aspects of French colonialism. In this spirit, scholars met in Nha Trang on March 1 and 2, 1991, to evaluate the legacy of Yersin's half century in Indochina.[5] At the conference, Đặng Văn Vinh, a world-renowned rubber expert and former head of the Rubber Research Institute of Vietnam (RRIV), argued that "Yersin became an agronomist despite himself and heveaculture [rubber production] was for him a totally novel technique." Yersin himself admitted that he lacked "the training for this kind of sport" and that "the most mediocre French gardener would know a hundred times more than I do." Yet, Yersin's interest in agricultural matters is not surprising, given the arc and themes of his career and his commitment to France's colonial empire. Like his mentor Louis Pasteur, Yersin linked agriculture and medicine, drawing together production, research, and training in these fields in the colonies.[6]

Yersin established important connections among industry, government, and the scientific community in Indochina, according to Đặng Văn Vinh. Basic research needed to be justified in terms of its possible applications, which meant new cures, improved health, and increased production. As Vinh, who trained at the Suối Dầu plantation, put it: "Another achievement of growing hévéa at Suối Dầu lies in the tight collaboration between production and research, between the plantation and the laboratory. This fruitful collaboration was created by Yersin 90 years ago and today we are in the process of putting ourselves on this successful path." Vinh concluded that "natural rubber has become an industry at the forefront of our national economy."[7]

Đặng Văn Vinh did criticize what he saw as Yersin's complicity with the colonial project. For example, Vinh noted Yersin's failure to provide metal screens for worker housing on the Suối Dầu plantation. The spring and surrounding forests provided an ideal habitat for *Anopheles* mosquitoes, which transmitted malaria to the workers, and while Yersin spoke of the need for protection against malaria, he never pursued prevention research. Vinh also noted that political imperatives, rather than observations of *hevea brasiliensis* itself, formed the basis for some of the findings of Yersin and a collaborator, the chemist Georges Vernet. Vernet argued that *hevea* would not grow north of 15 degrees latitude, a finding that Vinh argued was motivated by the French colonial government's order to focus on *ficus elastica* as a source of latex in Tonkin.[8] Vinh failed to note, however, that Yersin probably introduced plague to Indochina. While the evidence is circumstantial, the first known outbreak occurred in Nha Trang in 1898, with multiple outbreaks in the surrounding population and among laboratory workers in subsequent years. Given that the role of fleas in transmission of plague had not yet been established, it is unlikely that effective precautions would have been taken against the spread of the disease.[9]

Yersin, the Pasteur Institute, and medical science abetted the exploitation that took place in French Indochina. *Rubber and the Making of Vietnam* explores this entanglement of science, commerce, and governance on Southeast Asian plantations. The book provides an ecological perspective to explain how rubber refashioned human societies, economies, and politics. It explores the ways that various human relationships to their environments have influenced their ideas about nature. Finally, it demonstrates the ambiguous effects that plantation rubber had on nationalism, decolonization, and nation building in colonial and postcolonial Vietnam.

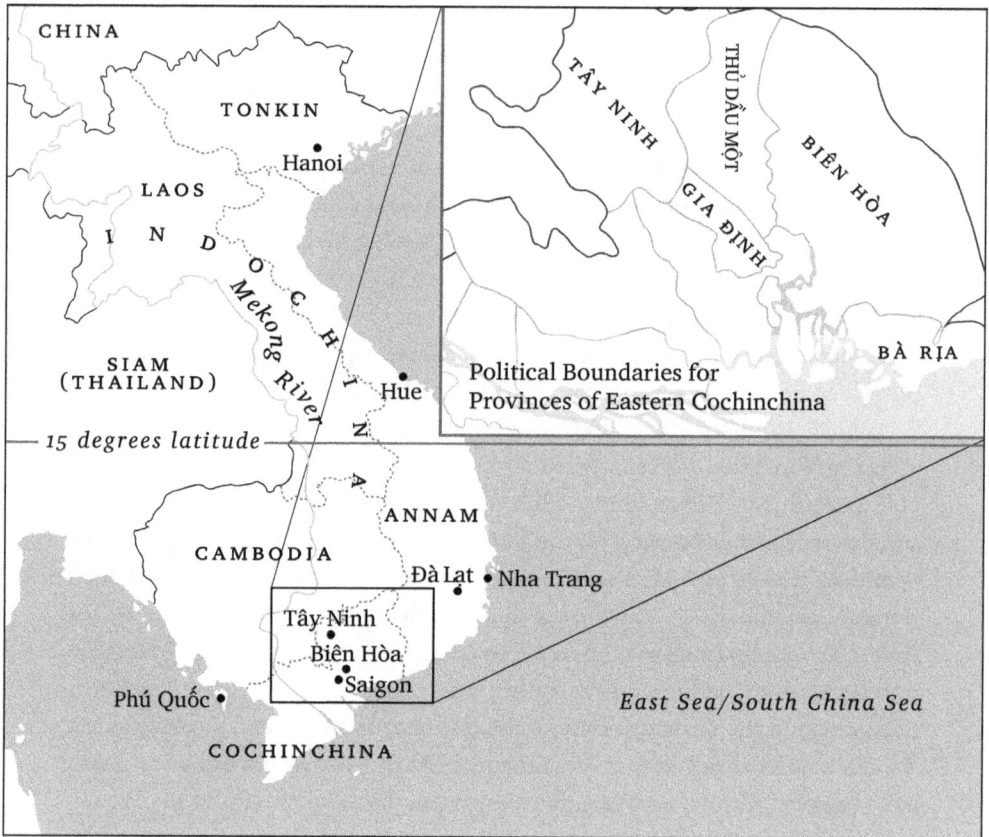

Map of Indochina. This map gives an overview of French Indochina and the major political boundaries. Note the 15 degrees north latitude that represented the imagined limit of rubber during the colonial era. Source maps: United States Central Intelligence Agency, *Indochina* 1985, 24 × 16 cm, Library of Congress, http:/hdl.loc.gov/loc.gmd /g8005.ct001585; ANOM Indochine nouveau fonds 1853/227; Brenier, *Essai d'atlas statistique*.

Where the Rubber Met the Road

Like all commodities, rubber has been defined by the ways and locations in which it has been produced, along with the people who have produced and consumed it. In 1876, an Englishman named Henry Wickham and his family accompanied 70,000 *hevea brasiliensis* seeds on a ship traveling from Brazil to England. Wickham then delivered these seeds to Kew Gardens, from where they were sent to different parts of the British empire, including Ceylon and Singapore. Voon Phin Keong argues that *hevea's* combined abilities to regenerate bark and to respond to more frequent tapping with greater amounts of latex has made it a favorite among Southeast Asian rubber producers. In its Amazonian habitat, *hevea* occurs in low densities and grows best in well-drained soils with an extended rainy season and frost-free climate at altitudes lower than 300 meters. As the historian Warren Dean has observed, *hevea* formed part of a long succession of seed and plant transfers, including maize, tobacco, coffee, sugar cane, and fruits, and has been among the most highly politicized plants in history.[10]

At the beginning of the nineteenth century, most Europeans considered rubber to be a curiosity known only to a few savants. What is now called "rubber" was most often called India-Rubber or *caoutchouc*. A wide range of plants, vines, and trees found in the tropics, a region with warm climates and abundant rainfall, secrete latex. Tappers collected latex by making incisions in the plant and letting this milky white fluid flow out. Latex is then transformed into rubber through various means, from slow smoking over fires to industrial-scale processes involving acid coagulation and large drying sheds. Humans have long used rubber: in the Amazon, rubber balls, shoes, and bottles were everyday objects; on the Malay peninsula, knife handles were constructed with gutta-percha, a hard, nonconducting substance closely related to rubber. In the 1830s and 1840s, the development of vulcanization—a process by which rubber could be stabilized at a larger range of temperatures—greatly increased its commercial value. As industrial demand for rubber's elasticity and imper-meability to gas and liquids grew, the material moved from peasant cosmolo-gies into a transnational capitalist system. Soon rubber became intimately associated with industrial society, with one rubber manufacturer calling it "the most astonishing and useful discovery of the nineteenth century, after the practical application of Steam and Electricity." In the 1890s, a 110-page price list for a Toronto-based gutta-percha and rubber manufacturing company included the following mundane items for sale: "Belting, Packing, Hose, Moulded Goods, Carriage Cloths, Drills, Mackintosh Clothing, Rubber Clothing, Tubing and all kinds of Rubber Goods for Mechanical purposes."[11]

In addition to industrial needs, new mass consumer items such as the bi-cycle stimulated increases in rubber consumption. Throughout the nine-teenth century, Kirkpatrick Macmillan and others steadily improved designs for a foot-peddled machine, or velocipede, promising faster travel. Its market was limited, however, by the lack of quality roads, particularly in the United States. Solid rubber tires existed but offered only slight comfort. In 1845, Robert William Thomson took out a patent on a pneumatic tire, but it was so unpromising that the idea sat virtually unnoticed until John Dunlop inde-pendently reinvented such a tire in 1888. Coinciding with the "safety bicycle," which included rear-wheel propulsion, front-wheel steering, reduced wheel size, and diamond frame, the pneumatic tire promised freedom for the masses. The following year, the Dunlop Rubber Company and the Michelin Company began to produce pneumatic tires and bicycle sales boomed, as the product reached frontier areas such as the U.S. Midwest.[12]

The growing price of rubber led to prodigious collection of wild latex starting in the 1850s, a time when most rubber was still gathered from latex-producing trees interspersed throughout forests in South America. Semi-independent workers called *seringueiro* walked paths connecting dozens of trees that they tapped with a small ax. They used heat and smoke to coagulate the harvested latex into balls that were sold to traders who shipped them down the tributaries of the Amazon. The lives of the *seringueiro* were difficult and often brutal. Indians, too, suffered heavily as they lost land and served as tappers and mercenaries in militias. In 1867, Brazil opened the Amazon up to international commerce, and between the 1860s and 1910, about 60 percent of the world's rubber came from this region. This period, which generated tremendous wealth, became known as the age of the rubber barons. Between 1890 and 1910 the price of rubber peaked at US$3 per pound. One symbol of this wealth was the ornate Opera House at Manáus, which was built between 1891 and 1896 at a cost of US$2 million (1890 U.S. dollars).[13]

The human and environmental costs of this period were high. Critics such as Roger Casement revealed the extreme violence that Julio Arana and other barons employed to control labor and enforce their monopolies. Foreign ventures were often just as deadly, with over 1,000 workers perishing between 1867 and 1882 in the doomed attempt to build the "Devil's Railway" in the Bolivian Amazon. In the middle of the nineteenth century, some observers thought that the supply of India rubber was "literally inexhaustible," but in Nicaragua between 1870 and 1900, *Castilla elastica* were virtually wiped out.[14]

From its inception, rubber production generated controversy and was promoted as a symbol of both modernity and oppression. Throughout the 1880s, workers in the United Kingdom and United States attempted to unionize and to challenge the harsh conditions and chemical-filled workplaces of rubber factories. In 1889, the House of Lords formed a select committee to investigate working conditions in the rubber industry, and some rubber manufacturers were fined for violating labor laws.[15] The worst abuses of the industry, however, took place in areas where latex was collected. In 1885, rubber gathering started in the Belgian Congo Free State, and just a few years later, E. D. Morel, Roger Casement, and others attempted to end the horrendous practices that were taking place. In his book on the King Leopold–owned Congo, Morel included various letters read aloud to the Manchester Geographical Society, including the following letter from October 18, 1892: "The frequent wars upon the natives undertaken without any cause by the State soldiers sent out to get rubber and ivory are depopulating the country. The soldiers find that the quickest and cheapest method is to raid villages, seize prisoners, and have them redeemed afterwards against ivory. At Boucoundje they took thirty prisoners, whom they released upon payment of ten tusks. Each agent of the State receives 1,000f. commission per ton of ivory secured, and 175f. per ton of rubber." Morel argued that rubber and ivory came out of the Congo while chains and guns went in, which was a potent symbol of trade during high imperialism.[16]

In the twentieth century, the automobile drove a surging desire for rubber. In 1895, there were only 350 cars in France, 70 in Germany, and almost none anywhere else; soon afterward the automobile became a mass-produced item. These automobiles required ever-growing amounts of rubber for their tires and other parts, a demand largely met by the rubber plantations of Southeast Asia. As latex production shifted from South American and African forests to Southeast Asian plantations, European capital and Asian labor quickly replaced varied landscapes with geometric plantations. In British Malaya, rubber expanded from 20,000 ha in 1905 to more than 200,000 ha in 1910 and 891,000 ha by 1921. In Indochina, the most rapid expansion took place from 1919 until 1929, when the area claimed for rubber rose to over 100,000 ha due to soaring world prices. By the end of the 1930s, most of the rubber exports from French Indochina came from vast *hevea* plantations that required a docile workforce consisting of Vietnamese peasants from northern and central French Indochina. The horrendous working conditions that existed on many plantations before 1940 caused scandalous rates of sickness and death, and many of these laborers joined the anticolonial movement as they struggled for better working conditions and higher wages.[17]

Global and Local

Plantations were key sites for the testing and implementation of different arrangements of power, from imperial domination, to anticolonial resistance, to nation-building programs. Almost every major imperial power and postcolonial nation-state in Southeast Asia was involved in the rubber industry. Violence was an integral part of this experience, and while *Rubber and the Making of Vietnam* attempts to convey some of the harshness of plantation life, it does not discuss the violence of rubber plantations to the extent of previous scholarship. This book moves away from narratives of victimization and the denials of power that such narratives entail. Rubber workers were not passive victims of the French colonial empire; most were able to exercise some sort of agency and employ "weapons of the weak" amid the daily brutality of plantation regimes.[18]

Early twentieth-century plantations were complex socio-technological systems that built on developments in plantation agriculture dating back over a millennium. After a period of growth around the Mediterranean Sea, the Portuguese started sugar production on the island of São Tomé in the sixteenth century. The mature sugar plantation complex, as the historian Philip Curtin called it, became the model for plantations in Africa and Asia in the nineteenth and twentieth centuries. It developed around three factors of production: land, labor, and capital. In the Americas, many Amerindians had perished because of European conquest, leaving large tracts of land available for plantations. A steady supply of slaves arrived from Africa to replace the Amerindians as a source of labor, and the growing demand for commodities of mass consumption in Europe, especially sugar, made capital available for investment. The nationalist revolutions that shook the Atlantic world at the end of the eighteenth century and the beginning of the nineteenth century seemed to put an end to the plantation complex, but it lasted well into the twentieth century. It is sometimes difficult to distinguish between contemporary multinational companies operating plantations and the previous plantation complex.[19]

Plantations served as the means to bring global rubber to the local road in Southeast Asia. The historian Frank Uekötter defines the global plantation as "an intellectual construct that serves as a vehicle for a discussion of the common challenges for plantation systems worldwide." Yet plantations have also been intensely local phenomena and have altered an infinite variety of environmental, social, legal, political, and economic conditions. As Anna Tsing has argued, the assimilation of global forms, such as science, biomedicine, and resistance, to specific places depends on friction in global connections. "Industrial rubber," she writes, "is made possible by the savagery of European conquest, the competitive passions of colonial botany, the resistance strategies of peasants . . . and much more that would not be evident from a teleology of industrial progress." In this way, plantations have been "mid-level places" where the global and the local have met, interacted, and reformed each other, without collapsing into an indistinct "glocal." Although historians have tended to treat the ways in which plantation agriculture has reformed local conditions like a bulldozer flattening all in its wake, plantations have been "delicate" mechanisms—a tumbler in a lock, as Stewart McCook puts it.

In this analogy, the different tumblers of the lock each represent a different factor, such as environments, markets, politics, knowledge, and labor. To operate well, all these factors have to be more or less aligned. This sensitivity of the plantation mechanism has encouraged planters and laborers to pay close attention to local environments. In the case of planters, such attention was a matter of profit; in the case of laborers, such attention was a matter of survival.[20]

Rubber and the Making of Vietnam focuses on embedded practices and the process in which ideas and actions came to be grounded in social and material environments. The production of rubber was not merely a social affair, and to keep the latex flowing, humans had to negotiate their relationship to nonhuman nature, ranging from the transplanted *hevea* tree to plasmodia-bearing mosquitoes. I analyze the ways in which certain environments and social configurations retain "lessons" of the past, thus enabling certain behaviors while constraining others. Plantation practices have been difficult to reform because they have been built into natural and social systems. "Above," global phenomena such as empire, capital, and science serve as frameworks for action and paper over the infinite array of unique circumstances. "Below" exist local phenomena including the colonial, the environmental, and the vernacular. Critical scholars often argue that a source and sign of the power of both empire and capital is the taking of local situations and making them commensurable and therefore amenable to a set of universal principles. This conversion always remains incomplete and masks the ability of changing local conditions to influence outcomes.

In addition to midlevel places, this book examines midlevel actors, especially those charged with re-creating the global plantation who had to grapple with local environments to create latex. These actors included planters, scientists, government officials, and workers who created and appropriated vernacular knowledge for their projects. Largely forgotten scientists such as Henry Morin, Paul Carton, Dang Van Du, and Đặng Văn Vinh did not generate ideas that shaped global practices, but rather took global ideas and attempted to apply them to local circumstances. Depending on the results, these actors could then make claims to have created global knowledge. In this process, the categories of global and local were contingent, dynamic, and constantly redefined by agents positioned as mediators in networks connecting the global and local.

To put these midlevel actors in their place(s), I think with and against social models such as actor-network theory that have proven useful in examining various assemblages such as plantations. The historian David Turnbull has defined an assemblage as an "amalgam of places, bodies, voices, skills, practices, technical devices, theories, social strategies and collective work that together constitute technoscientific knowledge/practices." As key enablers, and major beneficiaries, of colonization, experts in tropical medicine and tropical agriculture were also tied to its projects and concerns. They carried out research in agriculture and medicine both to improve native welfare and to strengthen the colonial project.[21]

Environment and Health

The production of latex from *hevea brasiliensis* in French Indochina started in the 1890s. One of the first attempts to grow *hevea* took place at the experimental field site outside Bên Cát near the provincial town of Thủ Dầu Một. In 1897, this site became the Ông Yểm experimental station, the first of its kind in French Indochina. In eastern Cochinchina and Cambodia, extractive practices similar to those of the Amazon did not develop despite colonial conquest and intrusion and the exclusion of foreign capital. Instead, plantation managers hired wage labor at low wages and with terrible working conditions, and invested in the means of production.[22] The literature on these rubber plantations can be divided into two broad camps: (1) critical scholars and revolutionaries who condemn plantations as sites of exploitation of both humans and nature and (2) planters, government officials, and experts who believe that their work led to economic and social development. While both perspectives reflect certain experiences of plantation life, they do not adequately explain the trajectories and multiple consequences of plantations and their associated agricultural and medical institutions. Few accounts of tropical plantations have taken into consideration how humans and nature formed, and were formed by, plantation agriculture. Plantations have been an iconic form of both slave societies and free peoples, and we cannot understand twentieth-century Vietnamese and Cambodian history without understanding the environmental and health effects of plantations.[23]

Plantations unified industrial agriculture and medicine, both conceptually and in practice, on the Indochinese peninsula and had profound effects on the environment and human health. Lenore Manderson and Laurence Monnais have shown the role that medicine played in increasing exploitation, and my book extends their findings by thinking about environmental change together with transformations in health. Planters, tappers, and others involved with plantations had to develop a colonial know-how, including new links forged between the environment and health that enabled the production of

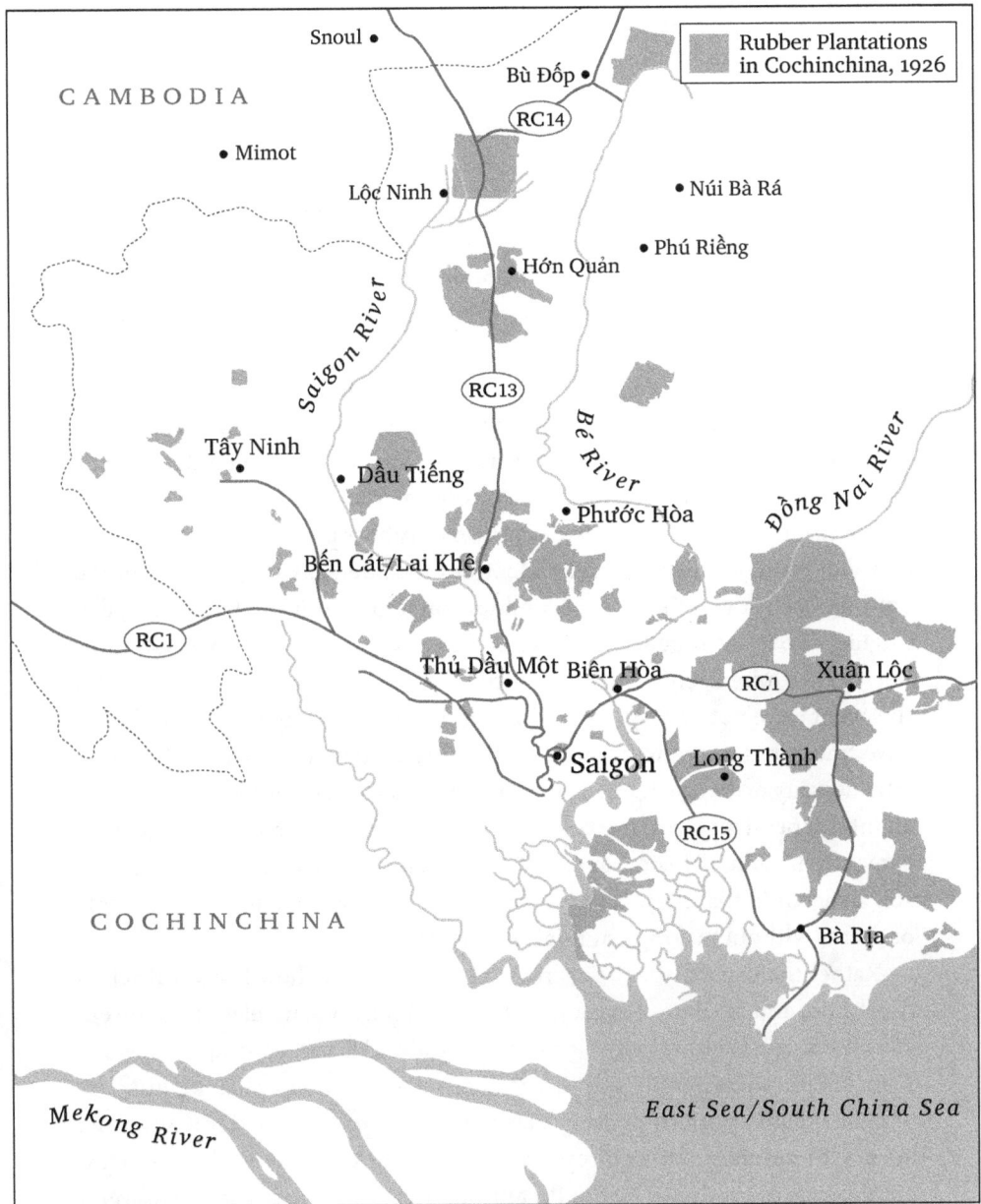

Map of the rubber plantations in Cochinchina, 1920s. This map gives an overview of the rubber plantation region after the rubber boom of the 1920s. Note the geometric areas of plantations, which had not yet crossed over into Cambodia. Source maps: SPCI, *Annuaire du Syndicat*, 1931.

rubber. At the same time, outbreaks of malaria and malnutrition diseases exposed social and political structures as struggles took place over where to place blame for diseased bodies. Discussions that took place in newspapers and journals about plantation regimes also shaped these landscapes of health. In narrating the history of rubber production, *Rubber and the Making of Vietnam* highlights the shifting political economy of scientific knowledge production about agriculture and medicine. Instead of viewing transformations in environments and bodies as separate, this book examines commonalities and differences across these locations, thus reformulating plantation histories from an ecological perspective.[24]

The concept of agency is integral to notions of human health and the environment. In a 2003 article, Walter Johnson critiques social historians who use a nineteenth-century liberal notion of "agency" in the context of the institution of slavery. Johnson's analysis of the terms "humanity," "agency," and "resistance" is useful for my discussion of rubber plantations in Cambodia and Việt Nam. Johnson helpfully identifies the liberal assumption about the equivalence of these three terms. Building on Johnson's insight, this book distinguishes between agency and agents, and actors and intentionality. Simply put, agents have agency, or the ability to affect historical processes, and if the agent were not present, or not present in a specific way, events would not have transpired in the way they did. This broad definition is agnostic about intentionality of the agent. As David Shaw wrote in *History and Theory*, "Agency is an interdependent structure or dynamic in which neither self nor intention is required." Thus, a historian can call a human, a rubber tree, a mosquito, or soil an agent in that they, individually or collectively, can affect the outcome of events without making claims about intent.[25]

Rather than strictly define agency, this book follows how historical actors thought of and attributed agency, and for what purpose, in relation to the environment and health. In addition to the question, "Can the mosquito speak?" I document what people said for it/them and what was at stake in those words. Most twentieth-century scientists were not comfortable granting agency to animals, but anthropomorphism snuck its way back in. For example, images produced by the Pasteur Institute in the 1920s and 1930s reinforced a human-like agency to mosquitoes by depicting them as individuals rather than in swarms. Such agency displaced the responsibility for horrendous rates of malaria onto "the" mosquito. Others, because of different social positions or divergent ways of visualizing health, mapped agency onto plantations differently. Workers often spoke of their competition with trees for plantation resources while the rural health movement of the 1930s helped critics of plantations put responsibility back on planters. This agency compelled Michelin and a few other plantations to build hospitals that remained understaffed and underutilized due to company labor policies.[26]

If agency has shifted during modernity, when and where was the modern? Scholars in the social sciences and humanities have generally applied the terms "modern" and "modernity" to the nineteenth- and twentieth-century societies that made rapid advances in science and technology. Recent work on colonial modernity, however, has shown how continuity can be disguised by the rhetoric of novelty and change by the notion of tradition. Frederick Cooper has spelled out the many uses (and abuses) of the terms of modernity, and he makes clear that modernism, an ideology that made a fetish of the modern, was not simply based on scientific logic but also on nonscientific rationalities and emotions.[27] Bruno Latour has gone further, arguing that "we have never been modern," meaning that modernity, at least in the West, was not a world-changing rupture but rather offered a mere "extension of practices, slight accelerations in the circulation of knowledge, a tiny extension of societies, miniscule increases in the number of actors, small modifications of old beliefs." The term used by the French to translate plantation, *đồn điền*, referred to a Nguyen dynasty (1802–1945) strategy to open new lands to Vietnamese settlement and industrial agriculture in southern Vietnam and Cambodia. Thus, *đồn điền* both pointed to and depended on processes already at work in the region.[28]

French colonial modernity did bring about quantitative, if not qualitative, changes in the relationships among land, the state, health, and agriculture during French encroachment. *Rubber and the Making of Vietnam* examines how the collection of ideas, practices, institutions, and technologies called modern has operated on and through nature and culture. It also analyzes "how historical actors have embraced various definitions of the modern to make political claims and envision different futures."[29] Just as with agency, various actors associated with rubber plantations each had their own ideas of modernity. The French, and later Vietnamese, planters and officials, who wished to portray rubber plantations as modern, argued that they involved scientific logic, industrial production and expanding markets, the use of machines, accelerated transportation and communication, and an emphasis on a future-oriented concept of time. Critics of plantations often portrayed them as nonmodern and argued that plantations collapsed the division between public and private, and reinforced social hierarchies such as class, gender, and race. They believed that they were neither constitutional, representative, nor bureaucratic; instead, they relied on arbitrary terror and kept the consumption of workers to a minimum. "Mid-level or intermediate analytical concepts" that are commonly viewed as constitutive of modernity, as Lynn Thomas argues, help to avoid unwieldy generalizations. Relying on such concepts also allows discussion of the ways in which actors viewed rubber plantations as simultaneously both modern and nonmodern. As Martin Murray finds in his study of plantation labor regimes, "market/non-market and free/coerced relations are inextricably entwined."[30]

Rubber and Nation

Not long after graduating from the Special School of Agriculture and Forestry of Indochina(ESASI) in Hà Nội in 1941, Đặng Văn Vinh joined the Việt Minh in the fight against the French, the RVN, and the Americans. Vinh's trajectory from colonial subject to national citizen raises the question, What tools and vocabulary did rubber plantations provide for would-be nation builders? To explore the specifics of environment, health, and labor in southern Indochina, *Rubber and the Making of Vietnam* adopts a narrative arc that moves from the French colonial empire to the postcolonial Vietnamese nation. It shows how shifting alignments of science, private capital, and government affected the colonial projects of economic development and the civilizing mission and the postcolonial projects of nation building. The colonial situation encouraged medical doctors and agronomists to fashion new connections between their disciplines and to link environment, health, and knowledge production. These discoveries relied just as much on interimperial as on intraimperial networks and exchanges, and Southeast Asian and global experts and laborers continue to create, spread, and contest rubber knowledge. Because of their demand for agricultural and medical infrastructures, and a disciplined, skilled labor force, plantations helped spark the industrialization of southeastern Vietnam. With *đổi mới*, or renovation, of the 1990s, transnational corporations once again became important agents and the plantation form of rubber production remained a solution to the limits imposed by Vietnamese social and natural environments, regardless of the ideological predilections of state leaders.

To challenge standard narratives of empire and decolonization, I spent several years gathering documents in English, French, and Vietnamese from multiple archives and libraries in North America, Europe, and Southeast Asia. These documents include government reports, planters' correspondence and memoirs, rare medical and agricultural journals, and visual materials. They show that colonial agronomists and medical researchers treated plantations as experimental sites and used knowledge generated there in their efforts to control Indochina's peoples and plants, efforts that continued into postcolonial society. In addition, I located and translated rare workers' memoirs and communist documents and conducted oral histories with former plantation workers to gain insight into diverse Vietnamese perspectives. Combined with over a century of secondary scholarship on the rubber industry and conversations with Vietnamese currently working in the industry, this material has allowed me to analyze multiple perspectives on the formation and maintenance of plantations and explore on-the-ground realities that are invisible in most histories of rubber. These realities helped define nation making in colonial and postcolonial contexts as rubber experts and laborers contributed to the material and mental networks that structured Southeast Asia throughout the twentieth century.

Profits motivated the construction of, and then utilized, imperial and national networks of knowledge production that linked French Indochina to France and to other colonial territories around the world, and placed plantations at the heart of efforts to discipline the tropics. Chapter 1 examines the introduction of *hevea* to Indochinese environments during the late nineteenth and early twentieth centuries, a process that involved laws marking the physical and intellectual boundaries between forests and plantation agriculture. It begins with a discussion of the southeast region of Việt Nam and the study of nature in Indochina during the nineteenth century. New understandings of human and nonhuman natures enabled the production of commodities such as rubber, and rubber production for global consumption, in turn, helped reformulate the coproduction of human and nonhuman natures in local places. The chapter lays down a baseline for evaluating later transformations in environment and health as plantation agriculture replaced biological diverse habitats with much simpler ecologies.

Plantation regimes encouraged knowledge production about plant and disease ecologies and the relationship among organisms and their environments more generally. More detailed knowledge about newly introduced plant species, plant and human diseases, and their shared environments was a key ingredient of better, more profitable management of rubber plantations. Chapter 2 explores the process by which agronomy came to support the burgeoning rubber industry after rubber arrived in Indochina in 1897. The French colonial government was not the first to encourage agricultural improvement on the Indochinese peninsula, but the qualitative and quantitative investment that it made in these projects set it apart from previous states. Encouraged by the success of their British and Dutch neighbors, French planters envisioned turning biologically and culturally diverse landscapes into neat rows of *hevea*. Plantation agriculture also played an important role in defining the political and intellectual scope of the science of ecology in Indo-china, encouraging agronomists to direct their energies toward transnational businesses and the colonial project. The process of integrating the efforts of scientists, officials, and planters was not always smooth, however, and this chapter highlights the conflicts and tensions generated by a political economy of plantation agriculture.

Plantation environments emerged in the 1920s and 1930s as humans, animals, and machines cleared the land to plant rubber. After World War I, colonial administrative policy, ecological necessity, and economic logic converged to promote Vietnamese migration to meet plantation demands for labor. Greater mobility by ship and by road meant that peasants from the north were brought to plantations, where they sometimes displaced previous inhabitants. These workers helped carry out the deforestation that created the limpid, sunny streams in which the mosquito species associated with malaria in the region bred. More than the lick of the whip, it was malaria, beriberi, and horrible living conditions that resulted in the illness and deaths of thousands of plantation workers. These outbreaks, along with the more famous cases of abuse, provided much fodder for opponents of imperialism, French and Vietnamese alike. Even as medical doctors recognized the poor health of plantation workers, they found it more plausible to blame workers' moral failings and culture rather than the imperial system. By placing the human suffering of laborers in the context of changing disease environments, chapter 3 furthers the investigations concerning the relationships among science, business, and government. As with agricultural science, industry played a key role in creating medical institutions and knowledge in Indochina during the colonial period and, partly because of this role, economic concerns trumped humanitarian impulses, at least until the 1930s.

Because plantations relied on a massive labor force, they required extensive medical studies of human biology and diseases. Researchers at the Pasteur Institute carried out numerous studies of mosquitoes and plasmodia, and to a lesser extent other pathogens, among plantation workers. Race served as an important analytic category for these researchers even as anthropologists were beginning to question the coherence of racial categories. Chapter 4 investigates the type of racialized society that the architects of industrial agriculture imagined they were creating. It also discusses the interactions in Indochina between the burgeoning tropical sciences and government and transnational capital, drawing on human disease environments to focus on how "rubber science" was applied to the surrounding countryside. If plantations were microcosms of the global colonial society, they were also laboratories where solutions to colonial problems were worked out. Tropical agronomy, geography, and medicine, linked by an ecological view of climates and soils, helped naturalize racial distinctions for the colonizers. Yet the colonial subjects who were the targets of these projects did not act in ways that race makers expected. While these subjects could not control the discourse of race, they could appropriate it for their own ends, and they attempted to do so before the outbreak of World War II.

Rubber plantations structured the violent transition from empire to nation-state during nearly thirty years of conflict on the Indochinese peninsula. Chapter 5 focuses on the struggle over plantations that took place in Cochinchina and Cambodia between 1945 and 1954. During the First Indochina War, plantation environments served as a key military battleground. In the fighting that took place immediately after the end of World War II, many plantation workers, encouraged by the anticolonial Việt Minh, attacked the rubber trees as symbols of hated colonial-era abuse. Slogans placing the culpability of worker suffering on trees show how plantation workers often treated the trees themselves as enemies. Despite their colonial origins, plantation environments were important material and symbolic landscapes for those seeking to build postcolonial Vietnamese nations. French planters claimed to struggle heroically against nature, Vietnamese workers saw themselves as struggling against both nature and human exploitation, and anticolonial activists articulated struggles against imperial power structures. Industrial agriculture such as rubber was vital to nation-building projects, and by the early 1950s, Vietnamese planners began to envision a time when plantations would form a part of a national economy.

Planners from the United States and the RVN, or South Vietnam, initially looked to rubber as an important source of income for the Republic and as a way to create a class of Vietnamese smallholders, or individuals who owned modestly sized rubber plots. Chapter 6 considers the attempts to create this group, and discusses the fate of rubber plantations during the First Republic of Vietnam from 1954 to 1963. It demonstrates the persistence of development ideologies that valued plantations over smallholder production and formal science over informal modes of knowledge, showing the power of the ideology of modernity during the process of decolonization. This chapter also examines the participation, or absence thereof, of Vietnamese in the rubber industry. Between 1955 and 1965, planters benefited from relatively peaceful conditions and Vietnamese smallholders began to take part in more significant numbers in the industry. Despite Ngô Đình Diệm's nationalism, the costs of production, the arrival of the U.S. military, and the lack of practical support meant that for the most part, only well-off Vietnamese could benefit from the expanding industry. This nonadoption of technology shows the limits of postcolonial development and raises questions about the movements of scientific and technological knowledge. It also suggests that the exclusion of most Vietnamese from the industry was a product of both colonial and postcolonial modernity.

The extreme violence brought to bear on the Vietnamese society and environment by the American war machine during the 1960s meant that measures taken by South Vietnamese leaders ended up sustaining plantation production. Ironically, the communist insurgency also benefited from plantations, which continued to serve as a valuable source of material and recruits. Meanwhile, North Vietnamese rubber experts worked to extend the range of *hevea* into more northern latitudes so that latex could flow in the socialist world. Chapter 7 extends the history of rubber to 1975 to show the ways that memories of colonialism continued to structure thoughts and behavior regarding rubber, and suggests why human-environment interactions on the plantations of post-1975 socialist Vietnam often resembled those of their colonial predecessors. This chapter focuses on the degree to which colonial discourse as materialized on plantations was subverted by various actors and revisits the historiography of the Vietnam War by adopting the lens of environmental history to show the unexpected consequences of plantation agriculture. Finally, it considers how the post–World War II development of "synthetic" rubber affected the actions of those associated with "natural" rubber plantations.

Industrial plantations have had some of the most significant impacts on the surface of the earth, and while natural rubber is no longer hegemonic, palm oil, soybeans, maize, and coffee production, each with its own nexus of human and nonhuman agents, continue to have major impacts on the environment and human health. The conclusion briefly analyzes post-1975 memories of colonial and national plantations as participants use the memory of rubber production to negotiate their relationship to each other and to the politics of Vietnamese history in the present. Planters' associations in France recall heroic times, the Communist Party celebrates the heroic contributions of rubber workers to the socialist revolution, and some workers use memories of colonial efficiency to critique present socialist mismanagement. Many Laotian and Cambodian farmers, and their allies, decry Vietnamese "colonialism" that is associated with the expansion of Vietnamese rubber company interests into the territory of neighboring nations, thus calling into question the continuing role of tropical commodities in shaping Southeast Asian lives.

FEW TRACES ARE LEFT of the rubber plantations that once girded the outskirts of Sài Gòn. Yet their imprint is everywhere, from roads and old railroad tracks to hospitals and dispensaries. My narrative about environment and human health in southern Vietnam does not assume a fixed end, as humans have continually adapted to evolving landscapes of labor and health. A Japanese animated film, *Pom Poko*, captures this ambiguity. Its central characters, a group of badger-like creatures called tanuki, must adapt to life in a Japan that is rapidly urbanizing. These tanuki can shape-shift, and in a climatic point in the film, they use this ability to re-enchant the neighboring urban landscape to change human behavior. The humans and the tanuki share a hybrid (nature/culture) landscape, albeit a highly urbanized one, in which the groups live side by side, sometimes interacting, sometimes not.[31]

Technological change is not deterministic, and human choices do matter in what technologies are adopted, how they are deployed, and what effects they have. Environments, too, are not deterministic and histories depend on assemblages of material and symbolic natures. As Julia Thomas notes, "The political lessons of this nature (as ideological concept and as physical environment) are a rejection of modernity's promises of perfection, but also an awed sense of our responsibility to choose among the multiple, competing, incommensurable values that modernity has developed."[32] This book explores the lessons of a few of the choices that those such as Yersin, Vinh, and rubber workers have made about their relationships to nature and to each other.

Part I
Red Earth, Gray Earth

CHAPTER ONE

Civilizing Latex

All around her was the hostile jumble of tall bamboo, the forest and its threats, solitude.
—François de Tessan, *Dans l'Asie qui s'éveille*

On all four sides was the ancient, untouched forest, empty of bird calls or the cold cries of the gibbons. Everywhere nothing but weeds, dust, and thorns.
—Trần Tử Bình, *The Red Earth*

The road, here as everywhere else, is the best instrument of civilization.
—Lieutenant-gouverneur de la Cochinchine, 1922

From the rice deltas of Southeast Asia to the swampy southern coasts of the United States, the spread of industrial agriculture during the nineteenth century revolutionized the world. In Southeast Asia, European capital, Chinese trading networks, and indigenous labor converted the great river deltas of the Mekong, the Chaophraya, and the Irrawaddy from biodiverse swamps and mangrove forests into vast monocultures of rice and sugar. Mining, forestry, and fisheries likewise expanded in size and intensity of activity. Similar forces reconstituted Southeast Asia's upland forests as planters replaced them with monocrops of coffee, palm oil, tea, and rubber, and colonial governments cordoned them off as reserves and hunting parks, thus criminalizing common usage. Nature itself functioned as an agent of change and resistance, as disease environments imposed limits on human activity and roads and bridges sank into the quagmire of waterlogged soils.[1]

The Indochinese peninsula was not spared from this social and environmental change. In the eighteenth century, Nguyễn lords founded *đồn điền*, a Sino-Vietnamese term for agro-military settlements, near the Đồng Nai River to increase tax revenue and secure new territory. Chinese traders and Minh Hương in Biên Hòa and Mỹ Tho submitted to the rule of the Nguyễn, while the state equipped the outpost of Sài Gòn with a Vauban-style fort. Established in 1802, the Nguyễn dynasty sought further to colonize southern parts of Vietnam and to exert tighter administrative control over the region and its borders with Khmer lands. Gia Long, the first Nguyễn emperor, sent out emissaries to bring back smallpox vaccines not only to save lives but also to discipline Vietnamese bodies and their microbes. Although the Nguyễn dynasty focused more on consolidating its political hold on the nascent nation-state of Việt Nam than on spreading agricultural or medical improvements, government officials did import new crops and carry out agricultural experiments on maize, tobacco, and coffee. One Vietnamese official even grew a *ficus elastica* to produce latex at his residential garden.[2]

The arrival of global financial markets and European liberalism in the mid-nineteenth century also refashioned preindustrial views of nature and landscapes on the Indochinese peninsula. In 1858, a combined Franco-Spanish naval fleet bombardment of Đà Nẵng near the imperial capital of Huế hammered home the point that, to survive, states required the type of military and techno-scientific dominance that could be obtained only by further integrating human and nonhuman nature. The Nguyễn dynasty's response to this challenge was ineffectual, and the French proceeded to conquer the kingdoms of Việt Nam, Cambodia, and Laos. The French military and political subjugation of the peninsula accompanied its concurrent quest to bring Indochinese natures under its own form of control. While several institutions such as the botanical gardens of Sài Gòn, the Grall military hospital, and the semiprivate Pasteur Institute helped remake social and ecological relationships in southern Indochina, rubber plantations proved to be one of the most effective tools of such transformation.[3]

This chapter analyzes how a latex frontier was formed and "civilized" by written natural histories and European conquest that allowed planters to establish rubber plantations on the Indochinese peninsula. Frank Uekötter has suggested that "learning by doing" informed early plantation efforts, and this was true of planters in Indochina as well.[4] While planters relied on a process of trial and error and imitation, however, knowledge of nature framed as "science" provided them with a sense of purpose and mission. Colonial states established the legal, economic, and transport infrastructure necessary for planters to appropriate land and plant rubber. The incorporation of Indochinese forests into networks of exploitation and conservation, the instituting of European land tenure regimes, and the setting up of new practices of mobility and immobility all paved the way for plantations to replace forests. Conflicts arose between planters and officials, but the colonial state was mostly eager to use the forested regions of eastern Cochinchina for industrial agriculture. As planters and laborers cleared the southern forests, new biological and linguistic environments emerged on plantations, influencing in turn the surrounding region. These emerging environments replaced the existing habitats of both animals and non-*hevea* plants and more clearly delineated natural, social, and political borders. Finally, these hardening borders, along with texts of explorers, missionaries, and administrators, formed the basis for initial state approaches to the Montagnards, or uplanders, and influenced their interactions through-out the colonial period.[5]

Indochinese Natures

Curiosity about nonhuman nature in Southeast Asia predated the formation of colonial states; missionaries and scholars had long studied the airs, waters, soils, diseases, flora, and fauna of the region. The earliest textual attention came from China, where authors viewed the south as a source of spices and medical ingredients. Missionaries continued investigations into the plants and animals of southern Indochina, and Cristoforo Borri's seventeenth-century writing about the south-central Vietnamese region of Đàng Trong represents one of the earliest European-language descriptions of Indochinese nature. In 1790, the Jesuit missionary João de Loureiro (1710–96) published one of the first catalogues of plants in Cochinchina to follow the Linnaean system of classification, providing names in both Latin and Vietnamese. These pre-nineteenth-century publications often combined observations relevant to botany, agriculture, medicine, and ethnography, tasks eventually divided up among distinct scholarly disciplines.[6]

A Vietnamese court official, Trịnh Hoài Đức, wrote *Gia Định Thành Thông Chí*, perhaps the earliest scholarly work in any language to consider southeast-ern Vietnam as a coherent social and environmental region. Trịnh characterized Gia Định and Biên Hòa, the two easternmost provinces of the Lục tỉnh (six provinces), as generally unhealthy and sparsely populated, and focused more at-tention on the four rice-growing provinces of the west, which were more densely populated and economically vibrant. Đông Nam Bộ (ĐNB), a more recent term for an area that is roughly isomorphic with Gia Định and Biên Hòa, forms a fron-tier zone where the central highlands meet the southern deltas. The region has distinct dry and wet seasons and is watered by three main rivers: the Sài Gòn, the Đồng Nai, and the Bé. Humans inhabited the area during the Neolithic pe-riod, and connections to the Óc Eo civilization further to the south have been found in the Đồng Nai River basin. In the seventeenth century, Minh Hương, or Ming loyalists who fled China after the rise of the Qing, began to arrive in urban centers such as Biên Hòa, and, even though Trịnh's Viet-centric narra-tive omits their presence, the Stieng and other Mon-Khmer speakers had long inhabited the region. The region's location between the plains of the delta and the central highlands has made it an important trading point for delta and coastal products, such as rice and salt, and forest products, such as timber and wildlife. The vegetal cover on the undulating hills and sloping plateaus of the southeastern uplands included both dense and open forests, stands of bamboo, agricultural plots, and *tranh*, or lalang, grass, and environmental boundaries helped divide up a land with few firm political borders.[7]

When the Vietnamese government ceded the final three provinces of the Lục tỉnh to the French in 1867, and southern Vietnam became known as the colony of Cochinchina, a French view of the region's nature began to take shape. In 1876, the colonial government divided up the province of Biên Hòa into Biên Hòa, Bà Rịa, and Thủ Dầu Một (often in French these names were written as one word, Thủ Dầu Một becoming Thudaumot, or even Tudomot), which, together with Tây Ninh, produced most of Vietnam's rubber during the twentieth century. These political divisions were initially fluid and had little basis in on-the-ground facts. To destabilize older geographical units and increase the intellectual coherence of these provinces, the French colonial state commissioned numerous missions to map their living and nonliving topography. During the twentieth century, maps, statistics, and infrastructure related to industry, commerce, and agriculture in general and rubber plantations in particular gave this region added reality.[8]

The colonial government further imposed European knowledge systems on the flora and fauna of Cochinchina when it established a botanical garden in Sài Gòn in 1864, a mere five years after France took possession of the city. In 1865, Jean-Baptiste Louis Pierre (1833–1905), who was born on Réunion but moved to India for employment, became the garden's director, a position he held until 1877. Over those years, Pierre published an expansive, multivolume treatise called the *Flore forestière de la Cochinchine* as well as smaller works on specific plants. In 1878, a veteran of the navy artillery became the new director of the botanical garden and Sài Gòn's stud farm. One of the new director's first responsibilities was to organize Pierre's disorderly notes and publish a catalogue of the exotic and native species that had been cultivated at the garden. This initial annual report of the botanical garden, written in French and Vietnamese, also referenced the first phenological (i.e., agricultural and botanical) calendar, which was published in 1866.[9]

French colonialism transformed the preindustrial meanings of nature. A common term for "nonhuman nature" in Vietnamese is *thiên nhiên*, and the calques in Mandarin (tianran, 天然) and Japanese (tenzen, 天然) are characters drawn from the Taoist understanding of "a logical, cosmologically ordered Universe . . . that operates according to a Way." Another standard translation of nonhuman nature, especially as opposed to human invention, is *tự nhiên*.

The equivalents in Mandarin (ziran, 自然) and Japanese (shizen, 自然) are older characters meaning "the thing itself," "innately," or "automatically," "in the sense of without effort, or according to the innate characteristic of something or someone." In Vietnamese, *tự nhiên* also means "naturally," or "spontaneously," as in "be natural" or *tự nhiên đi*, a shade of meaning also suggested by the Japanese and Chinese terms. Finally, the nature of something, its essence, and human nature can be translated as *bản chất* and *bản tính*, respectively.[10]

The Vietnamese word for "environment," *môi trường*, is a more recent invention and can stand alone or be used in combinations such as *môi trường sinh thái* (the ecosystem), *môi trường acid* (the acid environment), or even *môi trường công tác thuận lợi* (the advantageous business environment). Lay terms for "nature" among Vietnamese farmers consisted of more concrete ideas, such as *trời đất* (heaven and earth, with the Sino-Vietnamese equivalent of *thiên địa*) and the names of animals, plants, and other "natural" objects. Even political terms were often rendered very concretely (e.g., *đất nước*, or earth and water, meaning country or nation). Terms such as *sinh vật học* (the study of living things) for biology, *sinh thái học* (the study of the relationship between living things and the environment) for ecology, and *môi trường* for environment were not introduced until the 1920s and 1930s. Those responsible for helping to spread, if not originally translate, such terms included agricultural agents, some of whom also articulated important criticisms of colonial agricultural practices.[11]

As elsewhere in Asia, the idea of "nature" played a role in modern political projects in Việt Nam, including the colonial and postcolonial states' obsession with natural resources, intellectual concerns about Vietnam's place in a social Darwinist universe, and the communist and noncommunist projects of anticolonial national identity. In other words, each constructed division between nature and culture associated with modernity drew on a set of power relations to sustain it. As Pam McElwee has shown, colonial and postcolonial governments in Việt Nam enacted environmental rule to legitimate their control over the land and the people. Officials and scientists claimed that they alone knew how to best manage nature, and that they alone would protect its resources.[12]

Forests to Manage

During and after colonial conquest, the French debated how best to exploit and conserve southern Indochina's forests. The peninsula is in the monsoon zone that stretches from near the equator into the foothills of the Himalayas, and a high degree of variation in latitude, altitude, and climate had created a highly diverse flora and fauna. French sources from the early twentieth century listed the surface area of Indochina as approximately 700,000 square kilometers, or about 70 million hectares. Of this area, trees covered about 300,000 square kilometers, or about 25 million hectares. The eastern provinces were heavily covered by forests; for example, one-half of Tây Ninh's 700,000 hectares was covered by forest, with only 16,000 hectares devoted to rice production at the beginning of the twentieth century. French naval officers viewed these lands as a reservoir of wood that could supply them with building materials, French botanists viewed these forests as a store of unknown and unclassified plant material, and foresters tried to balance goals of exploitation and conservation. Meanwhile, locals continued past activities and exploited new opportunities.[13]

The first step toward industrial agriculture came as Europeans measured and defined the legal life of forests, making legible the type and extent of the forest cover of Indochina. Questions of social control, rather than concern about natural conditions, dominated the colonial government's approach to forest management. This emphasis left the Forest Service in a weak position and led to conflicts over the conversion of Indochina's forests into land that was devoted to industrial agriculture. Even before plantations were first established in the 1910s, the Indochinese Forestry Service and the Agricultural and Commercial Services struggled over definitions of what constituted a "forest" and whether land set aside as reserves could be used to grow "forests" of *hevea*. A map from a 1930 file on colonization in the red earth region shows some of the major routes, population centers, upland peoples, and surviving forest reserves. Despite the areas closed to colonization, those promoting industrial agriculture were largely successful in their attempts to recast forests as "empty" and "wasteland." In account after account of plantation creation, the pre-plantation landscape was depicted as an unproductive, unpopulated region—depictions that were patently false.[14]

The French government initially took a practical approach and relied on local laws and customs to manage forests in a system of indirect rule. This approach was soon reversed, however, and the French installed checkpoints to control the sale of wood in the provinces of eastern Cochinchina. Instead of regulating large enterprises, these checkpoints were directed at small woodcutters whom the government blamed for the destruction of many trees. In 1875, the Cochinchinese government secured a measure of control by requiring annual permits for cutting. These permits, which cost 400 francs, authorized companies to employ unlimited numbers of woodcutters, and allowed them to cut as much wood as they liked. This loose regulatory regime had disastrous effects on the forests of Cochinchina, as admitted by the first conservator of forests in Indochina.[15]

In response to rapid forest degradation, the French colonial state passed two more *arrêtés*, or decrees, in 1891 and 1894. The *arrêté* of 1891 established forest reserves where exploitation was banned, and the first three reserves were established in Thủ Dầu Một in March 1892. The reserves created by these *arrêtés* resulted in fines and punishments for inappropriate cutting. Posters advertising these restrictions were published in Vietnamese and Chinese, which suggests that the French were not the targets of these fines. These reserves were monitored from Bên Cát, a village that later became an important site of rubber production. Under the Nguyễn dynasty, the area had belonged to the province of Biên Hòa, canton of Bình Lâm, or "peaceful forest." In the 1890s, the area was known for its small-scale wood production, and twenty-five kilometers up the Thi Tinh creek (*rivière, rach*) from Bên Cát, two foresters maintained a former military post that had been transformed into a forest guard station. These foresters had little authority, however, and observers complained that many cutters sold wood to Bên Cát sawyers without permits. By controlling the use of wood, these reserves, and others like them in eastern Cochinchina, affected the lives of many residents.[16]

The 1891 *arrêté* also established an ecological perspective in Cochinchina, a process that extended to Cambodia and Annam with *arrêtés* of the 1910s. First, it divided the colony's forests into two simple but useful regions of west and east. The western forests consisted mostly of mangrove forests, while the eastern region contained large hardwood trees, which the state considered more valuable. Like Nguyễn dynasty thinkers, French colonizers made distinctions between the rice-growing Mekong Delta and the higher, drier lands of the east. The 1891 *arrêté* clearly articulated the importance of the environment for living beings as well. Article 63, for example, listed the following as reasons to conserve the forests: (1) to keep soil on mountain slopes; (2) to protect soils against erosion and removal by rivers, streams, etc.; (3) to maintain springs and water courses; (4) to protect dunes and sand from ocean erosion; (5) to defend the frontier territory; and (6) to preserve public health. The first four reasons were closely related to the regulating effects that forests were supposed to have on local environments, or microclimates. The fifth reason viewed forests as a deterrent to attack, while the sixth saw the regulative effects of forests on tropical climates as beneficial for humans.[17]

The wording of the 1891 *arrêté* recalls the long tradition of scientific forestry in France, and though officials often prioritized other political and economic needs, foresters had their own vision of how the forests should be managed. Foresters used the burgeoning science of ecology, a term coined in 1866 by Ernst Haeckel, to give older traditions of forest management a new language that helped advance scientific views of proper management within a colonial state. These foresters emphasized the importance of forests for preserving a proper climate and hydrologic cycle. "To conserve the minimum forested regions necessary from the point of climatology and hydrology," argued the forest conservator, was a key function of his service.[18]

French scientists had developed theories about the effects of trees on climates from their previous observations of island colonies such as Mauritius and Réunion, and they brought this knowledge to bear on their work in Indochina. The French experience with trees in Algeria, Morocco, and Tunisia also factored into thinking about nature in Indochina. Foresters and administrators fretted about swidden, a farming practice that clears patches of forests with fire in a rotating system that cycles through several years. They believed that swidden, called *rẫy* in Vietnamese, had a role in deforestation, and looked for ways to tie mobile agriculturalists to place. Officials feared desiccation, or long-term drying of Indochina's climate, a concern that Diana Davis has shown in the context of Algeria was "based largely on inferences and deductions from partial evidence, and primarily advanced by those with economic or ideological interests in planting trees." By the end of the colonial period, some, though not all, scientists had begun to minimize the role of forests in determining rainfall levels at a regional, if not local, scale.[19]

The 1894 *arrêté* established the Indochinese Forest Service, a program that was motivated more by fiscal concerns than scientific forestry. The forest guards, often pulled from the ranks of former soldiers, had no real forestry training and largely served a policing function. Because the Forest Service derived its revenue from cutting royalties, its guards focused on collecting taxes, which left them little time to carry out conservation efforts and prevented them from setting trees apart for conservation. The 1894 *arrêté* also shifted the responsibility for forest damage away from transnational capital and large forest cutters to small-scale cutters and villagers, thus continuing a logic that blamed locals for destruction of forests. A ban on local wood harvesting was bolstered with another on swidden agriculture, a practice vital for upland peoples, and laws punishing those who started forest fires.[20] An *arrêté* issued three years later was supposed to place the Forest Service on a more "scientific" footing, from which it could claim a greater monopoly over the discourse of forests. Although the recruitment process began to test agents for knowledge of principles of forest management, those with scientific training remained in the minority and held higher-level positions that removed them from day-to-day management.[21] The high turnover of forest agents from Europe and the unwillingness of the colonial government to give responsibility to local agents ensured a continuing lack of expertise in the Forest Service.

Foreshadowing their approach to the rubber industry, administrators favored large European exploitations over small native cutters in Cochinchina. Indochina's forests contained a high diversity of species and lacked the type of timber that was valuable for economic activities like shipbuilding. The lack of roads and waterways also made the harvesting and shipment of timber prohibitively expensive. In 1914, Emile Baillaud, an associated member of the International Colonial Institute, wrote that "wood, and in particular those of construction and woodworking, is very expensive as a result of the difficulties of searching and transport (a lack of routes) and because of the continuing great distance that separates the market from the place of cutting." Given limited governmental manpower and an administrative reluctance to manage small woodcutters, large European-run enterprises became prevalent. One of the largest such companies in eastern Cochinchina was the Bienhoa industrielle et forestière (BIF), created in 1911, which possessed 30,000 hectares, some of which it used to grow rubber.[22]

Administrators who jealously guarded their budgets did not always see eye to eye. The administrator of Tây Ninh complained bitterly about the loss of revenue to the province that was caused by the moving of wood at night by train. He argued that the province was receiving less revenue even as woodcutting increased, in part because of the lack of oversight over large trees. Enterprises that exploited forests could also run into trouble with administrators when they crossed the border separating Việt Nam and Cambodia. Articles 31 and 32 of a 1903 *arrêté* placed control of forestlands at the local level, which led to a fight over cutting rights on the frontier between a Tây Ninh–based European named Huguenin and the French administrator of the Cambodian province of Soai Rieng. The administrator in Soai Rieng arrested cutters working for Huguenin, including three Vietnamese and thirty-six Cham, near the village of Mai-Khum-Mere. Huguenin had written two letters to the administrator demanding information and had appealed to the Tây Ninh administrator to intervene. Huguenin argued that both his permit to cut and those that he gave to his cutters were valid in Cambodia.[23]

Until 1910, an ecological perspective remained absent from forestry practices and, as the historian Frédéric Thomas has demonstrated, a centralized forest service was not well adapted to the environmental realities of Indochina. First, the minimum diameter of trees common in the four "pays" of the

Indochinese Union (excluding Laos) was assumed to be the same, with little regard for their different environments. Thomas argues that this perspective reflected a colonial emphasis on profitability rather than adaption to local conditions. Second, Thomas's examination of the spatial organization of the Forest Service across Indochina shows lacunae in its coverage. The Forest Service was not located in the richest forests, but rather at communication points, where wood could be taxed, and in forests that were already impoverished. This arrangement was the result of economic priorities rather than an ecological perspective. After the 1894 *arrêté* had eliminated the division of Cochinchina's forests into two zones, further changes introduced between 1905 and 1910 eliminated the western canton, replacing it with a central canton based in Sài Gòn. This left some of the great mangrove forests of the delta, such as U Minh forest, unmonitored. Finally, Thomas shows how the service was incapable of cooperating with locals to manage the forests, relying instead on capitalist enterprises that exploited forests and cut down trees. This situation led to lack of oversight, as most of the active woodcutters operated on a small scale.[24]

Albert Sarraut (1872–1962), who took the office of governor general in 1911, recognized some of the problems that plagued the management of forests. "The organization of a local Forest Service," Sarraut reasoned, "has been easily justified elsewhere by the fact that forests, as well as customs, differ from region to region." The differences across the Union, Sarraut continued, required a consideration of the variable conditions and a move away from a blanket approach to forestry. Sarraut employed an emphasis on local conditions to justify this move to local political control of the activities of forestry. Sarraut also referred to the climatic theories of the day, stating that forests played "by an intercontinental wave of great amplitude" a role in weather formation, and even discussed the significance of the "conservation of woods and forests." Although Sarraut deployed the discourse of ecology, he listened to economic demands and viewed conservation and local control as compatible with the economic needs of the colony.[25]

Sarraut's efforts at decentralization left local branches of the Forest Service to advocate for the interests of forests against *hevea* plantations and prefigured later conflicts among planters, administrators, and agricultural scientists. The tensions between cutters and the Forest Service also left the foresters largely unable to assert their authority over eastern Cochinchina as industrial agriculture moved into the region in the 1910s. The debate about the desirability of industrial agriculture revolved not around an ecological perspective but rather on political and social questions. These questions outweighed environmental concerns, and planters had to convince officials of their ability and right to exert authority over the people living in the region. Some officials were concerned about the outsized role played by planters in local affairs, but overall the colonial state's emphasis on economic development meant that planters were largely successful in their attempts to recast forests as "wasteland."[26]

Concessions to Agriculture

Would-be planters took the second step toward plantations when they agitated to rewrite land-ownership laws. At the beginning of the twentieth century, rubber planters looked to precedents and the legal framework developed for rice concessions in the Mekong Delta from the 1860s through the 1890s. As with the Nguyễn dynasty's forest policy, the French military government initially adopted the existing land regime as understood through readings of the *địa bộ*, or land register. French administrators largely used the Codes of Gia Long to understand Nguyễn policy, which had divided land into three types: individual, village, and state. Some French viewed these laws favorably, and in 1881 the director of the interior for Cochinchina wrote that these codes had encouraged settlement in the Mekong Delta by allowing settlers who farmed land and paid taxes for three years to claim the property. Other observers judged these laws as poorly thought out and punitive of colonization. Pierre Pasquier (1877–1934), the future governor general of Indochina (GGI), argued that the lack of a clear legal structure in the Mekong Delta arose from the constantly shifting terrain of land and water. Some officials attempted to reform and clarify the land regime, and in 1864 French laws applying to land ownership were adopted in Cochinchina. Yet these laws were never widely published and, as a Vietnamese lawyer in Paris in the 1930s observed, "custom replaces the law in many cases."[27]

An *arrêté* of August 1882 instituted the first significant reforms in land laws in Cochinchina, establishing a system of transferring land from the state to private hands. There were two ways to gain control of land. The first was through *gré à gré*, or private agreements, signed between the government and the buyer. The second was through *adjudication*, or the public auctioning of land to the highest bidder. In both cases the profits from the sale of land went to the state budget of Cochinchina. The government also had the option of giving land away for free. In practice, *terres rouges* was often given away in the hopes of encouraging agricultural development, while the peri-urban *terres gris* was sold. Concessions included the idea of granting limited rights over land in exchange for fulfilling certain duties and obligations. The stipulations for the land transfers of more than ten hectares were meant to encourage putting the land into *mise en valeur*, or development. These land grants, temporary at first, could become permanent after five years if the planter or enterprise cultivated the land. This basic system of land transfer stood, with some modification, until the land reforms of 1927 and 1928.[28]

Even after the formation of the Indochinese Union in 1887, land laws in eastern Cochinchina remained ill defined, leaving provincial administrators and the governor of Cochinchina (Goucoch) to decide colonization on a case-by-case basis. One of the main factors driving the interest of colonial officials in granting agriculture concessions was the potential of plantations to push economic development and increase tax revenues for a region with few major exports. In 1896, the Goucoch stated that the administration aimed "to favor by all of the means possible the progress of agriculture." French officials had been searching for economically profitable plants to grow in eastern Cochinchina since the late nineteenth century, but until *hevea* was grown in the 1910s, most of these efforts had failed.[29]

Reflecting the changing French priorities for the colony, the GGI Paul Doumer (1897–1902) adopted a policy of economic and social development and approved a 200 million franc loan aimed at promoting the *mise en valeur* of the colony. The reforms in land laws that took place under Doumer between 1899 and 1903 attempted to systematize the granting of concessions, but land sales remained a highly idiosyncratic affair. For instance, the Colonial Council supported a request for land by the Suzannah plantation (named after Suzanne Cazeau, the daughter of the plantation's owner), noting in its report that "the importance of the general costs that the company has already invested, from the beginning, proves that the society really desires to develop the lands that it requested to buy." In this way, individual administrators continued to determine the vague definition of "desire," which some planters argued caused too much uncertainty in the process.[30]

When it was formed in 1910, the planters' syndicate, initially called the Association of Rubber Planters of Cochinchina (APCC in French), waded into the battle over land concession policy (see chapter 2 for more details on the planters' syndicate). The APCC pushed planters' interests within the commission that was working on reforms, and on October 13, 1910, the Goucoch passed an *arrêté* that set up a system of land transfer aimed specifically at encouraging rubber production that allowed for free concessions, private sales, and public auctions. In addition, *terres rouges* lands that were not yet owned were set aside for the cultivation of rubber. The only restriction on these lands was that they be turned over for rubber growing, and the size of the land holding of any individual enterprise was essentially unlimited.[31]

Not all planters were satisfied with the new laws; some of them faced local resistance to plantations, and they urged the Goucoch to pressure Vietnamese authorities to agree to more concessions.[32] While planters recognized the danger of land speculation and admitted that some controls were necessary, ultimately they argued that the concession of land was too slow and rejected the idea of selling land through public bidding. Reporting on his conversations with the reform commission, André Crémazy, the president of the planters' association, wrote that the commission had started from the position of limiting land sales to 500 hectares, which the association had rejected, arguing that the limit on land holdings neutralized the ability of large societies, with their reserves of capital, to develop land. The commission had shown itself to be receptive to this argument and had allowed plantations to work around these restrictions by legalizing the control of multiple lots under one enterprise. Crémazy did not think that the colonial administration would follow the commission's already limited restrictions, and believed that the GGI would prefer to keep the former system of a combination of free concessions, private sales, and public auctions.[33]

Not all planters welcomed the new rules for land grants, and heated debates in 1912 resulted in the resignation of Octave Dupuy, a senior member of the planters' association. Dupuy had worked with the Comité to draft a petition calling for clearer regulations on the sale of land to avoid sudden changes in prices. This faction submitted the "Thiémonge" petition, which was addressed to the minister of colonies, in April 1912. Gaston Sipière and other planters disagreed with the measure, preferring the ambiguity of the local land commissions, which the planters felt they could manipulate.[34]

Despite internal debate, the planters' association effectively lobbied the administration, as evidenced by the *arrêté* issued by the GGI in December 1913. This *arrêté*, the most important reform before the late 1920s, gave very favorable terms to planters. Those with land grants up to 500 hectares, which later was one definition of a "smallholder," had to plant 10 percent of their land in *hevea* each year, with 50 percent planted at the end of five years. Those with grants over 500 hectares (i.e., estate holdings) were required to plant only 5 percent each year, reaching 50 percent by the end of ten years. The other 50 percent was completely left up to the owner's discretion. A tax of 2.75 piasters per hectare was charged on land planted in *hevea*, but the tax would not be imposed until the seventh year, and would be applied to only

10 percent of the land. This increased until the tenth year, when the tax was due on all land planted in *hevea*. For the other 50 percent of the land, a 0.48 percent tax per hectare started in the seventh year, payable in three installments starting in the eighth year. The purchase price of the land was paid in two installments, the first due twenty days after the sale, and the second due the following year. The price of land varied from 0.50 to 5 piasters per hectare, and foreigners could own land. These terms were some of the most favorable in Southeast Asia, according to the planters' association.[35]

Land concessions had generated conflict in the Mekong Delta, but provoked less immediate controversy in eastern Cochinchina. This was not because the land was "wasteland" or "empty"; as Marianne Boucheret and Mathieu Guérin have shown, groups such as the Stieng and Phnong had long occupied the region, and the records of land-granting committees reveal that they knew that the land was in use when they awarded it to applicants. Some officials expressed concerns about the consequences of such land grants, and a 1912 report by the Thủ Dầu Một administrator indicated skepticism about handing over too much state prerogative to private enterprises in relation to upland peoples living in Hớn Quản. The administrator complained that some planters were attempting to establish a state within a state and to ignore the general good in favor of specific financial interests. In the 1920s, the question of land claims in the region became especially politically sensitive as several rebellions by "moïs indépendants," such as the Phnong, took place. Legal challenges to land ownership also increased, as groups such as the Stieng requested rights to all the land east of the Kratié-Kampong-Cham road. The colonial administration was not prepared to grant this petition, as it would have eliminated most of the desirable agricultural land from European development, but potential unrest and a willingness to balance the needs of locals with those of capital encouraged the colonial administration to put some restraints on the territorial ambitions of plantations.[36]

The dossier for the land grant to the Suzannah plantation includes a 1904 report by the waters and forests warden that acknowledges the connection between swidden and human presence. The warden noted: "One finds indeed the presence of former *rays* that almost completely wiped out the forest, leaving only on average twenty or so trees per hectares." Earlier that year, the forester thought that the land being requested had recently been subjected to *rẫy*, an agricultural system that he did not condone. The local "moï" who had lived there, he thought, had left upon "the first appearance of Europeans in the region." The site of the Suzannah plantation was in fact still inhabited when it was sold, and Suzannah paid 15 piasters to the sixteen or so locals who were forced to leave their homes. The value of the piaster over time is difficult to calculate, as exchange rates were not recorded until 1913, but the Indochinese piaster in 1910 was worth roughly between 33 and 36 French francs, or 6.4–6.9 U.S. dollars (150–200 U.S. dollars in 2015). Perhaps a more relevant comparison is the salary of a rubber tapper, which was 0.40 piaster per day, meaning that 15 piasters represented about thirty-eight days of plantation work. This amount of money could not compensate for the villagers' loss of their livelihood and homes. While planters mostly received the swidden land they requested, Vietnamese authorities did retain some say in local governance, and village leaders successfully blocked the construction of a plantation village on land under their jurisdiction.[37]

Increasing French economic activity led to a growing state presence, and plantations offered the potential of political and social influence over frontier territories. In addition to uplands inhabitants, the control was aimed at groups of people who were viewed as operating outside the law, such as woodcutters. Turning these cutters into agriculturalists was part of the civilizing mission, and in a 1908 report, the administrator of Tây Ninh wrote to the Goucoch that "the tendency of the Tây Ninh woodcutter to transform into a farmer is increasing each day," a trend that was "very reassuring for the future of Tây Ninh." Plantations were in turn concerned about their security, and colonial state documents discuss perpetual attacks by "pirates," "brigands," and "bandits." In 1913, the colonial government placed plantations under the authority of the canton heads and village notables. In addition to regular tours of guards, the Biên Hòa administration set up a permanent police post at Xuân Lộc.[38] Attacks continued, especially on locals, and a 1917 Goucoch report noted that "in the isolated regions of Tây Ninh province, on the border with the Cambodian frontier where the vast stretches of forest make the surveillance and the pursuit of criminals very difficult, there have also been signs of some acts of piracy."[39] One of the key tools used by the colonial state and its police in such matters was the road.

Paths to Civilization

Since the mid-nineteenth century, rubber has become associated with some of the most potent symbols of modern mobility: the shoe, the bicycle, the automobile, and the airplane. In reciprocal fashion, the increasing numbers of these modes of transportation have driven a continual increase in the world market for natural and synthetic rubber. At a local level, establishing rubber plantations both depended on and encouraged the development of

"Mirage . . . no, reality!" This image shows the tight link between mobility and rubber. It was sponsored by the Institutes for Rubber Research in Indochina and France during the First Indochina War to sell the French public on the value of keeping its colony. Source: *France outre-mer, le monde colonial,* no. 256 (1951).

transportation and communication networks connecting eastern Cochinchina to the outside world. Indeed, plantations promoted certain forms of mobility while inhibiting others within a broader imperial framework. Various social groups used these changed forms of mobility and immobility to their advantage: planters sought to regulate mobility and immobility to gain profit; colonial officials attempted to manage flows of people within Indochina to better govern; and, even though they exercised little control over the system, Vietnamese workers attempted to take advantage of the opportunities offered by, and avoid the worst consequences of, new patterns of mobility and immobility.[40]

Footpaths and waterways that were already being used to transport goods and people were crucial for the creation of plantations in the early twentieth century. As the historian David del Testa has argued, "The French did not initially improve transportation but instead relied on existing footpaths and navigable rivers."[41] Immediately after colonial conquest, Biên Hòa had only 46 kilometers of main roads, yet it had 282 kilometers of smaller paths, with Bà Rịa and Tây Ninh exclusively serviced by 189 kilometers and 169 kilometers of minor routes, respectively. While autobuses ran along the main roads, the smaller ones allowed a "pioneering ethic" to operate on the landscape: many individuals, mostly Vietnamese, moved into previously inaccessible regions and enacted continual small changes that collectively resulted in deforestation. As the Colonial Council argued, such byways "permit the exploitation of uncultivated land, that bring civilization to the most remote villages."[42]

By 1880, Cochinchinese officials sought to increase the movement of goods, ideas, and people in the colony. As the region was home to many bandits, a lack of security was viewed as a key impediment to the development of an export economy. To make banditry more difficult, the colonial government began to extend the telegraph network, which at the time primarily served to connect Sài Gòn to the provincial capitals. Also in 1880, the Colonial Council considered a proposal for an automobile service that would take advantage of the newly built main road between Sài Gòn and Thủ Dầu Một, based in part on the already successful automobile run from Sài Gòn to the provincial capital of Biên Hòa. Such services led to a steady increase in traffic between urban centers by the end of the nineteenth century. By making it easier to police the countryside, exploit natural resources, and increase mobility, newly constructed roads such as those into the forests of Bà Rịa and Tây Ninh offered officials a way to bring *mise en valeur* and civilization to the region.[43]

Although some improvements to the transportation system had been undertaken, a great deal of work remained. In the 1910s, the Public Works

Department began a road-building project in the region, funded by a 90 million franc loan made by the Bank of Indochina, which was approved at the end of 1912. A significant part of this loan was slated to service Doumer's "railroad loan" of 1898, which was directed largely toward building the trans-Indochina railroad, although 9.5 million francs was also directed toward road building. This was not a large amount, but it represented one of the first significant efforts to expand the road system since the beginning of colonization.[44]

Much of the initial road-building effort was put into expanding the main east-west road that crossed the province of Biên Hòa and formed part of the famous Route coloniale 1 (RC 1) that ran north to south along Vietnam's coast. By 1915, the road system had expanded north into the interior of the province, serving the rubber plantations via a biweekly bus run in operation between Thủ Dầu Một town and Hớn Quản in the heart of the *terres rouges*. Maps made for construction projects show how push-pull interactions operated between roads and plantations. Since no plantation could operate without a means of bringing workers and supplies in and sending latex out, they tended to cluster around waterways, railroad lines, or already constructed routes. Meanwhile, more roads into red earth territory were paved to provide economic benefits to plantations.[45]

Although plantations sometimes encouraged road building, their lands could also impede route extension, and the construction of roads raised questions of authority and caused debates between planters and officials regarding public versus private ownership. The colonial government was concerned about such right-of-way conflicts, and the granting of concessions included clauses ensuring the public right to access plantation land via road or railroad. For example, in 1913 the provincial administrator of Biên Hòa approved the Gia Nhan Rubber Company's request to buy 68 hectares of land adjacent to Route Chesne. The only two restrictions were that (1) "free circulation would be maintained on the section of Route Chesne that formed the eastern limit of the current holdings of the Company," and (2) "there would be no obstacle to work done to repair Route Chesne."[46] Retaining the right of passage was also raised in a 1914 letter from the administrator of Hớn Quản, Tonarelli, to the administrator of Thủ Dầu Một. Tonarelli wrote: "One of the most important questions that has come up in the Hon Quan region is that of circulation around and within the plantations—immense blocks of thousands of hectares that will lay claim to nature, the soil, the labor fixed there and, in a general way, all of the commodities, past, present, and future that are only intended for the plantation; there can be constraints, customs, and rights to respect."[47] This quote offers a critique of the local power of plantations and planters to disrupt local life and to lay claim to many resources. Tonarelli continued that the planters did not necessarily ignore the laws of the administration, but stated that very few regulations applied to them. Not surprisingly, while pushing for greater autonomy in their treatment of workers, land planters rejected the idea of paying the costs of governing.[48]

Notwithstanding these conflicts, after World War I, a network of roads was extended to exploit the agricultural potential of the red earth region. By 1924, 600 kilometers of paved road had been developed in Biên Hòa, and by 1930 this number grew to 703 kilometers, including RC 20 to Đà Lạt, which was finished in 1929. Lonely outposts such as those at Phú Riềng and Núi Bà Rá were connected to urban centers in Thủ Dầu Một by RC 13 and 14 along with smaller routes. By the estimate of one author, the TP efforts after World War I constituted one-third of the total roads and buildings that had been constructed in the province. And according to one set of calculations, between 1917 and 1954 "the French built or improved over 9,000 kilometers of roads."[49]

The government's grandiose descriptions of its road-building projects must be taken with a grain of salt, as waterways remained the most important means of transport. One planter complained that when he arrived in the border region of Cambodia in 1927 to manage a rubber plantation, the roads were only halfway completed, with "Eiffel" bridges rusting uselessly beside ravines. This planter criticized what he saw as the TP's apathetic approach to road construction, although road-building efforts were often abandoned because of endemic malaria. Former workers spoke of the poor state of the roads such as RC 14 near the Vietnamese and Cambodian border at Lộc Ninh. Regardless of their quality, these capillary networks of paths, roads, and waterways increased the reach of state and nonstate agents into the forests of eastern Cochinchina.[50]

Time to Tap

For the twentieth-century plantations to function, new everyday relationships between humans and nonhumans had to be created. Following the daily motion of a tapper around a tree helps to imagine these diurnal patterns. Early in the morning during the rainy season, the relatively cool temperatures encouraged the latex to flow in the phloem located just under the outer bark of the tree. Around 6:00 A.M., just before sunrise, a tapper, usually male, made incisions a few millimeters deep in the bark, just enough to tap the latex passageways but not deep enough to sever the flow of nutrients and energy between the tree's roots and crown. After much experimentation, these cuts were made down and at an angle, in either one direction or both, across the tubes that carried the latex. Then the latex dripped down the grooves in the tree into a bowl. A single worker repeated this action as quickly as possible for between 300 and 500 trees per morning. By 11:00 A.M. the flow of latex slowed considerably, and this worker returned to tip the ceramic collecting bowl into large canisters. Early in the afternoon, the seasonal rain rolled in, and for thirty minutes or so, a deluge fell, soaking trees and bodies. Later, if the rain stopped for the day, a worker (often a woman or child) returned to weed, cut branches, and perform other "light" tasks.[51]

Regional climate patterns of temperature and rainfall set seasonal rhythms. In southern Vietnam, during the wet season (May to October), when the monsoon winds blew from the southwest, the latex of *hevea* flowed freely and workers focused on tapping trees. During the dry season (November to April), monsoon winds reversed direction and blew from the northeast. Many plantations concentrated their clearing work during the dry season, when latex flowed more slowly. The alternation of dry and wet seasons had important consequences. First, mosquitoes bred during both seasons, but were more numerous during the change in seasons. Second, the extended dry season limited the attack of parasites and fungi on *hevea*. Europeans were initially worried that the extended dry season would not allow the maximum growth of *hevea*. But, French planters argued, unlike the ever-wet Brazilian rainforests, Cochinchina's dry season eliminated many of the pests of *hevea*.[52]

Another tool to understand human and nonhuman relationships is ecology. Ecology in its current sense refers to both "the interactions that determine the distribution and abundance of organisms" and the academic discipline that studies these interactions.[53] By the time the French colonial empire began to crumble in the middle of the twentieth century, ecology had become a key element within the studies of agriculture and medicine. But at the beginning of the twentieth century, the word "ecology" rarely appeared in publications printed within the French colonial empire. While ecological concerns predated the formation of colonial states, and missionaries and local scholars had long studied the climate, waters, soils, flora, fauna, and diseases of the Indochinese peninsula, the science of ecology was not a major subject of institutional teaching or research in Indochina until the mid-1920s.

Ecology in the tropics tends to bring to mind a "wild profusion" of animals and plants, but scientists in French Indochina understood ecology through an economic lens; their studies focused on increasing the yield of plants and animals.[54] In the global scientific community at the turn of the twentieth century, ecology meant both "the science of communities" (called synecology) and "the environmental physiology of the individual organism" (called autecology).[55] Government scientists were often concerned with synecology, which was almost always referred to as "pure" or "theoretical," while scientists concerned more exclusively with *hevea* (or other export crops) tended to work on autecology problems, which could be considered either "pure" or "applied." Scientists focusing on rubber plantations were, however, interested in synecology in relation to diseases of *hevea* such as South American Leaf Blight. These diseases highlighted the importance of the interactions of organisms with each other, and the research began to veer away from the traditional focus on topography and general environmental factors. Profitability was another concept used to differentiate pure from applied science, and Arnold Sharples, a mycologist in the Department of Agriculture in Malaya and the Pathological Division of the Rubber Research Institute, drew a distinction between applied and pure science when he wrote that "in the former, profits must be clearly visible, while, in the latter, the completion of a line of research work is the primary consideration, and whether or not profitable issues may be the ultimate outcome does not affect the question."[56]

The term "ecological perspective" can be used to identify the precursors to what later became ecology's objects of study.[57] An ecological perspective also provided a way to understand colonialism as specific material and symbolic flows of things and ideas through systems. This general awareness of connections in nature and society became an increasingly important way for government officials and planters to view the colonial situation.[58] During the First Indochina War, the Vietnamese and French begin to imagine the modernity of rubber plantations in ways that extended past the colonial. Colonial agriculture and medicine were reframed by socialism or development. So too were ecologies of plantations. The term "ecological modernity," used by the historian Ian Miller in his work on the Tokyo Imperial Zoo, is useful here to describe human and nonhuman relations on plantations that cross the colonial and postcolonial divide. Miller has characterized ecological modernity in Japan as a process that is "persistently concerned with Japan's place in the modern world and the place of flora, fauna, and other natural products within that modernity." Likewise, rubber planters, the Việt Minh, government officials, experts, and laborers all articulated different visions of ecological modernity and simultaneously enacted the modernization of ecology, a "double process of intellectual separation and social transformation."[59]

Cultivation on plantations during the twentieth century altered *hevea's* spatial and temporal ecology in contradictory ways. Spatially, monocultures simplified the tree's surroundings, while at the scale of individual trees, grafting and crossings increased diversity. Temporally, the overlapping yet distinct time units of years, seasons, and days defined the plantation experience, and while these periods varied across the world, a general plantation rhythm persisted. The importance of years was often tied to the global operation of markets and to labor concerns. During the clearing and planting stages of plantation creation, capital raised by anonymous societies or invested by individual planters sat idle and earned no return on investment, which encouraged French capital initially to flow to plantations in British Malaya and the Dutch East Indies. During the first half of the twentieth century, planters faced the biological fact that *hevea* required a six- to seven-year waiting period before tapping could begin. During this time, young trees were vulnerable to attack from animals such as elephants and faced the constant threat of fire. This precarity encouraged planters to push workers hard to clear and plant as quickly as possible. Early in a plantation's life cycle, death and disease were widespread among workers and managers; planters often viewed labor as expendable and saw little reason to invest in their physical well-being. But as the trees came to maturity, more highly skilled workers became a priority, and the yearly production of rubber depended on a content, trained workforce that could consistently tap latex without destroying trees. Companies gradually sought ways to retain workers, and the reproduction of the workforce became a bigger concern.[60]

Like other perennial export crops such as coffee and tea, *hevea* provided object lessons in linear historical development, competing with cyclical conceptions of space and time. Unlike the annual cycles of rice agriculture, the perennial *hevea* tree lent itself to a narrative of birth, growth, maturity, and decay on the scale of decades, which planters understood to include four main phases: clearing, planting, tapping, and regeneration. As one plantation observer put it in 1969, "The generations of *Hevea* still march closely with the generations of men who produced them."[61]

While none of the individual developments related to industrial agriculture were novel, their combined effects represented what J. R. McNeill has termed "something new under the sun."[62] The evolving characteristics of individual plants and the networks in which they were embedded became crucial as organisms threatened to escape the meanings that human groups had assigned to them. Although most rubber growers treated *hevea* trees as if they were just machines with inputs and outputs, plants such as *hevea* are more accurately described as "gardens in the machine," a designation that precludes equating *hevea*, and agricultural technologies more generally, with machines.

European managers sought to establish a homogenous space, but their gardens sometimes responded to the social and environmental conditions of Cochinchina and Cambodia in unexpected ways.[63] Above all, planters envisioned environments in relation to *hevea* trees and the processes that supported and threatened them. The introduction of plantation agriculture also encouraged colonial officials to use their reach to start new forms of governing people on the frontier, including the Montagnards, who later played a key role in the creation of plantations.

Montagnards to Govern

The term "Montagnards" refers to a diverse assortment of people, often grouped by anthropologists under the term Austro-Asiatic, who live in the Indochinese uplands. Some scholars argue that the Stieng, Phnong, and others in this region had moved down the Annamese Cordillera, or Trường Sơn Mountain Range, while some postulate an oceanic origin. If these groups came from the north, then they also most likely brought malaria *falciparum* with them on their journey. Although these groups formed highly variable relationships with the lowland Kinh, or Việt, and Khmer states before colonial conquest, the southern uplands had long been a crossroads for lowland armies, giving most groups their experience with hostile centralized states.[64] Early contact between the French and Montagnards in the uplands was often carried out through religious missionaries, explorers, and other men-on-the-spot outside of official channels. The first areas open to French contact in eastern Cochinchina were those close to the Sài Gòn, the Đồng Nai, and the Bé Rivers. Still, the area contained many unexplored pockets, and as late as 1882, the Bé River, according to an early French explorer, remained unknown.[65]

The mobility of the Montagnards complicated their integration into a concessionary framework of land ownership. Many conflicts between plantation land claims and Montagnard land use arose because of the incompatibility of the two systems. At the time, the Montagnards still had room to maneuver, and the settling of the international boundary between Cambodia and Cochinchina in the early 1900s did not stop groups from finding ways to cross this frontier. As the anthropologist Ian Baird suggests in the context of the border between Laos and Cambodia, groups such as the Mon-Khmer speaking Brao still contest and at times subvert state functions of the border.[66] The French colonial government viewed the problem of the Montagnards as more political than economic, and implemented few incentives to introduce agricultural and land reform.

In January 1903, Ernest Outrey, the administrator of Thủ Dầu Một province, wrote of the need to explore the surrounding region. The same year, the
French established a military post at Hớn Quản.[67] Colonial geographers and
explorers such as Auguste Pavie had already defined the broad outlines of the
topography, peoples, and the borders of Indochina in various publications
such as *Excursions et Reconnaissances*, a journal running from 1879 to 1890. Yet,
the area where Biên Hòa, Thủ Dầu Một, Haut-Donnai (in Annam), Cambodia, and Laos met resembled many other internal frontiers and remained a
lacuna in the administration's knowledge. The French administration was
generally content to leave these remote spots alone until tighter integration
made better knowledge urgent, and from the newly opened Hớn Quản post
the colonial government press-ganged the "moï" into building roads and at-
tending schools. This region also neighbored the territory of the "moïs indépendent" or "moïs insoumis," whose inhabitants refused to recognize French
authority and supposedly made their living through piracy and slave raids
on the "Annamite" villages and *srocs* (Khmer word for "settlement") near the
border with Cambodia.[68]

A largely forgotten exploratory mission represented the type of attempt
that was made by the French to engage with the "moïs indépendent." In a letter to
the Goucoch in March 1903, Outrey outlined his plans for a mission of renewed diplomacy. He suggested that the same man who had explored the region almost twenty years earlier would be ideal. When the Goucoch rejected
this choice, Outrey suggested the inspector of agriculture for Cochinchina
with the assistance of a forest guard. The Goucoch chose instead an explorer,
Paul Patté, following the recommendations of his acquaintances. Patté viewed
his mission as a human intervention, one of developing knowledge about
people, writing that even as knowing the land was important, his survey mission was one of "political and not just topographical exploration. The goal
was to know man, the terrain was of secondary importance."[69]

Patté's mission was an utter disaster. Among other setbacks, his inter-
preter died under mysterious circumstances, and reports noted an increase in
"turmoil" among the independent villages, which was attributed to Patté's
handling of affairs. In a damning letter from 1905, Outrey concluded that the
undertaking was "run with such inexperience and by a mission head so little
prepared and so little concerned about accuracy, that it could do nothing to
increase our influence."[70] Yet, Patté's failure was not simply the result of his
poor leadership skills. Many villages along the right bank of the Bé River had
recently been decimated by a smallpox epidemic, and an outbreak of rinderpest had eliminated many of the area's cattle. Long-standing social tensions

predating Patté's arrival also worked against him. The "Srai Kaman" (or pirates as the French called them) saw the European presence as threatening their livelihood of banditry and slave raiding. Patté's mission, which was small and carried explicit instructions to act peacefully, was not the first to be denied access to the region; a previous mission had been rebuffed twenty years ago by these same villages.

Outrey, among others, placed the blame for unrest primarily on Patté, but resistance to French rule resulted from various factors, according to an official report from mid-1904. One was contact with Patté's mission, but the second was proximity to the hill Jumbra (also called Yumbra or Núi Bà Rá). Resistance to state rule was a well-established reality, and this report mentioned the role of specific leaders such as Chhrap, Loït, Ngeurt (of Buton), and le tong Kouit (of Phuoc Lê), along with the "Penongs" east of Kratié. Outrey used the results of Patté's mission to proclaim that the era of missions was over, arguing that French influence "would be more surely extended through regular tours made by functionaries assigned to outposts and in constant contact with them [locals], than by costly missions run by explorers who were more worried about their fame than about the practical results that they should reach." In other words, regularized bureaucratic contact would now be advanced as colonial policy vis-à-vis the Montagnards.[71]

The era of exploration came to an end in the 1900s, even as certain villages continued to successfully resist French incursion. In a 1909 report on Kinh penetration of the "moï" region near Hớn Quản and Bù Đốp, the Colonial Council noted the problems caused by the "moïs indépendants" from the border, and a 1936 article notes little "pacification" work done between 1911 and 1925.[72] In August 1914, the feared Phnong chief Pa-Trang-Loeng (Pou Tran Lung), who had carried out a series of devastating attacks on French posts near Kratié, killed Henri Maître (1883–1914), a famous French explorer who only a few years earlier had published a definitive work on the peoples of the south-central highlands. The French colonial state did not renew its efforts to extend government rule to the countryside until the second rubber boom in the mid-1920s. Around that time, the Goucoch wrote that the 30,000–35,000 Montagnards living in the far reaches of the provinces of Bà Rịa, Biên Hòa, and Thủ Dầu Một in Cochinchina were settled on the land, working on plantations, and "in contact with our civilization."[73] Colonial administrators described the Montagnards as "calm and docile," and the Colonial Council reported that the "moïs" of Hớn Quản were voluntarily working on public works projects and plantations. Even the "Stiengs insoumis" were said to be coming to Bù Đốp for "civilization."[74]

Planters portrayed the Montagnards and their use of fire as an existential threat to plantations, but the Montagnards played a key role in establishing and maintaining plantations by providing necessary labor. In the nineteenth and twentieth centuries, European planters, experts, and colonial officials valued fire differently, depending on whether it was used to clear land for plantations or for swidden agriculture, reflecting the dichotomy between "indigenous" and "European" agriculture. Many colonial experts, who viewed fire as environmentally harmful, supported a crusade against swidden. One colonial agronomist held that "settled farming belonged to a higher plane than shifting cultivation or hunting and gathering." Rather than view swidden as a viable agricultural adaptation to low population density, these experts often blamed it as the cause for diminished numbers of Montagnard settlements. Many Europeans also viewed swidden as harmful for the soil, and, largely due to concerns about Cochinchina's forests, the colonial government outlawed the practice in 1894.[75] Yet European agricultural experts also chastised plantations for their own destructive practices. For their part, plantation owners, who coveted swidden land, railed against nonplantation use of fire, while often employing uplanders to clear land with swidden techniques. Seeking a balance between economic development and political stability, colonial officials adopted an ambivalent position, sometimes arguing for the necessity of fire for Montagnard lifestyles and sometimes condemning it as a threat to plantations.

Most Montagnards were unaware of the laws that forbade swidden, and the practice continued. Some French colonial officials recognized swidden as a necessary source of food for the upland peoples and accepted its practice among the Montagnards. In a 1904 letter to the district head of the Cochinchinese Forest Service, the temporary head of the Indochinese Forest Service argued against the strict application of punishments for this practice, particularly the seizing of the harvest, as such punishments served only to drive the Montagnards away from the French outposts and further into inaccessible forest reserves. He suggested that "the fixing of the Moïs on the land, is certainly the solution that best meets our humanitarian ideas and our peaceful penetration." This forester even suggested an agricultural system that was used in France as a possible model for settling the Montagnards and preserving forest reserves.[76]

Resistance to swidden stiffened when European desire for land and the creation of plantations increased pressure to suppress fire. While knowledge of forest-clearing techniques made upland groups invaluable during the initial stages of plantation creation, as the cover of the *Annales de Planteurs de Caoutchouc de l'Indochine (APCI)* suggests, many planters and officials viewed swidden as a threat. In 1911, a planter discussed the dangers of the fires that "moïs" used to follow their "lazy" way of agriculture. He admitted that these fires helped clear *tranh* and provided range for large mammals to roam and to be hunted, but argued that Montagnard fires endangered the plantations. A similar situation for the central highlands was noted in a 1932 report from the Annam Chamber of Commerce and Agriculture. The report's author was troubled by the fact that fires "lit in successive lines and sometimes over a distance of more than 2 km by 100s of moys [*sic*], are above all practiced in the regions where French colonization has been settled and that the other regions are little touched by the fires."[77] Ironically, environmental changes introduced by colonialism itself played a large role in starting and sustaining fires. After planters oversaw the cutting down of forests and the draining of marshes, fires began to burn hotter and faster as soils dried out, and the detritus of *tranh*, felled trees, and monocultures of *hevea* all provided excellent fuel. A head forest officer argued that "the damage from burning was due as much to plantation owners and their workers . . . as to the indigenous groups."[78]

Fire was a key tool in creating plantations and during the rubber boom of the 1920s, and the question of clearing for plantations generated two schools of thought that are nicely summarized by Christophe Bonneuil. The first method, employed on the Suzannah plantation in Cochinchina, consisted of removing stumps mechanically and completely clearing the land of *tranh*. This method, called "intensive culture," required high initial inputs of nonhuman energy such as tractors or, more commonly, large herds of cattle, at times almost one buffalo per hectare of rubber. The second method, called the "Malay method," mimicked swidden and entailed cutting trees and then burning the ground without taking out the stumps. This system required intensive labor during the planting phase, as workers had to remove the remaining root systems of *tranh* with pickaxes. This technique, imported by planters such as the Belgian financier and plantation owner Adrien Hallet (1867–1925), was employed on the fertile red earth soils and allowed for a more exclusive devotion of resources to latex production. After the clearing of the land with fire, soils were exposed to the climate, and chances of erosion and laterization, or the formation of a stone-like substance, increased. Although some disparagingly compared the method of "our neighbors in the East [the British]" to the method of the "moïs," this system won out during standardization of plantation practices in French Indochina during the 1920s.[79]

Throughout the 1930s, planters continued to ignore the similarities between clearing practices and swidden, and officials often depicted swidden as irresponsible and detrimental to French economic interests. In 1931, the

Montagnard worker on the cover of *APCI* in early 1918. This cover offers a romanticized version of the Montagnards and the rubber industry and it shows how the industry valued their labor. Rather than tapping, they often carried out brush and tree clearing and performed guard duties. Source: *APCI* 1918, held in ANOM.

agronomist Yves Henry (1875–1966), who had transferred from his post in Africa to serve as head of the Department of Agriculture, Breeding, and Forestry, condemned swidden as destructive of the soil. "This system of cultivation," he wrote, "has the consequence of impoverishment of the soil by continuous cultivation and the denudation that have considerably diminished the value of this land capital." French colonization, according to Henry, had worked to save the soil quality by settling agriculturalists and reserving forests. By contrast, forestry commissioner Consigny admitted the impossibility of eliminating swidden and acknowledged the source of food that it provided for the Montagnards, even as he consistently opposed brush fires, regardless of cause.[80]

Scientific opinion about swidden did not begin to shift until the 1940s and 1950s, with authors showing either the relatively minor impact of this technique or how it was a rational approach given a sparse population and sloping agricultural land. An exhaustive study of French Indochina's forests published in 1952 held that the open forests of the south were created by climate and soils and not the human use of fire.[81] The tropical geographer Pierre Gourou came to a similar conclusion, writing that swidden could be well adapted for certain environments. Gourou did foresee problems if the highlands became more densely populated but, like Consigny, understood the dilemma in trying to eliminate the *rẫy* system. "The *ray* is, very often, the sole means of existence of mountain populations; even though the latter are not numerous it is impossible to deprive them of their principal source of food because the *rays* are ruinous to the future of Indochinese economy; such an attempt would not be humane, and it would be ridiculous because unenforceable. *Ray* can disappear only if the peoples who practice it are given means of existence by other agricultural methods. . . . Such an attitude, purely negative and of policing nature, would be intolerable."[82] This change in Gourou's thinking reflected his shift from colonial to tropical geography. In his history of the rubber industry in Việt Nam, the expert Đặng Văn Vinh went further, offering an appreciative account of the bamboo-clearing skills of the Montagnards, making a development case for these ethnic minorities in postcolonial Vietnam.[83]

Fluid Borders

Before the colonial era, the geographic border between Cambodia and Cochinchina had been fluid, and while the colonial government sought to clearly demarcate its boundaries, differences remained ill defined on the ground. Early colonial soil and climatological surveys further undermined political borders and gave visual representations of a borderless region.[84] On the Cambodian side of the political border, rubber production began relatively late in comparison with neighboring regions such as British Malaya (1890s) and Cochinchina (1910s). As in Cochinchina, almost all the rubber in Cambodia came from plantations of *hevea brasiliensis* trees. The first plantation in Cambodia was established in Kampot in 1911, but the most significant extension of plantation land happened in the middle of the 1920s during the second rubber boom, which was the result of the Stevenson plan of 1922 that restricted rubber production in the British Empire. Cambodian plantations went from processing 346 tons of rubber in 1931 to more than 13,000 tons in 1937 as trees planted in the late 1920s began to produce latex in the early 1930s. The late date of this expansion in production meant that plantations in Cambodia benefited from a framework for land, labor, and transport that had been established in Cochinchina.[85]

Although rubber was being harvested in small pockets of Cambodia, most of the production took place in the three provinces of Kompong Cham, Kompong Thom, and Kratié. For the Khmer, Kinh, and other locals, the space of northeastern Cambodia was full of meaning. For the upland peoples, a system of *chamcar* was in place long before the land became of interest to potential *hevea* growers. The Stieng and other local groups had carried on trading relationships with Chinese merchants, trading forest products for rice and salt. According to one French official, the name "Kompong Cham" had possibly originated from the area's past strategic importance as a staging area for armies from Angkor, Ayudhya (Siam), and Huế (Việt Nam). For local Khmer residents, the small wat, or temple, outside Snoul, the site of a rubber plantation, suggested worlds of significance. These meanings were mostly lost on the Europeans who moved and thought in colonial networks that connected the region to Sài Gòn and global commodities markets.[86]

The process of creating plantations required the establishment of racial and gender divisions that were new to local society. Plantation populations were large enough to affect the surrounding Cambodian society, and the large Kinh migration meant a changing ethnic composition, with a tenfold increase in absolute numbers and an increasingly strong Vietnamese community in the region. A 1904 census carried out in the district in Kratié, where Snoul is located, counted 1,131 "Annamites," or Vietnamese in a general sense, a little over 1 percent of the total population of 85,349. Estimates for the entire province of Kompong Cham in the late 1930s counted 17,000 Kinh, of which 14,000 lived on plantations, just over 3.5 percent of the 463,000 residents of the province. While the increase may not seem significant, the first percentage comes from a district that later was heavily covered in *hevea* trees, while the second percentage includes regions that did not produce latex.[87]

One difference between plantations in Cochinchina and those in Cambodia is that conflicts arising between Kinh and Khmer were blamed on the historical intrusion of Kinh into Khmer land. As a 1923 report states, "The Annamite is for the Cambodian an age-old enemy."[88] While some Khmer may have resented Vietnamese incursion, this framing of the conflict depended on a certain understanding of history. Peaceful mixing had taken place in the border region of Cambodia and Cochinchina, and the Goucoch stated in 1915 that the peaceful expansion of the Vietnamese into the Phnong lands around Kratié should be encouraged. He wrote positively of a measure that "would encourage and protect infiltration by Annamese who, in crossing the Song-Be [river] already, have imperceptibly spread into the groupings of the moïs indépendants. There is, as much from a political as from an economic viewpoint, a great benefit in encouraging this attempt at peaceful penetration."[89] Still, the plantations' demand for Vietnamese workers created resentment among the Phuong, Khmer, and others due to precolonial conflicts. Centuries earlier, slave raids on the Kinh had been carried out by the Montagnards, and literary exhortations from the sixteenth century called for the Kinh to face the Montagnards (Da Vach) in central Vietnam. Plantations often employed Montagnards as overseers and guards, and though plantation workers I talked to recalled harmonious relations between Montagnards and Kinh, this may have more to do with the Communist Party's project of promoting interethnic solidarity than with historical conditions.

On the Cochinchina side of the border, rubber growers and the government renewed their penetration efforts during the 1920s. After roads such as RC 14, which extended to Lộc Ninh on the border with Cambodia by the early 1920s, had been built, the government commissioned maps that showed the detailed topography of contact. Official French presence was extended with a military outpost established on Núi Bà Rá. Although the Montagnards were treated administratively at a local level in the same way as the Kinh as late as the 1910s, the incursion of industrial agriculture and the resulting violence among the Montagnards, Kinh, and French caused the colonial government to adopt special policies vis-à-vis the upland regions over which it exercised some control.[90] The Montagnards were given preferential treatment via taxes; they were responsible for corvée labor while the Kinh often paid their taxes in piaster. More significantly, Montagnard lands in this region, like forests, were given the status of reservations, where no concessions would be granted.

The government continued to exercise caution, and individual colonial officials were reluctant to grant concessions in regions of continued trouble. One of the most important changes in concessionary laws was a reduction in the size of concessions and a restriction on free land. Until 1927, the GGI had the right to grant concessions of up to 15,000 hectares that came free or at a nominal price for private capital. In 1928, a law reduced the maximum size of the GGI's grants to 4,000 hectares and eliminated the possibility of free concessions for enterprises by regulating the actions of local land commissions that controlled smaller concessions. This law stipulated that concessions under 300 hectares, which were available only to individual planters, could be sold or given away for free while those over 300 hectares, for either individuals or enterprises, had to be sold. The law also required that land sales be conducted through public auction and be published in the official journal, with private agreements granted on a case-by-case basis only. These new articles meant that cheap concessions, granted out of the public's eye, were to become, in theory, the exception and not the rule.[91]

The 1928 law also emphasized the rights of those who had been farming the land before the plantations, a guarantee that was especially relevant to the land claims of the Phnong, Stieng, and other groups in the region. In the reserves of the "moïs" and the "muong," land concessions were to be granted as "baux," or as leases instead of sales. This law also included restrictions controlling what happened on concessions. In 1924, for example, the Société des Plantations de Kratié requested 6,000 hectares, the maximum allowed for such claims, of prime *hevea* growing land in Cambodia. Based on administrative calculations, the 200 families living in the area, who practiced swidden agriculture on chamcar plots, needed approximately 2,400 hectares to complete their six-year agricultural cycles. The administration could grant only 4,000 hectares of the 6,000 hectares originally requested by the Société des Plantations de Kratié. Article 17 of the 1927 Terms and Conditions granting land ownership to the Société stipulated that residents who had been growing crops on the concession land before the signing of the contract could not be evicted unless they were given land elsewhere. While the Société continued to hold out hopes for claiming at least part of the 2,000 hectares, the land law of 1928 and a 100,000 piaster loan granted by the administration to help the Société through the rubber crisis of the early 1930s effectively ended such ambitions.[92]

For the most part, these reforms did not change plantation practices, and large enterprises came to dominate the industry, especially after the economic crisis of the 1930s. In response to a concession request made by

Michelin near Phú Riềng in 1927 in a region that was not yet opened for colonization, the Goucoch wrote to the director of Michelin warning about possible trouble with local peoples ("moï insoumis") and disavowed responsibility for what happened if Michelin decided to construct a plantation there. Despite this warning, Michelin proceeded to set up the infamous Phú Riềng plantation, later the setting for important labor and anticolonial struggles.[93]

During the early 1930s, the colonial government undertook efforts to normalize political relations and reaffirm borders, sending Auguste Morère to Phú Riềng to deal with tensions among the Montagnards, the colonial government, and plantations. During the three years that Morère, with his wife and young son, lived in the region, he faced questions about both the development of rubber plantations and simmering tensions between the French and the Stieng. Morère seemed to genuinely care for the people placed under his charge, and as an administrator-ethnologist, he published a brief French-Stieng dictionary. Violence between the Montagnards and the French continued to erupt, however, and Morère's part in the colonial dialogue was "interrupted" in 1932 when he was killed by a group of Stieng.[94] His death was widely reported and, in the climate of communist fear of the 1930s, it was rumored that some "Annamites" had been in the region inciting the "moïs insoumis" to kill Morère. It is not clear whether anticolonial activists had indeed been in the region, but Morère's death points to the tensions created by continuing colonial incursion, whether by the Kinh or the French, into the uplands. Morère's death did not halt colonial administration of the highlands, and in January 1936, the Goucoch, the *résident supérieur* of upper Donnaï (Đồng Nai), and the chief of Thủ Dầu Một met in Đà Lạt for the first Interprovincial Conference for Moï Regions.[95]

Plantations could trouble gender boundaries as well. While most plantations were directed by French men, French women, too, directed them. The most famous of these women was Janie Bertin Rivière de la Souchère, known as the princess of rubber, who received the Légion d'honneur and the cross of the Officier du Mérite agricole, among many other awards. In her role as plantation manager, she was responsible for the health of around one hundred Montagnard laborers. She went on to serve as the first and perhaps only "female labor inspector" (the term employed at the time), as well as the vice president and then honorary president of the SPCI. The lives of such women made for useful colonial propaganda, and even though de la Souchère was one of the only women members of the SPCI, the union and colonial officials played up her role in the rubber industry. In the 1922 text published for the Légion d'honneur, for example, the French minister of colonies wrote: "Her work is not only an agricultural work, it is also a truly political work, because if she knew how to overcome the natural difficulties, of a virgin soil, she also knew how to attach the natives to herself so that she is the 'Good Protector' of the Long Than [*sic*] region and through her the natives love France."[96]

IMPERIAL HOPES FOR INDUSTRIAL agriculture encouraged a reordering of nature and society in eastern Cochinchina and Cambodia. More intense management of the forests, the creation of a concessionary framework, and the building of roads to take advantage of global markets helped create the environmental conditions and interconnections necessary for plantations. Transformed ways of conceiving of nature helped the colonial state retain ideological control over environments. New animal, plant, and human landscapes evolved to replace the open and primary growth forests cleared for plantations. These landscapes included roads; herds of cattle; monocultures of *hevea*, coffee, and coconuts; clean weeding; and a steady introduction of people, with material to feed, clothe, and house them, all brought together to produce latex. Like the coffee plantations in the central highlands, both *hevea* and its planters attempted to put down material and symbolic roots in Indochina.[97] For others, the introduction of industrial agriculture caused deep anxieties, and the process of becoming "native" for rubber was contested and incomplete even in 1954. In part because the introduction of rubber technology was associated with colonialism, this transfer was never neutral for those involved.

Previous successes of scientific knowledge in controlling nature promoted a blind faith among some administrators in the ability of scientists to provide solutions to problems involved in colonial expansion. Yet, the colonial state was not omnipotent, and "challenging both colonialists' self-representation and the anticolonialist historiography that overestimated the success of European control over African lives," Christophe Bonneuil notes, "many historians have recently underlined the epistemic and political weakness of the colonial state." The next chapter reconsiders the relationship between the state and its experts and the capacity of the colonial state to refashion society and nature.[98]

Part II
Forests without Birds

CHAPTER TWO

Cultivating Science

Even here, in Indochina, the necessity to bring to the natives the aid of European science for regulating and increasing yield has imposed itself for a long time on the mind of those who have been interested in the development of an essentially agricultural country.

—Maspéro, 1918

The problem of science vis-à-vis our compatriots is really new and infinitely hard. Is it possibly because of this difficulty that the popularization of science in our region has been very slow?

—*Khoa Học Phổ Thông*, 1934

The introduction of new cultures following European methods is wished for to show the population the French agricultural sciences. Up to now, one cultivates the earth according to the principles left by the ancestors, without searching for how and why. Thus, one cultivates in order not to die of hunger.

—Nguyên Tân Biên, Commission Guernut, 1930s

On a Sunday in the spring of 1933, a group of agricultural engineers traveled to the gray earth soils of southern Indochina. Among other places, the engineers visited the Ông Yêm research station experimental field, where their guides showed them the agricultural history of the past four decades. This experimental field was one of the earliest in Indochina, and occupied a band of homogenous gray earth that sloped down from RC 13 to a small stream. Halfway down the slope, water emerged from a layer of impermeable clay that allowed for irrigation. As the participants walked across the field, their guides recounted the history of the *mise en valeur*, or economic development, projects that had taken place there. On the oldest part of the station, investigations had been made into various crops, including jute, manioc, peanuts, potatoes, and pepper, at a time when *hevea*, tea, and coffee did not yet dominate agricultural exports. As they walked, the engineers saw magnificent *ficus elastica* trees covered in tapping scars, which served as physical reminders of early studies on rubber. In the forests surrounding the station, *Manihot gloziovii*, *Manihot piauhyensis*, *Palaquium obovatum*, and *Castilloa* vines stood silent witness to past research into rubber production.[1]

In 1897, a naval pharmacist sent *hevea brasiliensis* seeds to Ông Yêm from the botanical garden in Sài Gòn. These seeds were planted near the river in marshy ground due to its apparent similarities to the Amazonian environment, but the saplings were soon moved to higher ground. They remained there for the next few decades to serve as test sites for tapping techniques and to provide training for would-be tappers. The Practical School of Agriculture (l'Ecole pratique d'agriculture [EPA]), later the farm-school of Bên Cát, was established on the opposite side of RC 13 in 1916. Students at Bên Cát worked in the morning and studied in the afternoon in preparation for plantation employment and an important role in rural life and the modern economy. The engineers' voyage over the gray soils of Ông Yêm, Bên Cát, and the rubber plantations highlights the importance of botanical gardens and research stations, which were key tools for colonizing, and their history opens onto the successes and failures of the French empire in Indochina.[2]

This chapter discusses how individuals, organizations, companies, and states established and spread the intellectual infrastructure and institutions needed to produce rubber. It builds on frameworks that have been articulated by historians and anthropologists to analyze the movement of techno-science in terms of global networks, markets, exchanges, and assemblages. A close examination of select episodes during the formation of rubber science illuminates the ways in which the interests of government officials, scientists, and planters both overlapped and clashed. Institutions in France were largely unable to centralize this science, thus demonstrating the need to qualify estimations of the strength of colony-metropole techno-scientific links in the French empire. Instead, rubber depended on trans-colonial connections and the growth of a regional identity centered on agricultural production. The efforts of individual planters and joint-stock companies were vital to the manufacture of rubber, highlighting the importance of private initiative in the generation of scientific and technical knowledge.[3]

Within the colonial system, however, the production and circulation of rubber science restricted the ways that Vietnamese could participate in the industry. French education teetered between the goals of altering "native mentality" and increasing agricultural exports through *mise en valeur*. This education system also trained agricultural agents to serve as mediators between the colonial government and the rural population. The French shunted Vietnamese, Cambodians, and other colonized peoples into positions as technical assistants and extension agents. While these agents popularized agricultural science, including traditional techniques and concepts such as grafting, the colonial state did not provide any meaningful support to Vietnamese rubber growers. Despite the official rhetoric of the late 1930s touting the extension of rubber, this technology never spread deeply in the countryside.

Production

The early period in the generation and movement of rubber knowledge in Indochina, which spanned from the late 1890s through 1910, was a time of great hope, and hype, as high prices garnered by rubber planters in Southeast Asia during the first rubber boom of the 1900s encouraged predictions of fantastic futures for the material. Government officials played key roles as financial boosters, as individual planters were ill equipped to deal with the challenges of growing rubber, in particular the five- to seven-year waiting period between planting and production. The French minister of colonies claimed that "the culture of rubber was of great interest to most our overseas colonies."[4] Agricultural journals in both the metropole and the colony were full of companies advertising the sale of the seeds of latex-producing plants. This period was also marked by a lack of coherent policy that eventually left the development of rubber dependent on individual planter initiative.[5]

The government-funded botanical research that did take place involved exploration, experimentation, and information gathering, as officials sought to generate an inventory of latex-producing plants throughout the French empire. At the time, *hevea* was just one among many plants being considered for cultivation in Indochina, and botanical gardens helped planters to consider these alternatives. Over a year before the naval pharmacist sent his samples to Sài Gòn, Léon Pierre Joseph Josselme, a professor and powerful planter living in Cochinchina, wrote to the GGI to request information and samples of *Manihot Gloziovii*, or *Céara* rubber. By the end of 1897, he had received his seeds and answers to his questionnaire from the head of the arboretum in Libreville, French Congo. According to the geographer Charles Robequain, the late-developing rubber plantations in Indochina also "benefited continually from the experiments of the large neighboring producers in British Malaya and Netherlands Indies."[6] In this way, botanical gardens across the region served as nodes in the networks of knowledge transfer, as rubber research necessitated communication of ideas and personal travel between the colonies of the different imperial powers. In the early 1900s, Edmond Carle, an agricultural engineer based in Sài Gòn, went to Singapore to buy 1,000 *ficus elastica* cuttings for planters in Tonkin. On his trip, Carle visited Henry Nicholas Ridley (1855–1956), the famous director of the Singapore botanical garden, and used the opportunity to learn about the *ficus* and *hevea* plantations on the

Malaysian peninsula. In his report, Carle discussed growing and tapping techniques, ranging from how to choose appropriate land to issues of plant disease and labor relations.[7]

The failure to establish plantation production of rubber in Tonkin and northern Annam suggests the limits of government agency. In 1900, trade statistics from Marseille listed Tonkin as exporting close to eighty-nine tons of forest-harvested rubber, piquing the interest of northern officials.[8] The resident superior of Tonkin (RST) sent out a letter to the provincial heads introducing a new method of extracting rubber from the skin of vines that was recommended by the colonial garden in Nogent-sur-Marne, France. In his letter, the RST also forwarded detailed instructions for the collection of all rubber-producing plants in the region, which he offered to subsidize using Tonkin's budget.[9] The prospects for the plantation growth of *ficus* were particularly hopeful. Like *hevea*, *ficus* is a tree that can be tapped with "V" incisions channeling the rubber into cups. Although not all government officials agreed, the head of the Service of Agriculture in Annam argued that these trees had good latex yield and economic potential.[10]

By 1905, early tests of *ficus* had yielded few promising results, but the metropolitan government continued to push for the development of rubber in northern Indochina. The minister of colonies, who wrote to the GGI late in the summer of 1906, acknowledged the "mediocre results" of *ficus* tests reported in the *Bulletin économique de l'Indochine* (*BEI*) but included in his letter further instructions for the propagation of the plant by cuttings. The head of the botanical garden in Hà Nội mentioned in his comments on the minister of colonies' letter that the suggested methods were the same as those employed in Indochina for years, which had provided meager results. The head suggested that those tests used *ficus* from Indochina and that later tests on *ficus* from Singapore and the Buitenzorg research station on Java might yield more promising outcomes. This suggestion was followed by continued requests from planters interested in *ficus* and by analysis of samples, reports, and instructions on planting issuing from Nogent-Sur-Marne and elsewhere in the imperial circuits of knowledge. Despite these efforts, plantation production never rivaled the level of rubber collection achieved earlier.[11]

Although government officials were willing to send experts on study missions and maintain a network of botanical gardens, they balked at direct financial investment in rubber. The colonial government did not want to appear to favor any single group of planters, and officials were reluctant to be seen funding knowledge production that was tied too closely to one plant. In 1904–5, one planter wrote a series of letters to the head of the Department of

Agriculture, Forests, and Commerce (AFC) about the possibility of obtaining samples of *ficus* from a well-known Chinese seller in Malacca. The official agreed to the planter's request and proposed to defray the costs of transport using government funds. The RST refused, however, arguing that "despite all of the interest that the introduction of a new culture presents for the colony, it is impossible for the local budget to contribute."[12]

Growing emphasis on laboratory studies at the beginning of the twentieth century encouraged centralized control of knowledge about rubber.[13] Metropolitan scientists sought to become the hub for the rubber sciences as researchers in France envisioned themselves at the center of a classic Basallian model of center and periphery. In this relationship, the metropole would receive raw data, in the form of answers to questionnaires and plant samples fixed in alcohol, and return processed knowledge contained in advances in growing and processing techniques.[14] In this spirit, the minister of colonies announced in 1910 the creation of a special office for the study of rubber whose primary functions were to collect and store knowledge related to, and direct experiments analyzing, the growth of rubber plants, the harvesting of latex, and the industrial manufacture of products. This office, which had established models in London and Amsterdam, was intended to be a focal point for relations with other countries involved in the rubber industry and a reference point for those in the French empire interested in cultivating latex-producing plants.[15]

Even when laboratories remained "in the field," they faced significant hurdles to knowledge creation. At the Nha Trang branch of the Pasteur Institute, Alexandre Yersin, famous for his research on plague, grew *hevea* seeds he had received from Sài Gòn. Yet data that were gathered under the conditions of sandy coastal soils with frequent exposure to high winds, which bent growing trees, yielded few insights into the gray and red soils elsewhere in Indochina. Likewise, seeds sent to the agricultural station in Ông Yêm yielded little insight, and a 1908 report from the Agriculture Service in Cochinchina mentioned problems with record keeping. Notwithstanding interest expressed by the colonial government in the development of local research, little financial support for these early experiments meant that French planters continued to send abroad for information.[16]

Yet, metropolitan France was attempting to take control of the rubber industry without adequate resources. In 1910, the same year that the rubber bureau was formed, the minister of colonies sent out instructions for the collection and preparation of latex-related plant samples from throughout its empire. These instructions took up five pages, and read more like botanic

collecting forms than political instructions. In describing the method for preparing samples, the form instructed administrators that "a complete botanic
sample should include: root fragments, stems, branches, leaves, flowers, fruit,
and seeds" and that these "collected samples should be accompanied by a
card indicating the scientific and indigenous name, if possible."[17] Local colonial administrators balked at the amount, the accuracy, and the precision of
information required by the minister. The head of the local Service of Agriculture and Commerce in Cambodia stated that due to a lack of funds, it was
impossible to satisfy the Colonial Bureau's "very complicated" demands. The
minister's request to collect samples from all latex-producing plants in the
protectorate would have required "an agent sufficiently robust" to organize an
"exploration mission" to areas in Cambodia that were still "unhealthy, almost
deserted, and even dangerous to cross."[18]

The fate of the *Agricultural Bulletin of the Indochinese Scientific Institute* (*Bulletin agricole de l'institut scientifique de l'Indochine*) further demonstrates the
lack of colonial government funding for agricultural research. Founded in 1919,
the *Agricultural Bulletin* was designed to report the results of the newly formed
Indochinese Scientific Institute (Institut scientifique de l'Indochine). Led by a
French expert on tropical agriculture, Auguste Chevalier (1873–1956), its research largely focused on agricultural activities. From the beginning of its existence, officials discussed whether to fold the *Agricultural Bulletin* into the
BEI, Indochina's premier economic publication. Georges Vernet, the first editor of the *Agricultural Bulletin*, argued that the journal should remain separate
from the *BEI*, since the *Agricultural Bulletin* served as a focal point for scientific contact with France and other countries. He also said that the planters of
Cochinchina wanted a monthly publication printed in Sài Gòn to keep up to
date on scientific advances. The *Agricultural Bulletin*, unlike the *BEI*, offered
space for "experimental sciences" as opposed to applied sciences. These arguments did not sway the government, however, and it decided to fold the
specialized publication into the *BEI* in 1923.[19] Vietnamese language publications, both official and private, that dealt with agricultural issues were equally
unstable. These journals, which included the bilingual *Cochinchine Agricole*
(1927–30) and the monolingual *Defending Agriculture*, or *Vệ Nông Báo* (1923–33),
tended to be short-lived. Private efforts in Vietnamese could sometimes be
sustained, but their existence was threatened if their directors came under
suspicion of trying to subvert the colonial system. Even *Nông Cổ Mín Đàm*,
sponsored in part by European planters, was placed under investigation, and
a major contributor, Gilbert Trần Chánh Chiếu, accused of using the journal
to advance political causes.[20]

Parisian researchers did not contribute to rubber science until the signing of the 1934 International Rubber Regulation Agreement. That year, the major rubber producers met in London in yet another attempt to buffer the large fluctuations in world prices. The London agreement put many restrictions on rubber-producing countries, including caps on yield, area of land cleared and replanted, and percentage of grafted trees (some French officials used the agreement to argue that no room was left for Vietnamese rubber planters). To enforce these agreements, and assess fines, the signatories established the International Rubber Regulatory Committee (CIRC). Each country, in turn, was responsible for setting up a representative organization to communicate with the CIRC and ensure compliance with the terms of the agreement. In 1935, the Rubber Planters' Union in Indochina (UPCI), the metropolitan representative of Indochinese planters, founded the French Rubber Institute (IFC) to fulfill these obligations.[21] In addition to its role of ensuring treaty compliance, the IFC also contributed to the scientific work of the International Committee of Research (Comité international de la recherche), the scientific arm of the CIRC. In fact, the laboratory set up by the IFC in 1938 signaled the first time that the systematic study of rubber was undertaken in France. Slated to be housed in the organic chemistry section of the Collège de France, the IFC addressed industrial questions, such as properties of latex and the prevention of rubber degradation, during its first year of regular operation. The IFC also oversaw the formation of a center on rubber documentation, a longtime goal of many in the rubber industry.[22]

With little direct financial support from the government until the 1930s, rubber production relied on individual planter initiative and the investment of capital from within the colony itself. Even when prices rose as high as thirty francs per kilogram during the first rubber boom of the early 1900s, double the average of fifteen francs per kilogram of later times, French capital preferred to go into already established plantations, in part because Indochina possessed its own currency, the piaster, which was not pegged to the French franc until 1930.[23] The success of rubber production in Cochinchina ultimately depended on the initiative of individual planters serving as trans-colonial actors, and in 1910, planters in the south formed the APCC. The founding of the APCC, which later changed its geographic scope from Cochinchina to Indochina to include Annam and Cambodia, marked an important milestone in the creation of a rubber industry in Indochina. By 1910, both individual planters and company-owned plantations were growing rubber; the first two such plantations were Suzannah and Xa Trạch, which was started as a joint-stock company in 1907. During the 1910s, rubber plantations, mostly but not completely

European owned, began to spread slowly out from Sài Gòn, sometimes following, sometimes pushing roads and railways into the hinterland.[24]

The APCC's members included judges, lawyers, doctors, engineers, and others from high-level colonial society, and the organization often successfully influenced government policy. French, Vietnamese, and Chinese rubber-growing efforts at this point were not dissimilar, and joining the association depended on one's personal connections and social and political position. The association was technically open to Vietnamese and Chinese planters, but for reasons of race and class few such planters were admitted to the group. As a few French-owned companies gradually came to dominate production of rubber, non-European producers, seeing few social or economic reasons to try to join the association, faded into the background.[25]

With the creation of the APCC, privately funded books and journals began to carry crucial information regarding rubber. The most important source of information about rubber was the publication of the Association of Rubber Planters of Indochina (APCI), which ran from 1911 to 1943. Initially called the *Annals of the Rubber Planters of Indochina (APCI)*, the journal changed its name in 1918 to the *Indochinese Rubber Planters' Journal (BSPCI)* and published articles on a range of topics including science, law, economics, politics, and gossip, as planters hashed out issues that mattered to the industry and discussed how to put theoretical knowledge into practice. In the early years, the *BSPCI*, like the colonial government, relied heavily on information gathered from neighboring rubber-producing countries. For example, in its first year of publication, the *APCI* reproduced the complete report of Georges Vernet's government-sponsored mission to the Malaysian peninsula, as well as a study conducted by its own member, Octave Dupuy, a civil engineer, in the same region.[26] But the publishers of the *APCI/BSPCI* also recognized the need to develop a "science of rubber" specifically adapted to southern Indochina: "On the other hand, one can admit that in addition to general knowledge given by books and used in all lands, each region is liable to have information of practical interest limited to that region. The cultivation of rubber in Cochinchina and Annam, for example, will give rise, without doubt, to new theories, special studies, beneficial discoveries, especially for the planters, present and future, of this land. A body of work is thus set up, that of the science of the cultivation of rubber in Indochina; this will be a communal work, offering the best returns, if each is prepared to supply, day after day, his contribution."[27] According to the president of the APCI, knowledge was location specific, and the study of rubber in Cochinchina and Annam should give rise to new, localized theories that were useful to the community of planters engaged in the business of rubber.

A brief but heated debate over a law that regulated the importation of *hevea* seeds into Indochina further revealed tensions among planters involving the creation and dissemination of scientific knowledge. This debate also highlighted the importance of regional links and the degree to which plant material related to the rubber industry would be subject to regulation. Of concern were the fungal pathogens that threatened the plantations of the British and Dutch colonies, including *Fomes semitostus, Diplodia rapax* (*Massee*), and *Corticium javanicum*. Following the recommendations of an industry chemist and several planters, the president of the Chamber of Agriculture wrote a letter to the GGI in support of import regulations. As a result, in June 1910, the GGI banned the importation of *hevea* plants and required *hevea* seeds to undergo the same disinfection procedures used for other seeds imported from abroad.[28]

Not all planters agreed, and during the following year a debate over conflicting experimental results dragged on. Finally, in July 1911, the GGI compromised, maintaining the ban on the importation of trees while lifting the requirement that seeds must be disinfected. The GGI concluded that the costs and potential losses related to disinfection outweighed the possible risk of introducing diseases, and that the inconclusive experimental results were exploited to override the concerns of many scientists and other planters. However, the GGI refrained from arbitrating for over a year, as the profits from rubber were not yet key to balancing the colonial budget.[29]

Other diseases continued to pose a threat to rubber. The best-known *hevea* disease, South American Leaf Blight, effectively prevented all attempts to create rubber plantations in the Amazon, including the ill-fated Fordlandia. Many other diseases endangered *hevea*, and the growing research in plant pathology, or the study of plant diseases, helped colonial experts learn how to better exert control over nonhuman nature. These diseases opened the door for researchers such as Thomas Petch, who published a monograph about *hevea* biology in 1911 and a decade later expanded and rereleased one of its sections in a volume called *The Diseases and Pests of the Rubber Tree*.[30]

To combat disease, Petch suggested that managers and researchers begin by acquiring general knowledge about the environment, then zeroing in on specific plantation pathologies. When agronomists could not identify individual pathogens as causing stress on the trees, they should draw on knowledge of climates, soils, flora, and fauna, and their connections in order to reason about the maladies of monocultures. Petch advocated for prevention over remediation and argued that "the pathologist should be consulted beforehand, not five or six years afterwards when some disease has already appeared." These preventive measures involved general sanitation. For example, Petch warned against overcrowding of *hevea*, which would lead to diminished rubber yields through both stress and the easy transmission of disease. Initially, plantation practice had been to plant trees at 100–200 per acre or 500 per hectare, and Petch spoke of thinning trees to 150 per hectare or less. He also emphasized the dangers of "jungle stumps"—tree stumps that were left to rot because removal was expensive. During this process, these stumps became carriers of many of the diseases that then spread to *hevea* either through fructification, the spread of spores, or through fungus hyphae and direct root contact.[31]

Agricultural scientists such as Petch turned to their colleagues in the medical sciences for models of how to understand and treat plant diseases. The form and content of *Diseases* resemble medical texts about human ailments, and the book is arranged by the part of the tree under attack as well as by disease and its causes and effects, which were graphically described by Petch. In Petch's introductory chapter on general sanitation, he wrote of the "visitations" that characterized previous understandings of plant diseases and how experts had only recently realized "that plant diseases are as inevitable as those of men and animals." Furthermore, once diseases arose, the steps recommended often resembled coercive public health measures. If a disease was minor and did not threaten to spread to other trees, then "tree surgery" could be performed (i.e., the diseased part could be cut out and the resulting wound treated). If, however, diseases threatened to be epidemic, more drastic measures were required. The treatment for *Fomes lignosus*, a pathology earlier known as *Fomes semitostus*, was one of the targets of these drastic measures. *Fomes* was one of the most feared root diseases of Southeast Asia because it moved quickly and could spread widely before above-ground signs signaled its existence. Not only were the visibly diseased trees dug up and their roots removed, but planters also had to identify apparently healthy trees that had been infected. If the *Fomes* had spread widely enough, the infected ground was covered in lime and forked, with more lime mixed in. Some experts even recommended digging trenches up to a meter deep and covering them in lime. Though not always effective, these measures were analogous to steps taken to quarantine human communities during outbreaks of plague or cholera.[32]

Plantations also offered productive places for researchers to theorize about plant diseases. Major advances in the study of rubber pathologies were made at the beginning of the 1920s thanks to a large degree to years of experience with plantation-grown *hevea*. In Petch's 1921 edition of *Diseases*, he noted the rapid development of plant pathology, including knowledge about leaf blight. The botanist Auguste Chevalier stated that much was still unknown about *hevea's* biology, but his report on the 1921 International Exposition on Rubber and Other Tropical Products in London listed eighteen known diseases of this plant.[33] Furthermore, by the 1920s, rubber researchers were making contributions to basic science. Hugo de Vries, who worked for rubber companies in the 1910s and 1920s, continued his study of soil science and fertilizers after he left the plantation, while his son made important discoveries in genetics. In 1925, the Dutch scientist Alfred Steinmann published the next major text on rubber diseases, which joined Petch's monograph as an industry standard. Steinmann discussed diseases caused by identifiable parasites but also noted that *hevea* could suffer from environmental stressors. He mentioned in particular the harm to trees that could result from worn-out soils, their nutrients leeched by tropical rains and previous crops. Other damage to the tree could result from overexploitation of *hevea's* bark. The solutions for these diseases varied from application of pesticides such as Paris Green to the use of fertilizers to restore soil fertility. Yet, to a certain extent, diseases were viewed as unavoidable; Emile de Wildeman, a Belgian expert on rubber, argued that diseases inevitably appear on monoculture plantations.[34]

By the mid-1930s, more was known about the diseases of *hevea* but confusion in the field remained, threatening the relationship between planters and colonial experts. The British mycologist Arnold Sharples decried the inconsistencies in naming that had cropped up in the field of plant pathology, which were caused by disputes such as Petch and Van Overeem's back-and-forth over the naming of *Fomes lignosus*, also known as White-Root disease. According to Petch's 1921 publication, *Fomes lignosus* was the same disease as *Fomes semitostus*. In 1923, however, Van Overeem challenged this classification and proposed the name "*Rigidoporus microporus*, Swartz. Van O.," which would place *Fomes lignosus* in a more general class of fungus. Sharples lamented that such inconsistencies led to fruitless discussion, and "little can be gained," he claimed, "by constant disagreement over the correct systematic position and name of an organism causing a common plant disease, and it certainly leads to lack of confidence between planters and the pathological workers [*sic*] who are called upon for advice."[35]

The rubber tree was not the only plant that occupied plantations, and once they had been established, weeding became an essential daily task. Experts who advocated for clean weeding, which involved the elimination of all unwanted plants from between the rows of rubber trees, called on science to bolster their position. Some argued that the risk of fire was reduced by taking out the weeds that could later turn into combustible material. Other experts believed that superfluous vegetation could rob rubber of the "force de végétation," a vague concept that held that environmental conditions could support only so much vegetable matter. This concept is analogous to the idea of "carrying capacity," in which given environmental conditions of water, light, and space can sustain a certain amount of life. In the case of plantations, *tranh* (Lalang grass) stole nutrients from rubber, and various methods were recommended for clearing weeds, including hoeing and arsenate.[36]

The opponents of clean weeding also applied science to bolster their positions. H. N. Ridley of the Singapore botanical gardens advocated leaving some undergrowth as a means of retaining moisture and for use as green manure. The French agronomist Yves-Marie Henry argued that clean weeding resulted in the loss of humus.[37] The chemist Georges Vernet identified five types of possible intercropping (i.e., growing plants among saplings): (1) cash crops such as coffee, which produced product earlier than rubber trees; (2) cash crops such as cocoa and quinine, which produced their product at the same time as rubber; (3) plants that were meant to protect rubber against sun and wind; (4) food crops to offset the cost of labor and running the plantation; and (5) plants meant to block out weeds, to develop the soil, and to prevent erosion. To leave enough nutrients for rubber, Vernet recommended spacing of five to six meters and no fertilizer for new plantations on rich soil. When nutrients were lacking, planters could leave the products of forest clearing on the soil to decompose slowly, but this method had the drawback of increasing the possibility of diseases. For poorer soils, or for other situations calling for fertilizer, Vernet suggested that each individual case be considered. He quoted his professor, who wrote: "It is necessary to ask of the plants if they are happy to eat the fertilizer."[38]

The desire for legibility—the act of simplifying to effect intervention—extended beyond state projects to businesses and Christophe Bonneuil has shown how aesthetics dominated discussions of clean weeding on rubber plantations.[39] A 1933 report about the Canque plantation illustrates the slippage between science and morality when discussing well-weeded spaces, as even agricultural scientists themselves praised planters for keeping a neat appearance, regardless of the cost or benefit for latex production. R. Caty, an alumnus of the agricultural school in Grignon and head of the genetic laboratory of the Indochinese Bureau of Rice at the time, wrote that Canque was chosen for its visually perfect presentation, including its clean weeding, which gave the observer the impression that the owner cared for his operation. While Canque's old tree stock gave only modest yields of 507 kilograms per hectare, Caty emphasized that even though deep in the throes of a world economic crisis, good maintenance and appearance were necessary. Thus, planters turned to experts for advice regarding *hevea* disease, and they ignored it regarding clean weeding.[40]

In the British colonial empire, Arnold Sharples argued for a balanced approach to clean weeding. On some types of grounds, including flat and swampy grounds, this system could be useful, Sharples stated, but on undulating inland hills, clean weeding should be avoided due to erosion problems. His take on clean weeding fit into his overall approach to the application of the principles of "forestry methods" to plantations. These methods, Sharples argued, presented three main advantages: (1) shading soil for temperature reasons; (2) preventing soil erosion; and (3) providing humus. Sharples argued that only "natural" undergrowth was to be allowed, as other types of undergrowth could lead to increased diseases, and noted the tension between scientific methods and the nonscientific techniques commonly used among planters.[41]

Animals too played a key role in plantation practices, and they were both victims of and crucial contributors to plantation landscapes. Indochina is rich in wildlife, with a mix of fauna and flora that resulted from the immigration of organisms from adjacent areas. In 1928–29, the Kelley-Roosevelt expedition of Chicago's Field Museum, headed by Harold Coolidge (1904–85), later president of the International Union for the Conservation of Nature, traveled to northwest Indochina because of its rich animal and bird life. Plantation managers likewise noted different types of wildlife in their memoirs, and Raoul Chollet, the manager of the Snoul plantation in Cambodia, repeatedly told stories of animal encounters.[42] The populations of wild animals decreased during the colonial era, and while hunting in eastern Cochinchina killed some tigers, panthers, buffalo, boar, and elephants, the biggest factor in their disappearance from the region was the expansion of industrial agriculture. Early in the colonial period, the Goucoch offered a reward for the killing of big cats, as they were considered "pests" for agriculture. Almost all the money was rewarded to those in the provinces of eastern Cochinchina, and when this item was cut from the colonial budget, the Planters' Syndicate fought hard to have the reward reinstated.[43] Planters also targeted elephants, as herds could trample and eat newly planted rubber trees. The opening of remote regions to European settlement lessened this "threat." As a 1910 monograph of Thủ Dầu Một notes, large mammals had disappeared with the introduction of roads, and by the 1920s, the colonial government instituted rules limiting hunting that had originated in French West Africa.[44]

Rubber affected domesticated animals as well, as planters turned to animal husbandry for both food and power.[45] Herds of cattle helped to reduce plantation dependence on outside sources of meat, to clear and plow, to transport material, and to fertilize the land. Colonial experts encouraged plantations to raise cattle, with Suzannah and Anloc having a combined 2,200 cattle for 2,700 hectares planted in 1920. These cattle were also thought to be useful for protecting workers from malaria by serving as an alternate source of blood for mosquitoes.[46] Yet herds were difficult to maintain, as epidemic diseases ravaged the ungulates of the region and placed immense pressure on the biodiversity of surrounding lands.[47] When rinderpest decimated plantation herds in 1922, 1923, and again in 1929, plantation owners began to value veterinary medicine, which had appeared in the colony in 1911.[48] Machines were briefly considered as potential replacements for cattle, but they were implemented relatively late, and were often drawn by horses well into the 1930s. Although tractors did appear on the plantations, a 1912 poster selling the "Imperaton" still shows a metal machine being drawn by horses, with the driver sitting on the machine.[49]

Circulation

Although little metropolitan capital was invested in rubber before the 1920s, the situation changed radically after 1922, when the British-led Stevenson plan sought to limit global rubber production. This plan led to the second rubber boom and signaled the coming of age of the industry in Indochina. The land rush, which was particularly intense in the 1920s, focused on the *terres rouges*, and French capital was feverish about the prospect of massive windfalls. An initial step for the colonial boosters, or those promoting the colonization of Indochina, was to define a coherent object of investment. Maps from the time provide examples of how rubber planting became synonymous with *terres rouges*. Even though numerous rubber growers owned land outside the *terres rouges* zone, most rubber production maps show only the large plantations in this region. To attract investment, boosters had to convince capital in France and elsewhere that the land would repay investment and that officials would employ concessionary regimes to construct a specific political and economic approach across the provinces of Bà Rịa, Biên Hòa, and Thủ Dầu Một. Two characteristics defined the region "scientifically": the soil and the climate.[50]

A healthy climate, for both people and plants, was key to selling land. A directory of planters produced for a 1913 rubber exposition extolled the healthy climate of Cochinchina, noting that unlike earlier times, colonial officials and planters could now spend their whole lives in the colonies and remain healthy.[51] More importantly for investment, the climate of Cochinchina was represented as beneficial for *hevea*. Cochinchina's marked dry season, which was nonexistent in the Amazon, was a potential drawback, an impression that was still cited in the 1970s as a drawback to the rubber industry in Indochina.[52] French colonial boosters fought back. In the 1913 issue of the planters' journal *Annales*, which was produced for an international rubber exposition, boosters argued that experience had shown that the extended dry season slowed the growth of parasites, leaving monocultures of *hevea* in the region relatively free of the diseases facing plantations elsewhere.[53] Rubber plantations in Indochina did in fact confront common diseases, and Colin Barlow speculated in the 1970s that long boat journeys, rather than climate, had protected Southeast Asia from leaf blight caused by *Dothidela ulei* that had stymied all attempts, including Henry Ford's, to create plantations in South America.[54]

Some rubber experts downplayed the importance of minor differences in climate and argued that soil was the most important factor in determining good rubber growth. The focus of the concessionary rush was *terres gris* and *terres rouges*, terms which were as much political, economic, and cultural constructions as they were classifications of the physical and chemical properties of soils. These characteristics, however, played an important role in promoting *terres gris* and *terres rouges* as valuable commodities, encouraging investment, and managing the land. The first region, *terres gris*, was located for the most part close to Sài Gòn. These soils were sedimentary in origin and contained a high percentage of sand. This sand content meant that these soils drained well and were less exposed to malaria, as the habitats favored by *Anopheles* mosquitoes, carriers of malaria plasmodia, were less common. With a high silicate content, these soils also had a lower cation-exchange capacity (i.e., the ability to store positively charged nutrients such as potassium, phosphorus, etc.) than the *terres rouges* and were therefore more easily exhausted by the growth of plants with a high nutrient demand, such as tobacco. Often these soils were used for wet-rice agriculture, which recycled many of the essential micro- and macronutrients.[55]

As late as the 1920s, much of the lands called *terres rouges* were poorly charted and had few roads, making it impossible to draw detailed soil maps of the region. To engender trust in the region, promotors of colonization turned to science and laboratory analysis. The importance of chemical and physical analysis for selling land and building trust among faraway investors is evidenced by the inclusion of a chemical analysis table in the 1913 *Annales*. This table defines a sample of *terres rouges* soil as high in clay and sand content, with a significant portion of organic material. This composition would give the soil both good drainage characteristics and an ability to hold minerals. In

— 13 —

RÉPUBLIQUE FRANÇAISE

GOUVERNEMENT DE LA COCHINCHINE

Laboratoire de Chimie

BULLETIN D'ANALYSE N° 604

Composition pour mille parties de terres brutes sèches........................ 100° cent.

Analyse physique

Gravier ferrugineux........................	7,15
Sable...................................	478,40
Argile..................................	438,19
Carbonate de chaux........................	6,16
Matières organiques........................	67,72
Humus.................................	2,38
	1.000,00

Analyse chimique

Azote..................................	1,81
Acide phosphorique........................	6,70
Potasse.................................	1,02
Chaux..................................	3,44
Magnésie................................	2,49

Saigon, le 12 août 1912.

Vu :

Le *Directeur,*   Le *Chimiste principal du Laboratoire,*
MORANGE.   BUSSY.

Soil analysis in 1913 issue of *APCI*. This soil analysis was meant in part to help planters figure out how best to fertilize rubber trees. It was also meant to sell the red and gray earth soils to prospective planters. Source: *APCI* 1913, held in ANOM.

addition, the soils were rich in minerals such as nitrogen (*azote*), phosphorus (*acide phosphorique*), potassium (*potasse*), calcium (*chaux*), and magnesium (*magnésie*). This objective, quantitative laboratory analysis was complemented by subjective, qualitative accounts of non-French travelers who testified to the soil's fertility to counter claims being made by skeptics.[56]

The increase in rubber plantations created a demand for more detailed studies of the soils of the region. The first comprehensive study of the scattered soil analyses of the *terres rouges* region was written by the agronomist Yves-Marie Henry. In a 1931 monograph, he argued that most peoples' perceptions of the soils of the *terres rouges*, the main plantation region, were either based on ignorance or influenced by financial speculation. Newly constructed roads and recently drawn maps enabled Henry to offer an understanding of soils that was both comprehensive and detailed. He directed his writing at both planters and agricultural scientists, with the dual goals of helping planters and advancing knowledge about important types of soils. Henry's study was representative, if ambitious, for its time in two respects. First, he attempted to describe soils, climates, and vegetation in their interactions. Second, he sought to ensure the profitability of plantations.[57]

Physically, according to Henry, the *terres rouges* should be more accurately called basaltic soil since it was formed from decomposing basaltic rock that had been created during a recent period of volcanic activity. Henry divided this soil into two types: soil still containing large basaltic particles (called "bombs," larger than two millimeters) and soil that derived from ash and volcanic mud. This activity had helped form the central highlands but it continued down into the undulating lands of Thủ Dầu Một, Biên Hòa, and Bà Rịa. This soil's origin and its subsequent use by the Stieng, Phnong, and other groups for swidden agriculture had left a large range of fertility. In keeping with the economic focus of the book, Henry ventured a quick guide to soil fertility. The black soils were the richest, as their color came from a high content of humus, while those with the lightest color, up to a yellow, were quite infertile and not good for commercial agriculture.[58]

Henry's study also signaled a shift to more detailed biochemical knowledge of soils involving, in addition to the mineral levels, extended discussion of concepts such as C/N (carbon to nitrogen ratio, indicating relative amounts of organic matter), NPK (nitrogen, phosphorus, and potassium, nutrients essential for plants), soil pH (a measure of soil acidity), and colloidal activity (related to the total ion exchange capacity of a soil).[59] This knowledge was vital for planters who had to retain the productivity of the soil through the use of fertilizers, both "green" (i.e., through plant decomposition and animal manure) and synthetic. Planters were warned that they would have to carry out their own detailed studies for further soil treatment. Henry's work sought to convey the physical, chemical, and biological characteristics of soils through simple to understand means, like a colored chart meant to help with quick identification of soils. Henry also divided the basaltic soils into massifs, including the massif of Haut et Bas Songbé (Sông Bé), Bas Dongnai (Đồng Nai), and Bas Mekong in Cambodia, regions particularly attractive to rubber planters.[60]

Henry did note drawbacks to these soils. Some of the area was relatively flat, but much was formed in gently rolling hills, with occasional volcanic peaks jutting up suddenly from the surrounding countryside. This landscape could suffer from erosion. In addition, some of the *terres rouges* could form laterite, a hard, brick-like layer of soil that made growing crops very difficult (this same material had been used to make monuments such as Mỹ Sơn and Angkor). Henry noted in the Lower Mekong soils the presence of an upward movement of water that had leeched iron and aluminum toward the surface to form oolites that could harden into a solid surface.[61] Heavy rainfalls and leaching had also led to a high level of soil acidity. Finally, as chapter 3 will show, red earth soils were blamed for malaria outbreaks that struck down many rubber workers. Some medical doctors pointed to the physical and chemical characteristics of the soil, while others viewed the red earth as a synecdoche for several factors that kept workers in ill health.

The soils of the Kompongcham-Mimot massif in colonial Cambodia, for example, were rich in nutrients and resistant against laterization. But with the forest cover removed, followed by the practice of clean weeding, Henry warned, these soils could rapidly lose their nutrients and humus. Theories about desiccation that linked forest clearing to changes in regional climates also influenced Henry's thinking, and he noted the drastic reduction in rainfall (up to 50 percent) in some places. Henry showed, for instance, how rainfall levels on Mimot, a plantation on the Kompongcham-Mimot massif, went from 3,256 millimeters in 1927 right after the beginning of deforestation to 1,638 millimeters in 1929 at the end of this process.[62] In the 1940s, Gourou argued against the ability of forests to affect rainfall. "In a country of violent and prolonged rains like French Indochina," he continued, "the retaining capacity of leaves, trunks, and roots is rapidly exceeded, and the rain passes through the forest as if it did not exist."[63] Henry's and Gourou's theories about desiccation differed in part because they focused on different environmental scales. Henry noted changes in rainfall at the microscale of plantations, while Gourou wrote of an Indochina-wide scale.

In addition to soils, Yves Henry wrote about agricultural practices. In 1925, he argued that "studies [on native agriculture] have rarely led to a practical plan for progressive improvement mainly because they have not been consistent and coordinated, nor always submitted to scientific discipline. This discipline alone," Henry continued, "enables a unity of doctrine, which is the source of all productive action."[64] Henry harbored a growing interest in local knowledge that could be termed "vernacular science." Yet, native agricultural knowledge continued to be assigned an inferior place among agricultural scientists. While planters and experts attempted to correct their own misconceptions about the environment, and sometimes expressed concern for native welfare, they viewed native agriculture as a threat to plantations and either ignored or denigrated these practices. This trend was only exacerbated by the further privatization of knowledge production, and it was not until the 1930s that a few human geographers begin to take seriously the techniques of indigenous farmers.[65]

Henry's entanglements with the colonial project also led to controversy. John Kleinen, for example, notes that Pierre Gourou and some subsequent scholars questioned Henry's data and conclusions, which were based in part on studies undertaken during a time of rural unrest. Henry's collusion with the colonial state after the Nghệ Tĩnh uprisings of 1930–31 further called into question his credibility as an independent expert.[66] Ironically, liberal supporters of colonial business attacked Henry from the right. In 1932, an anonymous pamphlet called *The Agricultural Scandals in Indochina* (*Les scandales de l'agriculture en Indochine*) criticized Henry in his role as head of the AFC. While the main thrust of the publication (and a related series of articles in the pro-business *Presse indochinoise*) was that Henry had appropriated the department's resources to advance his career, the charges contained in *Scandals* raised broader questions about the proper role of the AFC by calling for a decentralization of its powers.[67] The insinuations against Henry continued in the pages of *L'Eveil économique de l'Indochine*, where he was condemned as a Bolshevist who sought to create a welfare state for his role in creating agricultural banks in Indochina.[68]

Internationally, organizations such as the Parisian Association for Colonial Sciences (Association colonies sciences), established in 1925, brought environmental research done in French Indochina to the attention of scientists in the metropole. The founders of the Association included Auguste Chevalier, Albert Calmette, and other colonial technicians, administrators, and members of the colonial economy and political lobby who viewed science as useful for economic development (Girolamo Azzi was later an honorary member). The Association was partially funded by large banks and companies, and many rubber planters subscribed to this group. Its two purposes, according to the *BSPCI*, were (1) to put in contact those related to the "development of colonial soil" and (2) to be a documentation center "on the plant and animal products of our colonies." To better coordinate efforts between imperial and local governments, the Association organized conferences, published information, and formed two subcommittees, one to collect soil samples and produce maps and another to study colonial plant parasites and diseases.[69]

Making agronomy more widely available in Indochina was another part of the colonial project. Colonial experts and officials paid little attention to "native" agriculture before World War I, focusing instead on "European" agriculture, defined largely through its focus on export and "modern" methods. These experts and officials attempted to force Vietnamese, Cambodian, and other farmers to adopt their "modern" methods. The mutually exclusive categories of native and European agriculture both required detailed knowledge of local environmental and social conditions to decrease social disruption and increase yield. Vietnamese farmers employed well-adapted techniques for growing fruits, maize, and, above all, rice; yet, until the 1930s, agronomists, geographers, and other experts did not take local practices and knowledge, or what Helen Tilley terms "vernacular science," seriously. By the end of the 1930s, critics recognized the failures of French-directed agriculture and the colonial system to deliver on its promises of improvement articulated in programs such as *Pháp-Việt Đề Huê*, or Franco-Vietnamese collaboration.[70] In response, reform-minded French and Vietnamese politicians called for political reform aimed at improving health and economic conditions. In 1937, the Indochinese Democratic Party, headed by the medical doctor Nguyễn Văn Thinh, called for improved hygiene education and a more fully realized Native Medical Service (Assistance médicale indigène [AMI]). The party also discussed the importance of popularizing scientific research into agricultural products, including those for export, such as rice, rubber, and maize.[71]

Once the latex began flowing, factories, replanting, and grafting took center stage on plantations. By 1935, rubber trees that had been planted on 1,196 hectares of Snoul's land in Cambodia in the late 1920s were producing latex, which was processed into rubber in a temporary-use factory, then shipped out on newly constructed roads. This factory used tanks to mix the latex with formic acid to coagulate the colloidal suspension. The resulting ribbons of rubber were then rolled into sheets, cut into meter-long sections, processed into crepe, dried, smoked, and packed to be sent to the port of Sài Gòn, where they were shipped to markets in Singapore, France, and the United States. In 1935, ten years after the land contract had been signed, ninety-six metric tons of rubber had been manufactured. After a permanent factory was constructed in 1937, Raoul Chollet reflected somewhat wistfully that from then on "the operation is bound to be nothing more than routine work."[72] Throughout the colonial period, the planters and officials envisioned a limited, if vital, role for the Vietnamese and the other inhabitants of Indochina that involved work with the hands and not the head.

Extension

Like other imperial powers at the turn of the twentieth century, the French looked to education as a key component of their civilizing mission. Colonial

officials also viewed education as a way of carrying out *mise en valeur* by training personnel for economic undertakings such as rubber plantations.[73] A
1904 report discussing the opening of a *collège* (at that time referring to any
higher level of training) at Bên Cát near Ông Yêm, designed to create Vietnamese assistants for government employment, helps shed light on educational aims in Indochina. This *collège*, the report's author noted, would have a
physics room and a chemistry laboratory to meet the requirements of a scientific education. "It appeared useful," the author continued, "to try, by the diffusion of elementary scientific knowledge and by the usage of experimental
methods, to modify the brain and the mentality of the native in developing
within him to a greater degree his intellectual powers."[74] French authorities
began to look seriously at educational reorganization in the wake of anticolonial uprisings that took place in 1908, and were connected to schoolteachers in
Tonkin and Annam. The trauma of World War I encouraged imperial powers
to adopt an educational policy that spoke of native development as much as
of imperial gain.[75]

Discourse connecting science with an ability to improve life in Indochina
continued during the first half of the twentieth century, and according to
many observers the countryside was the area most in need of improvement.
Although a strong educational system remained a low priority for a tightfisted
colonial government, agricultural programs in both the south and the north
were opened to train Vietnamese technical agents.[76] The Farm-School of Bên
Cát was charged with two tasks: to train assistants to help run the Indochinese
agricultural economy, and to transform Vietnamese farming practices and improve native welfare.[77] The newly trained Vietnamese assistants did more than
just aid French-trained engineers; they were essential to the transfer of scientific agricultural methods to Vietnamese farmers. Both French colonial officials and the Vietnamese elite expressed a sense of trusteeship, which Tania
Li defines as "the intent which is expressed, by one source of agency, to develop the capacities of another," vis-à-vis farmers in Indochina.[78]

Attendance at Bên Cát remained steady during the early years of its existence, with about thirty students enrolled, though the school met neither the
needs of Vietnamese families nor the goals of the colonial administration.
Many elites were reluctant to send their children to schools such as Bên Cát
in part because of the perception that students were not learning agricultural
skills—that they were trained only in how to become secretaries. For their
offspring to gain success in the colonial society, they needed to be trained as
agricultural engineers in Montpellier or Paris. The children who came from
poor families were rarely prepared to attend higher levels of education, which

left only the local small property-owning class as a base for students.[79] Furthermore, in 1904 the colonial government had also established at Ông Yêm a penal agricultural colony for wayward children and orphans that was directed by an agricultural agent from the research station. It is likely that the penal colony, the research station, and the farm school became closely associated for Vietnamese living in Sài Gòn.[80]

Most French explanations for the lack of Vietnamese interest in agricultural education centered on an assumed cultural aversion to manual labor—a complaint that the French levied whenever Vietnamese did not go along with French schemes. Gail Kelly has argued that an emphasis on manual labor was meant to keep Vietnamese backward, though, as has been pointed out elsewhere, general education in France (at least from the end of the nineteenth century) also extolled the virtues of manual labor, suggesting at least some overlap between pedagogy in Indochina and France.[81] Of the fifty-nine students who graduated before 1925, only two went to work for rubber plantations.[82] In response to this failure, the school changed its name and function, becoming the Farm-School of Bên Cát. Students were no longer guaranteed jobs after graduating and had to find their own work, often with nearby rubber plantations. In 1934, the Goucoch, Pierre-André Michel Pagès (1893–1980), changed the name of the school once again to the Practical School of Agriculture and Forestry of Bên Cát. Some observers perceived improvement in the school curriculum, which included more cooperation with local enterprises.[83]

The Advanced School of Agriculture and Forestry of Indochina (Ecole supérieure d'agriculture et de sylviculture de l'Indochine [ESASI]), which opened in Hà Nội in 1918, was situated near the city's botanical garden and, like Bên Cát, was designed to train agricultural technicians to work for government services and modernize rural agricultural techniques. From 1919 to 1921, Charles Lemarié headed ESASI while serving as the director of agriculture in Tonkin. In 1925, after years of disappointing attendance levels, the school was reorganized, partly as a response to the boom in agricultural production. In the north, science was viewed as an important adjunct to economic growth, but the ESASI was not immune to outside pressure. In 1928, the school was reorganized again, but in 1935 it was closed due to strains in the colonial budget.[84] In 1938, partly in response to Vietnamese political agitation during the Popular Front period, the ESASI was reopened under a new name as the Special School of Agriculture and Forestry in Indochina. The French had recognized the need to offer more educational and career opportunities to the children of Vietnamese elite, many of who were becoming disillusioned and susceptible to subversive ideologies like communism. According to the head of the Department of Agriculture, Breeding, and Forestry, the school, which enrolled only two female students, "marks a new stage on the road of progress of the educated youth of Indochina."[85]

The ESASI did not reopen without controversy, and its goals were questioned by an alumni association of the former ESASI, which was formed in 1935 (l'Association amicale des anciens élèves de l'École supérieure d'agriculture et de sylviculture de l'Indochine). Spearheading this attack was Vũ Đình Đại, who had graduated in 1936 and in 1938 became head of the Forestry Division at Bảo Hà in Tonkin. In two articles appearing in the summer of 1938 in *L'Effort indochinois*, a left-leaning weekly journal that was written in French by Vietnamese, Đại articulated doubts held by other members of the alumni association.[86] Đại discussed the love of Vietnamese, especially the young, of study and his optimism at having a new school. But for those already working for the government, there seemed to be limitations as to how far they could advance, and the ESASI was being remade for the young without addressing the needs of those already working. With overtones of class and generational struggles, Đại's letter expressed the frustration of those who had dedicated their lives to working for the colonial government, only to find themselves stuck in dead-end jobs. He wrote:

> There is nothing more discouraging for a man of good will than to see himself condemned to remain in a situation that he is capable of improving if the Administration would give him the means to do so. While the civil servants of the other administrative services have the hope of being able to improve their lot, why is not the government opening the door to become an Indochinese engineer to the past graduates of ESASI? The government would do well to institute a competition that would permit those who are successful to attain a higher post, or to admit them to the Special School under the above conditions in order to allow them to deepen their studies and, in leaving, to create a better situation.[87]

In the spirit of cooperation promoted by the Popular Front government, Đại pointed to the closing cultural gap between Vietnamese and French. "Our rulers are not unaware," Đại continued, "that the more a native enters into western culture, the smaller is the gap that separates Annamites and French. Raise the intellectual level of the Annamese people; and there you have it, the best way to help the French and the natives, who live side by side, to understand and love each other."[88]

Interest in science was certainly not new to Indochina in the 1920s, and already by the late nineteenth century Vietnamese intellectuals were reading about and discussing the role of science and technology in the world, particularly in China and Japan. For Confucian-trained literati at the beginning of the twentieth century, whose knowledge of the ancient texts was becoming obsolete, science offered a powerful means to participate in broader intellectual and political communities. Armed with scientific knowledge, literati could join in anticolonial struggles such as those carried out by the Modernization Society, Duy Tân Hội, whose purpose "was to encourage Vietnam to follow Japan's path in adopting Western science and technology to throw off Western domination." After the French crackdown on Vietnamese-run educational institutions such as the Đông Kinh Nghĩa Thục in 1908, those who sought to strengthen Vietnamese society were pushed toward popularizing science through publishing, and in 1913 Phan Văn Trường wrote about the usefulness of Latin script in "the teaching of the masses by popularizing, through the means of works in *quốc ngữ*, useful literary and scientific knowledge." But while Trường appreciated the potentials of the script, it still needed "to be endowed with a technical and scientific vocabulary."[89]

Popularizing efforts of the 1920s included journals and pamphlets introducing Vietnamese readers to scientific ideas. The venues for this popularization ranged from periodicals addressing a nonspecialized audience, such as *Phụ Nữ Tân Văn*, to periodicals devoted exclusively to science, such as *Khoa Học Tạp Chí*. These journals also offered an inroad to the countryside for intellectuals looking to connect with their people. One of the first successful efforts to spread science to the countryside came from a publication meant for owners of small businesses, *The Voice of Friendship*, or *Hữu Thanh* (1921–24), the journal of the Association of the Industrial and Commercial Employees of Tonkin (Hội Bắc Kỳ Công Thương Đồng Nghiệp, which had about 2,000 members in 1923). The editor of the journal, Nguyễn Huy Hợi, owned a small shop dealing in bicycles and other modes of transportation. Although the journal dealt mostly with business-related topics, discussions of topics such as experimentation and education relevant to agriculture also appeared. Historian Alexander Woodside characterizes this association as "not a proletarian organization, but rather an expanded, semi-modernized mutual aid society for the Vietnamese small business and service classes." The association, as Woodside points out, served to link the countryside and the city, and thus its journal was a way of popularizing science outside the urban centers.[90]

The French viewed themselves as diffusing science, while Vietnamese intellectuals worked to actively appropriate scientific knowledge. Vietnamese agricultural agents trained in colonial schools served as key mediators in networks connecting elite and peasant, urban and rural, French and Vietnamese.

These mediators contributed many books, pamphlets, and articles dealing with science that they either translated from French or wrote directly in Vietnamese. Vietnamese produced the content of these journals, but they were sponsored and published by the colonial government. The flow of information was multidirectional, and for the French, these agents worked as liaisons with the countryside; they were translators of not just the natural world but also the social world, and technicians occupied the privileged position of traveling the socially constructed division between nature and culture. The agents benefited from their position and could apply their knowledge of the village world to science.[91]

These technical agents drew some of their strength from their ability to translate the language of the countryside for a French audience. Able to speak both French and Vietnamese, these agents traversed these worlds and brought knowledge and material from each. For Vietnamese society to compete in the modern world, its people would have to be aware of at least the basics of science, and many authors expressed genuine concern for the plight of the peasants. Their jobs as agricultural technicians brought them into prolonged contact with the countryside, and many, even those who had grown up in urban settings, came to possess a profound sympathy and even respect for the people they called peasants. These observations led some agents to adopt more strident ideological positions that lay the blame for rural misery on French colonialism.[92]

Among the first generation of authors, those supporting the colonial order did not view "merely technical" knowledge as a threat, though many Vietnamese scientists, doctors, and engineers advocated the popularization of agricultural science in the name of nationalism. Afraid that the "Vietnamese race" was being threatened, these agents of diffusion attempted to increase the general level of scientific knowledge, which they saw as key to Vietnam's competitiveness in the modern world. Those attempting to popularize science believed that the introduction of science to Indochina had taken too long, and often expressed scientific determinism and an almost unbounded scientific optimism, seeing the introduction of science as a key to solving many of the ills of Vietnamese society. Although some authors wrote about the use of science in war and the dangers of chemical and bacteriological warfare, little doubt was expressed about the drawbacks of progress. "Taking nearby examples," one author wrote, "shows clearly enough that each race on this earth, every country, sooner or later, all must proceed on the path of progress, everyone must follow through to the end."[93]

Several popularizers, such as the agricultural engineer and constitutionalist Bùi Quang Chiêu (1872–1945), were actively involved in politics, while others limited themselves to writing opinion pieces. In the 1920s and 1930s, a less overtly anticolonial, though still politically active, generation of French-trained engineers, scientists, and physicians began to write and publish works about agricultural technologies in Vietnamese. One of their journals, *Khoa Học Tạp Chí* (not to be confused with Nguyễn Công Tiễu's later journal of the same name), appeared briefly in Sài Gòn between 1923 and 1926.[94] The next Sài Gòn-based journal that aimed to popularize science was *Khoa Học Phổ Thông* (1934–58), which was initially directed by Lâm Văn Vảng, a French-trained chemical engineer who was involved in agriculture. The contributors were often trained in France and could speak on the latest scientific developments and offer practical and specific advice dealing with topics such as grafting fruit and coffee plants, agricultural credit cooperatives, and tobacco production.[95] The journal advocated for changing farming practices, and almost every issue urged the adoption of new techniques. The journal also provided the latest information and argued that farming (*nông*) should be extended beyond rice to include corn, sugar cane, and other crops. The journal also addressed specialized problems such as fertilizer (whether chemical, animal, or plant-based) and pests (such as rats and insects). It discussed farming in relation to the science of ecology and considered the connections between living beings and climate.[96]

Respecting farmers was one of the journal's first objectives, and it did not presume to tell farmers how to do their jobs. In a supplementary section written in French, Nguyễn Háo Ca expressed an appreciation for local techniques and offered a resolution to the tension between "traditional" and "modern," writing that while French agriculture was scientific, Vietnamese agriculture was empirical (*empirique*) and based on centuries of observation. Citing the Vietnamese use of red ants to protect fruit trees, Ca argued that while French and Vietnamese methods often arrived at the same effect, the former were often costlier and the latter were often more adapted to local conditions. The solution, which paralleled one political platform, was a happy marriage between the two.[97] Some contributors, however, adopted a rhetoric that echoed earlier French writings about "timeless" Vietnamese agriculture. For example, the agricultural agent Nguyễn Văn Hưng wrote several articles about grafting. He argued that Vietnamese farmers did not know how to graft (*tháp* or *chắp*) plants, and understood only how to sow seeds. Vietnamese farmers had been recently paying attention to nipping (*chiết*), but this helped mostly with the fruiting process and did not improve the tree. He cited a recent tour of the region made by an official from Annam as evidence that many people growing fruit did not know how to graft, thus putting them at a disadvantage with foreign competitors.[98] Though evidence is scarce, the practice of grafting was probably already common among Vietnamese farmers, and the technique had been practiced in China for centuries. Although Nguyễn Văn Hưng was most likely aware of arguments made by Lâm Văn Vãng in *Khoa Học Phổ Thông* for the origins of sciences such as chemistry in China, many questions about the flow of agricultural techniques across the Indochinese peninsula remained.[99]

For other authors, such as the agricultural engineer Trần Thúc Kỳ, an intimate connection existed between the nation and its economy, industry, and science. He argued that the Vietnamese knew the meaning of science (*khoa học*) because of earlier popularization efforts. But a new attitude toward the purchase of domestic products would be required before science could be developed in Vietnam. Kỳ did not advocate for a boycott of foreign goods, as Indochina still needed to import a great deal of its products, but he did believe in purchasing local products whenever possible. Buying domestically would encourage the development of industry, and developing industry would in turn promote science.[100]

While many Vietnamese and French shared the common project of improving agricultural practices in Indochina, their practices indicated different goals. The French viewed investments in agricultural science as a way of increasing economic productivity and thus government revenue. Crops such as rice and rubber that provided export earnings and raw material received the lion's share of research money. Vietnamese, on the other hand, viewed investments in science as strengthening the nation. Both anticolonial and pro-French writers understood that the world was becoming more competitive and that the nation would survive only through the appropriation of science and technology, particularly in the countryside. Rubber production and consumption were critical to this survival, but officials and intellectuals failed to communicate new agricultural techniques for rubber (or any other crop) to Vietnamese farmers.[101]

Postcolonial observers of the rubber industry have critiqued the extension efforts during the colonial era. "The effective extension officer," observed Colin Barlow, "was a diffuser of ideas between the enterprises of commercial industry." Barlow writes critically of the work of agricultural extension officers of the 1930s, who blindly promoted the most "scientific" methods, meaning those methods associated with plantations. According to Barlow, "Their time was occupied in routine visits, and in 'telling' clients what they should do according to the current dictates of large-scale management."[102] Barlow argues that the situation changed in Malaya during the 1960s with the adoption of an emphasis on projects that involved direct application by farmers. This "project approach" was accompanied by a more efficient model of diffusion, as extension officers taught classes and provided advice. Extension officers who were the "same race" as clients were considered more sociologically appropriate. Psychological strategies such as completing projects quickly while enthusiasm was high, and engaging an effective community leader to avoid "catastrophic dissensions" proved highly effective. Larger-scale farmers were still more likely than smallholders to use extension services, particularly during wartime in the 1950s and 1960s, but the models also resulted in positive outcomes in smaller farms.[103]

Limitation

Colonial officials encouraged attempts to tailor education and agricultural technologies to the latex industry, a goal that served both ideological and practical ends. Associating a desire to improve rural conditions within the system of European-owned rubber plantations may seem perverse, given that these plantations often embodied the worst excesses of capitalism.[104] Yet plantations represented only one way to produce rubber, and the introduction of latex-secreting *hevea brasiliensis* fit, even if uncomfortably, within programs of native improvement as colonial officials envisioned ways that rubber could better the lives of Vietnamese peasants. Furthermore, the five-year development plans proposed by the colonial government at the end of the 1930s raised the possibility and desirability of encouraging the production of shoes, tires, and other industries based on rubber products.

In 1923, the journal *Nam Phong* published an article arguing that "with peasants, everything remained as it was fifty years ago, a time when nothing was known about Western civilization. Always the same stagnant pools from which people got drinking water, the same hovels built in mud or on dirty areas."[105] This rural stagnation was not uniform, as demonstrated by the formation of agricultural cooperatives, the rapid increases in industrial crops such as cotton, and other such innovations. The rate and character of change provide insights into the meanings of everyday agricultural technologies, and rubber plantations defined these meanings. *Hevea* had the potential for smallholder production because it was small enough and cheap enough for individual farmers to purchase and consume. Yet, during the colonial period rubber was limited to plantations and treated more like typical "big" colonial technologies such as railroads.[106] In contrast with rice, *hevea* technology was neither actively taken up by Vietnamese smallholders nor subject to much exchange between farmers and researchers. The political economy of rubber science played a key role in retarding the uptake of rubber production among non-French smallholders and thwarted improvement measures of colonial officials and Vietnamese intellectuals. In a time of intense regional and class divisions in Vietnamese society, the introduction of new latex-producing technology served only to widen these fissures.[107]

Judging from jobs-wanted advertisements, skilled plantation jobs were fairly attractive to graduates of the Indochinese school system.[108] By the late 1930s, however, the attempt to encourage Vietnamese smallholder production of latex had clearly failed.[109] Although quite a few Vietnamese grew *hevea*—over 50 percent of rubber operations were native-owned in 1937—these operations accounted for less than 10 percent of the cultivated surface area.[110] As Christophe Bonneuil concludes, "It was the Europeans who undertook the first [efforts] in this culture [of rubber] and who subsequently remained the masters. The Vietnamese economic elite started to assert themselves with regards to Chinese and French businesses, but this rather tentative movement had almost no effect on heveaculture."[111]

The Dutch in the Netherlands East Indies were confronted with many of the same choices as the French, such as whether to emphasize science and technology or social and fiscal policies in agrarian reform and whether to work through elites or directly with farmers to enact changes. When agricultural extension failed to address social unrest in Southeast Asia during the 1930s, many colonial officials turned to "pure science" to bolster the image of their state. While government officials in the Dutch East Indies supported pure science in part to appear "decent" in international opinion, in Indochina efforts by colonial experts to conduct "pure" scientific research met with resistance from the business lobby.[112] Returning to the "scandal" involving Yves Henry and the AFC, the author of *Scandals* argued that agricultural development depended on credit, technical and scientific studies, and direct improvements of agriculture. The role of the government was to create separate departments concerned with each of these fields and provide a loose linkage among localities. Henry's prizewinning work in rural economics did not convince everyone of his value; instead, *Scandals* emphasized the practical results achieved by Alexandre Yersin in both medicine and agriculture. Moreover, no centralized office existed to coordinate "pure" colonial scientific research in France until 1937 at the earliest, and the relationship between pure and applied remained problematic.[113]

In response to difficulties in creating rubber smallholders, some Vietnamese suggested trying to plant other crops. A 1935 article in the journal *Saigon*, repeating a message that had appeared one and a half years earlier, called for increased production of corn, cashew nuts, and soybeans to offset an overdependence on rubber and rice. Because the people of Bà Rịa, Biên Hòa, and Cap (Vũng Tàu) were too heavily invested in rubber, the crash in rubber prices of the 1930s had bankrupted many and caused the rubber gardens (*vườn cao su*) to be abandoned. Different solutions were considered and rejected, including increased production of rice, which was impractical in a land of forest and mountains; fruit, which grew well in the area but could not be exported for cash; and salt fields, which also did not produce wealth.[114]

In 1937, the Guernut commission investigation, and other reports on rubber, pointed to official anxiety about an absence of smallholders, a situation that had only worsened during the 1930s. The smallest plots of land in the Thủ Dầu Một province, which contained about one-third of the *hevea*-growing land in Indochina, were composed of less than five hectares, and rubber grown on those plots was grown to supplement family income from other agricultural crops.[115] These planters were relatively close to urban areas, as they did not have the resources for large clearing operations. They most likely used previously built roads to gain access to markets, were closer to urban health facilities, and did not have to deal with the explosion of malaria that followed land clearing in regions far from traditionally settled towns and villages. Their overhead costs were low and included the trees, which had been purchased from local sellers, a cutting knife, a container—and their individual labor.[116]

Issues of both race and class clearly impacted the spread of rubber-growing technologies. Many local notables and rich Vietnamese living in population centers owned small plantations. Employing a Vietnamese *cai* (*caporal*, corporal) and a few workers, these operations were more visible and their holdings were often listed in the *Annuaire*, a record of latex-producing operations in Indochina.[117] A forty-six-year-old planter, for example, lived in Bên Cát with his wife, six children, one son-in-law, two nephews, a chauffeur, and three servants. He earned a total income of 75,400 piasters, including 72,000 piasters from rubber, 1,000 piasters from renting rice fields, and 2,400 piasters from renting rooms.[118] Of the many expenses incurred by his 500-hectare plantation, 24,000 piasters were spent on labor, which involved about 150 "coolies." For the province of Thủ Dầu Một, this one grower's land and income composed about one-fifth of the total recorded for native growers.[119]

Vietnamese participation in latex production took place in a context of increasing native economic and political initiatives and a burgeoning sense of economic nationalism on the part of Vietnamese elite.[120] For some wealthy

Cochinchinese, rubber was a lucrative side business. Trương Văn Bền, who was born poor but became a rich entrepreneur and member of the Colonial Council, was involved in rubber production. Bền controlled several industrial operations, including rice mills and the Vietnam Oil and Soap Group. This factory depended on pressed plant materials for its operations, and in the 1930s agricultural engineers explored the possibility of *hevea* seeds as a source of oil for industrial purposes. Even middle-class Vietnamese often bought into *hevea*. Hồ Văn Lang, who was born in Sa Đéc in Cochinchina, worked as a secretary for the SPCI, and by 1926, he had made enough money through writing and publishing to buy forty hectares of rubber land.[121] Vietnamese plantation owners were relatively well off, and their economic position influenced their political stance. These politics could range from the conservative Nguyễn Văn Của, best known as the official government printer who supported the colonial administration, to Bền's more progressive, independent politics.

The concentration of production in the hands of a few wealthy Vietnamese ran counter to the desire of some French officials and Vietnamese elite to spread rubber to smallholders.[122] This was due in part to the strength of the planters' association,[123] which enabled planters to effectively lobby the government for loans during the 1930 rubber crisis. The association helped planters take advantage of technological improvements such as grafted trees, clones, and the capital to make use of these advances.[124] For poorer Vietnamese planters, the technical innovations involved in obtaining grafted stock were a barrier to entry. While such advances did not shield large plantations from fluctuations in the price of rubber on the world market, they did provide a buffer in times of economic instability.

For colonial officials in Indochina, problems included resettling the Red River Delta's surplus population, expanding the use of the underpopulated uplands, and reforming traditional farming techniques. The first of these problems was easier to solve with plantation rather than smallholder rubber, and officials encouraged solutions that were more useful to the former. Thus, limitations to smallholder planting were inherent in the practices of the programmers, those who the anthropologist Tania Li has argued "translate the will to improve into explicit programs."[125] In addition, according to Li, these midlevel actors must render solutions to problems in technical terms. In Indochina, officials used soil sciences, climatology, demographics, and allied natural and social sciences to help define technical approaches to defuse problems such as disease, agricultural backwardness, and poverty. These solutions were more available and more beneficial to larger-scale planters connected to colonial networks, including wealthy Francophone Vietnamese. Officials failed to meet objectives for native improvement in part because "the bounding and characterization of an 'intelligible field' appropriate for intervention anticipates the kinds of intervention that experts have to offer."[126] Changing ecologies also played a crucial role in the success of large-scale planters. Unlike the Dayaks of Borneo, who readily adopted *hevea*, Vietnamese could not draw on precolonial patterns of trading forest products such as gutta-percha.[127] Some interest in smallholder production remained, but for most Vietnamese, latex-producing *hevea* never became an everyday agricultural technology.[128]

Finally, rubber technologies played a role in the intense discussions in the late 1930s about whether Indochina should be industrialized. Agricultural agent Vũ Đình Đại argued that the industrialization of agriculture through private enterprise should be a priority. Đại stated that while a few Vietnamese trained as agricultural engineers in France, fewer returned, and those who did worked for the government, leaving almost no engineers to launch private businesses.[129] However, some colonial officials warned of the grave threat posed to France by the transfer of rubber-processing industries from Europe and the United States to the colonies "under the guise" of industrializing those countries. One official endorsed setting up a new political order in Asia to deal with these changes. In 1939, the chemical engineer Lâm Văn Vắng pointed out the difficulties in rubber processing that were created by the lack of trained personnel. He advocated for the initial training of these personnel through internship programs abroad, after which they could return to create industries at home. Because of World War II and the colonial system, even simple projects, like replacing metal and wooden tires with rubber ones in the Vietnamese and Cambodian countryside, made very little headway.[130] Such failures drove many Vietnamese agricultural agents to take part in nationalist movements and later to bring their knowledge of science and technology to the DRV.

A MIXTURE OF HUMANITARIAN impulses, a desire to govern, and greed motivated state officials to craft policies that encouraged increased agricultural output. A "scientific" approach to rubber meant one that benefited estates (i.e., large plantations) instead of smallholders. During this process, rubber production became an everyday technology for some participants in this industry. For plantation managers, rubber production became routine when their tasks became predictable. For workers, both European and Vietnamese, the everyday was embodied in physical relations to the means of production as they endured the discipline and violence imposed by making rubber for the global commodities market. As a result, Vietnamese tappers often had darker understandings of the everyday in relation to rubber than did their managers or European coworkers. For colonial officials, the trees represented a solution to the problem of "overcrowding" in the Red River Delta, and they wished to make rubber a daily, widespread fixture in the lives of peasants, whether as tappers, smallholder farmers, or workshop owners.[131]

Rubber grown on plantations in eastern Cochinchina and Cambodia also provides a lens through which to study how research was conducted and influenced by the changing political economy of French colonization, what was chosen as worthy of research, and how agricultural science projects were defined, structured, and funded. From the turn of the twentieth century until shortly before World War II, industrial agriculture served as a node for knowledge production, politics, and capitalism. The colonial government, private business, and imperial experts interacted to create the institutions and knowledge that existed in this region. Many government officials actively promoted rubber growing, but as with other colonial projects, they often lacked the means to complete them. Meanwhile, individual planters with access to government institutions helped sponsor science in the early stages of the rubber industry. Joint-stock companies subsequently exploited the existing foundations to take control of both the knowledge and the means of rubber production. As the next chapter demonstrates, the clearing of forests and the introduction of "unseasoned" bodies from Tonkin and Annam resulted in an explosion in rates of malaria.

CHAPTER THREE

Managing Disease

Malaria never takes on an epidemic nature.
—Rapport au Conseil colonial de Cochinchine, 1917

Malaria has increased in frequency and seriousness in Cochinchina from 1920 to 1928. This increase seems related to the rapid increase in labor imported to put into production several plantations.
—Undated government report

Because at that time there was forest, forest all around this rubber. At that time, because there was forest, the workers were often struck by malaria! [At that time, there were] many mosquitoes and flies!
— Đỗ Hữu Chuẩn, former rubber company director

In the 1910s and 1920s, thousands of workers migrated from the Red River Delta and coastal provinces such as Thái Bình or Quy Nhơn to eastern Cochinchina to seek employment in the newly established *hevea* plantations. The typical migrant was male and came from a village with scant rice-growing land or one that had recently been struck by a devastating storm or drought. While skeptical of the calls for rubber tappers, or perhaps tricked by a recruiter, this man had little choice to pass up the chance for a few years of food and possibly a bit of savings. He traveled to the closest recruitment center and was subject to medical examinations that included vaccinations for smallpox, cholera, and typhoid fever, and measurements of height and weight—although the labor recruiters knew many ways to cheat on the exam if the man was too sick to pass.[1]

Then the harrowing journey to the south began. He was shipped to a collection point (in either Hạ Lý or Hải Phòng in the north, or various places in the center) that consisted of wooden barracks with corrugated metal roofs. They were hot in the summer and offered little protection against the chill of the north in the winter. Rotten rice and spoiled dried fish were his only provisions, and he started to become anxious about his future employment. After three days of hunger on choppy seas, he arrived in Sài Gòn. He and his fellow mates were quickly herded into a depot near the Sài Gòn River, which was originally built for much smaller numbers of Chinese immigrants. There he was again exposed to the sun or baked under a metal roof, while guards delivered blows and kicks as he waited to be driven by truck to the plantation over rough roads. His journey could last several days and left him in miserable shape by the time he arrived on the plantation. Once there, he faced the lick of the whip, inadequate housing, and meager food. Of the many causes of ill health, perhaps the most detrimental was malaria transmitted by mosquitoes.[2]

At the same time, medical researchers were identifing the *Anopheles* mosquitoes in Indochina that transmitted the plasmodia, or parasitic protozoa, causing malaria. Clearing southern Indochinese forests for planting *hevea* helped generate the limpid, sun-drenched streams in which these species bred. When the immunologically-compromised migrants arrived, they were packed tightly into poorly constructed barracks, exposed to infected mosquitoes, and decimated. The increasing volume and speed of migration then helped spread the disease outside the confines of the plantations. The initial response of managers to the high, and costly, death rates of the 1920s was often simply to abandon a work site or wait a few years until the rates of the disease declined. And when plantation owners needed forests to be cleared, they hired Montagnards to do the heavy work, as they were supposedly less affected by malaria infections.

Initially, studies in mosquito biology and malaria prevention were not priorities for planters; because malaria was not seen as relevant to latex production, they had no incentive to create buffers between workers' bodies and local malaria disease environments. A few planters adopted housing improvements, mosquito nets, and landscape engineering measures, but only after researchers at the Pasteur Institute had convinced them that these initiatives would improve latex-producing systems. These environmental interventions were limited, however, and planters distributed quinine to their workers, albeit reluctantly and sporadically. By the 1930s, some of the larger plantations, in cooperation with the Pasteur Institute, had begun to construct nursing facilities, hire medical personnel, and distribute quinine while continuing to build drainage and carry out other engineering work.[3]

Up until the 1930s, medical researchers had characterized malaria in eastern Cochinchina as "endemic," or even "hyperendemic," but the experience of the 1920s occasioned explicit references to malaria "epidemics" as researchers attempted to place this disease in time and space. Such terms were political, and the choice to call malaria epidemic implied an unexpectedness that removed agency from planters and placed it on the mosquito (and plasmodium) itself. Planters further attempted to reassign responsibility for disease by raising arguments of racial susceptibility during discussions about the effectiveness of antimalaria measures and how these measures should be implemented. This chapter draws on recent scientific research into the dynamics of local anopheles populations to present a more detailed explanation of the mechanism of malaria transmission that places the epidemics within the context of changing local mosquito environments and the broader health conditions on plantations. While planters were caught unaware by particular outbreaks, they largely understood, and accepted, the risk of malaria among workers. Dense forest continued to surround many of the plantations well after the end of French colonialism in 1954, and these outposts of ecological modernity played a large role in changing the malarial disease environments in the wider region.[4]

This chapter also places malaria outbreaks within the changing political economy of Indochina. It focuses on the ways in which the human experience of disease was shaped by colonial labor regimes and the rapid expansion of capitalism. The workers' experience of health, the research of the Pasteur Institute, the responsibility of the colonial government for the health of its people, the demands of transnational businesses and international organizations, anticolonial politics, and local ecology all contributed to this history. Malaria, in turn, infected each of these realms of experience and helped bring about new subjectivities. Attempts to deal with malaria fell along what Gary Wilder has described as a spectrum of "political rationality," a term referring to a method of governing concentrated on managing populations that was concerned with the welfare of its subjects. Such a rationality enabled both planters and colonial officials "to promote socioeconomic individuality without creating legal and political individuals." In this bargain, administrators could promote native welfare through schools, hospitals, housing, and the creation of "Vietnamese" villages, while planters could continue to employ subjects without legal and political standing.[5]

Workers fought for improved welfare, which included control over their own health, and for expanded rights. In the 1910s, labor protests on plantations took on older forms of grievance submissions to the emperor; by the 1920s, Vietnamese laborers began to draw on resources provided by medical science to claim rights under the colonial government, and they often rejected planters' claims of improved laboring conditions. These resources included medical reports that showed the awareness among some government physicians that malaria epidemics among workers were largely the result of rubber plantations. The labor laws of 1927 were unable to resolve the tension between labor and capital, and worker protests that targeted the plantations with the worst living conditions continued into the 1930s.[6]

The money invested in finding prevention and treatment methods flowed from capital that could be tapped because it was threatened by diseases, meaning that malaria control measures were the result of the "marriage of health and agriculture." In other words, practical steps were taken to reduce the effects of this disease only when economic projects were in danger. When some engineers and physicians tried to move malaria treatment and prevention off colonial enclaves during the 1930s, these efforts failed, mostly due to lack of funds and government interest. In the 1930s, hopes for drug-based solutions were revived when synthetic substitutes for quinine, such as quinacrine, were developed, but these drugs did not find wide distribution until after 1954. By this time, French and Vietnamese physicians had drawn different conclusions from similar data about malaria.[7]

Diseases of Conquest

Malaria has a long history in Asia, and references in China to "a disease that accompanies intermittent fever" date from the first century A.D. Alexandre Yersin speculated that Vietnamese words ranging from "poison water," or *nước* (水) *độc* (毒), to "forest spirits" drew on understandings of nature like those expressed by the French concept of *terroir*. Yersin suggested that early translators of Vietnamese had imparted a literal sense to these words by connecting water to disease, a sense that was then reincorporated into common language. In the 1930s, French and Vietnamese researchers studied the physical and chemical properties of water that encouraged the breeding of malaria-transmitting mosquitoes. The term *bệnh sốt rét*, which describes the bodily symptoms of fevers and chills, marked malaria's separation from place.[8]

The Nguyễn court's interest in malaria, like that of the colonial state that followed, lay in the effect that this disease had on the stability of new agricultural and military settlements. Although royal ordinances of 1852 and 1864 dealing with the organization of settlements in the south made no specific reference to health, diseases were one important cause for the failure of many Vietnamese colonizing efforts. Almost no state medical institutions existed outside the courts, the Nguyễn rulers did not allow doctors to professionalize, few records from the Nguyễn dynasty discuss the health of local communities, and most people relied on local healers and other practitioners of folk medicine. Treatment for *nước độc* among Vietnamese peasants involved neither formal visits from practitioners of northern medicine, *thầy thuốc bắc*, nor large-scale campaigns to drive out demons that caused epidemics. Instead, the herbal medicines of *thuốc nam* and avoidance were the most likely methods of prevention and treatment. The noninterference of the state in matters of individual, family, and community health also explains, in part, the initial apathy of rural folk toward colonial state medicine. By reading this skepticism as resistance, the colonial doctors created antagonism instead of comprehension.[9]

The French and Spanish naval attack on Đà Nẵng in 1858 and the subsequent conquest of Vietnamese territory had serious repercussions for health conditions. Initially, French medical attention was focused on protecting French soldiers from diseases such as cholera and diarrhea. Few medical institutions were built by the French outside of the urban centers of Sài Gòn and Hà Nội, and, except for the native hospital called Chợ Quán near Chợ Lớn in Sài Gòn and the civilian hospital built by the Congregation of the Sisters of Saint-Paul of Chartres in 1872 in Mỹ Tho, almost none were built for Vietnamese patients. This was particularly true in eastern Cochinchina, where the Nguyễn dynasty had never established a foothold, limiting itself to colonization projects in the delta. With the breakdown of the Nguyễn state, even this limited presence in the delta ceased and the French colonial state took responsibility for providing medical care. Vaccination campaigns carried out against cholera and smallpox, occasional visits of health inspectors, and medical institutions run by religious associations such as the Société des Missions Etrangères de Paris and various sisterhoods, were the only contact locals had with biomedicine.[10]

When active combat subsided, military physicians started to turn their attention to the health conditions of the Vietnamese countryside. Hard statistical evidence regarding this topic is unavailable, but physicians' writings suggest that the initial period of colonization had a negative impact on the health of Vietnamese peasants. The disruptions caused by military violence threw food systems into disarray, and starvation often struck the countryside. Regular outbreaks of cholera, for example, were recorded in Cochinchina, with 1874 noted as a year of a particularly high number of native deaths. Cholera appeared with some regularity in Tonkin as well, but struck especially hard in 1885–88, 1895, and 1910, all years with seasons of low rainfall, when water sources became polluted and populations were already weakened by famine. Coastal populations and people in crowded cities were the most often devastated by cholera. During the last of these epidemics, at least 10,000 lost their lives in Indochina, including 5,000 in Tonkin alone. As Laurence Monnais has said of cholera, it "deserves its number one ranking among the deadliest peninsular diseases."[11]

In contrast to cholera, French physicians noted with surprise the relatively benign influence that malaria had on those living in the deltas, which challenged common medical understanding of the disease environments of deltas.

The endo-epidemic form of malaria in the Indochinese uplands, the worst of the four forms of the disease, was caused by *plasmodia falciparum*. Forests and malaria were linked in French medical thinking in part because of the association between fevers and the uplands that was common among many Vietnamese, and in part because of French military campaigns in the forested northern uplands. French medical doctors used malaria as an ethnographic tool, and one physician suggested that those on the coast associated malaria with uplands sorcerers who thwarted attempted invasions with fevers, while upland ethnicities saw malaria as a sign of evil water spirits that sucked the blood of their victims. Although significant levels of exchange and contact existed between the uplands and the lowlands, the disease had caused distinctions between the peoples of the two regions through the centuries.[12]

Malaria was no stranger to the lowlands, however, and in a treatise on European mortality in Cochinchina published in 1881, a naval physician described malaria as "reigning as master" in the region. Medical studies from the 1920s suggest that milder forms of the disease, most likely caused by another type of plasmodium called *vivax*, were endemic to places such as the river port of Châu Đốc. While agricultural development during the colonial period had reduced the number of mosquitoes, forested hillocks continued to be sites of infection. Often considered sacred, these hillocks were homes to temples that attracted pilgrims who spread the disease after being bitten. In 1938, the phrase "reigning as master" could still be applied to malaria in the delta province of Gò Công. Epidemics likely struck coastal communities with great frequency, especially after typhoons or tidal waves. French doctors divided incidents of malaria into three geographic zones: coasts, deltas, and uplands. More urban areas, such as Hà Nội and Sài Gòn, were essentially free of malaria. Some studies attempted to explain the difference among rates of malaria in the uplands and the deltas by the presence of forests, terraced rice paddies, or delta paddies, each searching for a reason for the difference in the type and extent of mosquito infestation.[13]

The 1890s, according to Albert Calmette, who helped found the first Pasteur Institute in Indochina, marked a key turning point in the trajectory of Western medicine in the colonies. Early Pasteur Institute interventions had isolated microbes that could be separated from the "terrain" and brought back to France for study. At the newly established Pasteur Institute in Sài Gòn, research could be carried out on "local" diseases such as malaria whose effects on Vietnamese had to be studied in situ. Calmette's contributions had broader political, social, and economic ambitions for the institute, which could contribute to the "undertaking of colonization, by making it eminently humanitarian and civilizing" and contribute to the economic health of the colony. These two interests existed in tension for, as Gerard Sasges has shown, Calmette's attempt to devise industrial processes to refine opium and ferment alcohol, thus enabling colonial state monopolies on both, did little to improve the health of locals.[14]

The initially low population densities of the red and gray earth regions of Cochinchina could not meet the planters' demand for labor. From the planters' perspective, the biggest problem facing the expansion of industrial agriculture of the 1910s was the lack of a sufficient workforce. Some planters sought to import labor from Java and other places outside French Indochina, but neither foreign governments nor the French colonial government was satisfied with this arrangement. Instead, their labor needs were filled by increased Vietnamese movement from north to south, enabled by road construction and shipping routes and a steady stream of migrants. At the time, mobile Vietnamese populations, especially southerners, retained some ability to leave plantations if they found the conditions too harsh or the pay too low. Planters began to call for regulations to gain more power vis-à-vis labor by bringing order to a chaotic situation in which they believed there was too much competition for labor.[15]

The *Mise en Valeur* of Malaria

The 1890s saw the opening of several institutions that intended to extend health care outside Sài Gòn, including hospitals offering free care for locals in provincial centers such as Biên Hòa. In 1897, Paul Doumer, newly arrived from France, aimed at *mise en valeur* through a policy of economic and social development and a 200-million-franc loan. Some of this money was slated for hospitals, and more medical personnel, both public and private, began to practice in Indochina, though no French-trained Vietnamese doctors were yet in place. By the 1890s, planters had also begun to consider eastern Cochinchina's potential for production, especially with respect to coffee, rubber, and other commercial crops. Like rice growers in the west, these colonialists faced a shortage of labor, and several solutions were suggested in response to these perceived labor difficulties. Some planters looked to Java and China for labor. At the end of 1896, however, the ninth reunion of the Committee of the European Planters Syndicate of Cochinchina (Comité du Syndicat des Planteurs Européens de Cochinchine) discussed the possibility of bringing workers from northern and central Indochina to the south. To regularize efforts to obtain labor from the north, the GGI issued a series of *arrêtés*, or decrees, concerning labor in 1896, 1909, and 1914, but the recruitment system remained chaotic.

Both planters wanting more control over their labor and administrators seeking to better control labor conditions found this situation undesirable.[16]

Until the 1910s, the number of Europeans working as full-time planters in the south remained small, with about fifty in 1907. In the early 1900s, the colonial government remained more committed to supporting rice growing in the Mekong Delta, as the immediate profits of rice and the potential difficulties of rubber plantation agriculture led the conservative government to favor the former. But the first plantations began to grow rubber around 1906–7 and were met by an increased demand for labor in the south. At first, local sources and informal recruitment filled most of the labor needs for the smaller plantations. There were attempts to import labor from Java, and later China, but migration and tax laws, as well as local opposition, especially in the case of Chinese labor, made these options unattractive.[17]

In contrast to previous colonization efforts, plantation agriculture was one of the first organized labor migrations to pay conscious attention to the role of health. It is difficult to know about the health of workers in the south during the 1910s because, with no labor inspectorate (inspection/inspecteur du travail), the government did not systematically collect information on plantation conditions. Not surprisingly, reports to the managing boards of plantations rarely mentioned the problem of labor health and plantation medicine; at that time, the agricultural sciences dominated corporate interests. In the case of Lộc Ninh, given the nature of the work, the plantation's distance from Sài Gòn, and later reports on worker health, it would not be surprising if morbidity and mortality rates were high. But as most *hevea* was grown close to areas long under cultivation, the health of workers seemed not to present such a great problem, for management at least.[18]

Provincial administrative reports give a general sense of conditions, and the creation of local boards of hygiene and sanitation in 1906 provided even more detailed coverage. Smallpox, cholera, and plague continued to dominate concerns of medical personnel and colonial officials. A cholera epidemic struck Cochinchina in 1908, resulting in 2,272 cases and 1,569 deaths and apparently followed the railroad line under construction.[19] In eastern Cochinchina, malaria was prevalent. Throughout 1911, the head of the Tây Ninh medical service reported treating many patients suffering from malaria, including several who came from the village of Kedol, located next to Núi Bà Đen and populated largely by Cambodians. During two days of consultations at the village, one doctor saw 114 patients, over one-half of whom had malaria.[20] In 1914, another doctor wrote of the region: "One can say that almost all of the general disease (consultations or hospitalizations) are malaria cases: the majority of villages are situated in the middle of or next to the forest, at places impenetrable, and certain marshy regions swarming with mosquitoes to the degree that each evening the natives are obliged to fill their huts with smoke in order to be able to protect themselves against these insects. This year all our effort will tend towards antimalaria prophylaxis as this disease is the scourge of the province."[21] Observations such as these were common, and to combat malaria, the colonial state created the State quinine service (quinine d'Etat) in 1909 with the purpose of providing quinine to the native population at first for free and then at prices below cost. Early accounts of the service were rather hopeful and reports to the Colonial council speak highly of the usefulness of quinine. Yet the service met with initial resistance. A physician in Tây Ninh reported resistance to quinine and noted the continued reliance on a "sorcerer" to cure malaria.[22]

Some attributed this resistance to quinine's bitter taste, but there were other problems. Even when Vietnamese expressed an interest in quinine, it was often unaffordable or just plain unavailable. Many administrators did not restock their province's depots because they did not want to pay the costs. Moreover, Vietnamese had doubts about the effectiveness of quinine and its costs relative to other medications. The costs were high in part because of a Dutch monopoly, and in response Yersin cultivated some quinine in Indochina. This did not stop the SPCI from reporting optimistically on the provision of quinine to plantation labor as part of a government service. Even though quinine was not seen as completely effective, other solutions were viewed as impractical. Officials argued that stagnant pools of water in the countryside could not be drained or oiled because they were also daily sources of water for the people and engineering solutions, outside of a few embankments in population centers, were fiscally impossible.[23]

Limited levels of quinine production meant that it was never a viable option for plantations. In 1921, the year of greatest distribution, there was only enough quinine available for ten centigrams (cg) per person per year. With forty cg (repeated multiple times during the malaria season) considered to be an effective dose, ten cg was woefully inadequate, even by the standards of the time. After several years of very discouraging results, the State quinine service was reorganized in 1921, but distribution of quinine remained limited to the confines of colonial enclaves such as plantations. In 1932, a League of Nations calculation showed that the projected need for prophylactic quinine in Indochina was greater than world production levels. Hopes for a drug-based solution were revived in the 1930s when synthetic substitutes for qui-

Growing capacity of the Native Medical Service

Year	Native hospitalizations in AMI, Cochinchina (deaths)	Quinine distributed by the state service in kg	
		Eastern provinces	Cochinchina, all provinces
1914	23,675	40.5	113
1916	31,086 (2,015)	n/a	126
1918	33,008 (2,957)	44	219
1920	45,647 (2,871)	n/a	341
1921	47,998 (2,787)	135.3	416

Note: Growing capacity of the Native Medical Service (AMI) to treat patients, along with the amount of quinine distributed by the state service for the beginning years of rubber plantations. The situation was in fact worse than this table shows, as all the depots in some provinces were closed and some provinces had several depots that dispensed no quinine.

Sources: *Rapport au CC*, 1910–1922; especially, *Rapport au CC*, 1915, 252–53; *Rapport au CC*, 1922, 279.

nine, such as quinacrine, were developed, but these drugs did not find wide distribution until after 1954.[24]

Established in 1905, the Native Medical Service (Assistance Médicale Indigène [AMI]) represented the creation of a civil sector of medical care, which up to that point had been dominated by military concerns. The AMI's services were concentrated in population centers such as Sài Gòn and the delta towns of Cần Thơ, Mỹ Tho, and Bến Tre, and the medical networks in the forested region of Cochinchina before the 1920s could not handle the influx of patients. Imported labor had a significant effect on the public health system, and a paucity of resources available to the countryside meant that plantations could easily overwhelm regional medical capacity. The plantations' demand for a network of local health centers pushed the government to create a few new clinics. For example, in Biên Hòa new medical posts were built in 1922 at Gia Rây (one nurse); 1927 at Phú Riềng (one nurse); 1928 at Long Thành (one Indochinese doctor, one nurse, two midwives, and four coolies) and at Xuân Lộc (one nurse); and 1929 at Tân Uyên (one nurse, one midwife, and one coolie). These medical posts were basic and had room for ten beds. In addition, they had one small pharmacy that may have provided quinine. Despite additional construction, by the late 1920s the resources of Biên Hòa to treat the rapidly growing population had become severely strained. The administrator who authored the 1930 monograph on Biên Hòa argued that the growth in population due to the "diverse agricultural, forestry, and industrial operations" of the province made the construction of a larger hospital essential.[25]

The initially low population densities of the red earth region could not meet the demand for labor, and, from the perspective of many planters, the biggest problem they faced during the plantation expansion of the 1910s was the difficulty of maintaining a sufficient workforce. The increasingly organized labor migration from north to south, enabled by road construction and shipping routes and motivated, at least in part, by the possibility of earning an income, brought a steady stream of migrants to the south. But this highly mobile Vietnamese population could just as easily leave, especially those from the south, and they often did when they found plantation conditions to be harsh and pay low. Planters did not respond by providing better working conditions but instead constantly sought to gain more power vis-à-vis Vietnamese labor.[26]

Health in the First Labor Code

In November 1917, after many years of French planters' demands for a steady supply of cheap labor, and growing concern about governing this population, the GGI established a commission to draw up regulations for agricultural and mining labor. This commission resulted in the first labor inspectorate for Cochinchina, one of the five "pays" or lands of French Indochina.[27] The president of the commission, who was also the inspector of political and administrative affairs, articulated the urgency of regulating imported labor and the government's role as both protector of the "natives" and facilitator of business. In an April 1918 letter, the president wrote:

> It [European colonization in eastern Cochinchina] is a revolution that is beginning and there will not only be *hevea* plantations, but also coffee culture, coconut palms, bananas, sugar cane, corn, etc. that will cover the red and gray earth. And all of that will not be done without imported labor that, I hope, will only be Indochinese. For the 1000s of employees that there are currently there must be substituted masses. This will be a real displacement of populations. There must be a very strict discipline imposed on all, employee and employer; penalties must be leveled on the one as on the other, for employees, fines, prison, days added to their contract, for the employers, large fines.[28]

The first labor code in Indochina, which went into effect only for Cochinchina, was announced on Armistice Day, November 11, 1918, perhaps to acknowledge the Indochinese contribution to France's war effort. This code provided some protection for labor, partly reflecting the influence of the

GGI, Albert Sarraut. Sarraut was a colonial reformer who promoted the policy of "Franco-Annamese collaboration." The GGI, however, was limited in his ability to affect the colonial machinery, and French economic interests at the lower levels of the administration played a major role in the process of law reform. This code was primarily designed to reduce friction between government and business, often at the expense of labor. In terms of food and hygienic conditions of the workers, medical science had not yet given answers to questions such as the minimum number of calories needed by workers or measures necessary to reduce the effects of malaria.[29]

The composition of the commission responsible for drawing up the 1918 labor laws shows the extent to which planter interests dominated the discussions. Of the fourteen-member commission, four were French directors of rubber plantations, and one was a rice planter. Five members were French colonial officials, three of whom had an interest in the economic development of the eastern provinces. Only two members were Vietnamese: one a *đốc phủ sứ*, a bureaucratic position, and the other an agricultural engineer. A French medical doctor and a secretary rounded out the commission. No labor representatives were present. Sarraut closely followed the commission's recommendations, often adopting its text verbatim, in a series of *arrêtés* that transformed the labor situation. The first law, enacted on November 9, 1918, required Vietnamese adults to carry identity cards. Although tax-paying cards were already in use, they were not sophisticated and were open to manipulation. The new identity cards included photographs of the holder, along with thumbprints, both of which helped employers keep better track of individual employees.[30]

By far the most significant changes in labor regulations came with the second *arrêté*, which was enacted on November 11 and created, among other things, the three-year contract for industrial agricultural labor. Planters had long bemoaned the fact that they could not legally bind Vietnamese workers for more than one year. The code of November 11 specified many penalties for workers, with the largest provided for resistance, "desertion," and sabotage of company property. The original draft of the labor code had included an article that called for the right for planters to hunt down workers who left the plantation, and to charge workers for the cost of the search. A vice president of the SPCI expressed the view of the planters when he "observed that the present regulations were inspired by the noble concern to put at the service of well-intentioned and active people [i.e., planters] the means to provide their effective contribution to the rise and economic development of the Colony."[31]

Vietnamese worker ID card. This card comes from the RVN period. It was similar to those used on plantations during the colonial era. Note the salary and life history information, including religion. On the back of this card is space for fingerprints. By the 1960s, there was a separate sheet listing health information, including number of children, results of annual medical check ups, and major diseases. Source: author's personal collection.

The legal expert on the commission, however, said that such an article reduced the workers to slavery. This expert noted that no law in France required the government to arrest absconding employees for their employers. He also noted that the practice of "work booklets" had been ended in nineteenth-century France because employers abused the system by marking a reason for termination, in effect blacklisting workers. According to the notes from the meeting, he "judged that the measures of article 33 are useless in law and vast in practice: bring back to an estate, individuals who have refused to work there, and, moreover, to force the employees to pay for the costs of their chase and capture, that is, in his [my] opinion, slavery." While the final law of 1918 included no provision for hunting down absconding workers, many provisions made "desertion" punishable by hefty fines and possible jail time. Left unstated was how such workers would be captured and brought to justice, or whether they would simply be brought back to the plantation for extralegal punishment.[32]

The government sought to defuse charges of exploitation by including provisions aimed at ensuring healthy living and working conditions. These conditions included housing, free medical care, adequate food, and protection for mothers. On November 23, the position of the local labor inspector was created for Cochinchina. This labor inspectorate had to strike a delicate balance among planters', workers', and governmental interests, and the commission president emphasized the need to hire a highly qualified person. Planters viewed the labor inspectorate as a possible tool to hunt down "deserting" employees. Two planters on the commission argued that "the labor inspectorate, in Cochinchina, should not be restricted to being a simple control agency, but that it should be endowed with a budget of which it should devote a part of these resources to the payment of the costs of search." In addition, planters argued that the labor inspectorate should report directly to the Goucoch, as they preferred to deal with the colonial administration rather than a potentially troublesome judicial system.[33]

Food was another key issue. The final version of the 1918 law did not spell out the amount of food to be provided by the plantation, leaving nutritional levels to the specific labor contract. Recruiters used this ambiguity as the basis for false promises of free food, and plantations were not required to provide daily rations of rice until the implementation of the 1927 labor code. Even then, rice was provided only to employees hired after the code was enacted. The only member of the commission to raise the issue of food was the agricultural engineer, newspaper publisher, and constitutionalist Bùi Quang Chiêu. In a short exchange, Chiêu argued that workers should have the right to sell any uneaten food received from their employers. One planter and the medical doctor disagreed with Chiêu, stating that malnutrition was a principal cause of ill health on the plantations, and arguing that workers could not be trusted to take care of their own health. "The ban [on worker resale of food] instituted by the fifth paragraph," they stated, "is an inevitable consequence of the obligation that is given to the planter to take care of the well-being and the health of his coolies." The 1918 law even provided for a fine of one to ten francs and one to three days in prison for reselling plantation food. This law raised many questions that remained unanswered over the next decade.[34]

Stamped by Rubber

After World War I ended, capital flooded into Indochina, but a drop in rubber prices after the wartime boom caused plantations to struggle. This period was especially hard on peasants because of increased taxes and high prices of basic goods. The poverty and landlessness promoted by colonialism helped create a floating population of paupers willing to travel south for plantation work. Many of these workers were in poor health upon arrival, and planters commonly complained about the state of the recruits from Tonkin and Annam. In 1918, Janie Bertin Rivière de la Souchère, later a "female labor inspector," grumbled about the quality of the recruits, calling them "rabble" (*racaille*). Physicians too repeated theories about the health of recruits, often without much evidence. For example, a medical doctor wrote in his 1928 report on the infamous Phú Riềng plantation, the setting for Trần Tử Bình's *The Red Earth*, about the "physical weakness of the contingent of workers. It seems that the recruitment from the beginning," he continued, "had emptied Tonkin of all of its undesirables, vagabonds, opium addicts, etc."[35]

During the early twentieth century, the main reasons for leaving Tonkin and Annam to work on the plantations were poverty, coercion and false promises, personal issues, and political motivation. Plantations in the south opened an economic opportunity and a chance to earn a living, and poverty was the most important factor driving peasants to seek plantation work. Yet not everyone left for purely economic reasons, as illustrated by the experience of the brokenhearted son of a poor northern peasant in Nam Cao's short story *Lão Hạc*. This mass migration paved the way for extensive abuses, and among the most infamous of colonial figures were the plantation recruiters, both Vietnamese and French. Often these recruiters could derive extensive profit from practices akin to human trafficking, which included garnishing portions of the advances of would-be laborers. Furthermore, recruitment companies attempted to cut as many costs as possible, making the journey from home to plantation a miserable, potentially deadly trip.[36]

Local resources were overtaxed due to the rapid growth in the plantation labor force and the underdeveloped health-care system. This situation caused tense discussions among plantations, health officials, and administrators about the financial costs of disease and colonization. Early in 1912, a series of letters between the Goucoch, the RST, and the GGI resolved the issue of responsibility for the hospitalization of northern migrants to the south. The Goucoch first told the GGI that, while providing hospital care for impoverished migrants from Tonkin and Annam was necessary on humanitarian grounds, the workers' home villages should cover the costs. The RST (whose budget depended on taxes from these villages) objected, arguing that Cochinchina and agricultural enterprises were benefiting economically from northern labor and that they should also bear the burden of caring for this labor. The RST further argued that if northern villages were responsible for hospital bills, they would not allow labor to leave—a disingenuous argument since villages were incapable of preventing migration. At the end of June, Albert Sarraut, the GGI, concluded that plantations would be responsible for the costs of workers who were contracted to serve in the south, while the village of residence would pay for all others. For those passing through Cochinchina for a few months on business and for vagabonds without papers, the "domicile de secours" or hometown as defined by a 1910 *arrêté*, would absorb the costs. For patients without identification, this decision resulted in a long process of determining their home and figuring out which location to bill.[37] The battle over the medical bills of Nguyễn Văn Đệ and Hồ Văn Lực from the 1920s illustrates how this distribution of responsibility between public and private sectors played out in practice.

Nguyễn Văn Đệ was born in the village of An Lac in Thái Bình in 1884. In January 1917, he arrived in Sài Gòn at age thirty-three to work for the Xa Trạch plantation. He labored for three years and was repatriated in January 1920. In the summer of 1923, he turned up in the native hospital of Chợ Quán, and on July 31, the hospital requested that the police department determine his identity so that his home village administration could be charged. The next day, the police replied that Đệ was a nineteen-year-old, originally from Tân Đệ village in Nam Định province and the son of Nguyễn Văn Lương and Trần Thị Nhạc. On August 3, the day that Đệ died, the Goucoch wrote to the RST asking for the proper reimbursement. A little over a month later, on September 5, the Goucoch received a response stating that the Resident at Nam Định could not locate Đệ in his records. The police were asked again to identify Đệ, and wrote back eight days later stating that they had no information. The Goucoch then turned to the immigration department, and on October 25 he received a letter with information about Đệ's true birthplace.

Hồ Văn Lực, forty-eight, was also recruited to work for Xa Trạch and arrived in Sài Gòn in September 1919. Lực's case was different; the government was seeking money from Xa Trạch to repatriate Lực to his home in Bình Thuận, Quảng Nam. Lực had been hospitalized immediately after his arrival in Sài Gòn and had received treatment at the native hospital for nine months, until June 1920. Xa Trạch had been paying for Lực's hospital stay, and, convinced that Lực's injury had been caused by the plantation, the government attempted to force Xa Trạch to continue to pay the bills. The Goucoch decided that the only choice was to repatriate Lực and make Xa Trạch pay the costs. Stibbe, the technical director, refused claiming that the society was responsible for repatriation costs only up to one year after an employee had left its service. Stibbe also sent copies of a series of letters in which he complained of the poor health of the recruits he had been receiving and his attempts to recoup money from the labor recruiters, particularly for the workers that had died shortly after arrival. On January 6, 1922, the delegated director of Xa Trạch had the last word, arguing successfully that Lực's injuries had been sustained before arriving at the plantation.[38]

Đệ's and Lực's cases show how the fate of workers could become a political football, passed back and forth between administrators and plantation owners. Workers like Đệ could work for three years on a plantation and be left in poverty afterward. Đệ's home village had to make payment after the plantation contract ended. For Lực, the Goucoch was willing to pursue Xa Trạch rather doggedly for costs. Such disagreements between government and business over labor arose in part because of the slow implementation of the regulations of the 1918 law. When in the spring of 1923 the GGI requested the Goucoch to report on the progress of this task, the Goucoch stated that the results varied greatly by province. The administrators of Thủ Dầu Một and Biên Hòa provinces reported that the regulations of the 1918 law were being applied with no complaints listed from either employers or employees. The administrator of Thủ Dầu Một added that "planters complained with reason to not be sufficiently protected against the coolies." The administrator continued to argue that employees ran away with ease to Sài Gòn, where they had better lives and higher pay than on plantations. The administrator of Tây Ninh province reported that only two planters had contracted employees, with only one from outside the province. Even the large plantations did not use contracted employees.[39]

The longest reply came from the administrator of Bà Rịa, a province with only "free," or noncontracted, labor, who were not covered by the terms of the 1918 law. Planters in the province still chafed under the regulations, a situation the administrator of Bà Rịa explained by quoting one planter anonymously who said that "when a contract is agreed to between two people of whom one is firmly decided to violate all the clauses and profit from all of the advantages and on the other hand, someone who is exposed enough to be constrained to carry out [the clauses] even in admitting that he tried not to, he will always be the dupe of the other." The planter argued that Vietnamese employees were constantly "cheating" French employers out of their money.[40]

The Goucoch reflected this planter's position in a July 1923 letter reporting on the implementation of the 1918 law. The Goucoch, who identified the anonymous speaker, continued by stating that "it is unfortunately known that the mentality of the employee is generally such as that described by the president of the Rubber Planters' Association." The Goucoch emphasized that some planters considered contract labor a waste of time, but the president's

Laborers and desertion rates of select plantations between January 1 and July 1, 1923.

Province	*Biên Hòa*		*Thủ Dầu Một*		*Hà Tiên*	
Plantation name	*La Souchère*	*Suzannah*	*Xa Cát*	*Lộc Ninh*	*Phú Quốc*	*Total*
Distance from Sài Gòn	51 km	70 km	104 km	135 km	850 km (sea)	
Day laborers	450 (est)	210	0	0	0	
Contract employees	246	205	395	3,117	94	6,693
"Deserters"	2 (1%)	11 (5%)	29 (7%)	92 (3%)	0	249 (4%)
Arrested "deserters"	1	0	0	23	0	33

Source: NAVN2 IIA.46/306, dossiers concernant la main d'œuvre agricole asiatique en Cochinchine, statuts, etc., 1922–25.

position could be read differently. His quote to the administrator of Bà Rịa continued: "I don't want to go as far as denying completely the effectiveness of the means of constraint that will sometimes by force bring back to the plantation an escaped employee. But I fear that under these conditions the cure is worse than the disease and that the employee who works unwillingly causes more pain and expense than his accomplished tasks can justify." This quote suggests that the system for bringing back runaway workers was violent and only somewhat effective. The larger problem was that reluctant employees were not effective employees. It is difficult to know why the Goucoch omitted this paragraph in his report to the GGI, but the hint of capturing runaway workers who were then kept on plantations against their will was probably not a reality that the Goucoch wished to convey.[41]

Judging from the numbers, the planters' complaints about runaway workers were not justified. Less than 4 percent of the workforce deserted in the first six months of 1923, a few of whom were recaptured. In the late 1920s, the reported numbers of broken contracts stabilized at about 11 percent of the workforce. These years included incredible death rates, and it is not surprising that many workers attempted to escape the horrible plantation conditions. Meanwhile, labor laws and an increased surveillance network were effectively reducing the chances of workers to escape, and the percentage of captured workers steadily increased, eventually reaching 45 percent. Desertion rates varied by plantation and did not always correspond with the rates of illness and death reported by the labor inspectorate. Some plantations with "kind" management and clean conditions had high desertion rates because they were closer to population centers. Plantations with greater European brutality and worse hygienic conditions, such as Phú Quốc Island off the coast of Cochinchina and Cambodia, were often located further away from existing settlements and were harder to escape.[42]

The labor inspectorate reports from this time do not describe plantation conditions as particularly poor, and offer few indications of death, disease, and brutality. Yet, these brief reports were based on preannounced visits of short duration. Testimony about abuse on the Phú Quốc Exploitation Company plantation places such labor inspectorate reports in a different light. The Phú Quốc plantation grew rubber, areca, and coconuts on 153 hectares of farmed land. The Vietnamese head of the local government on Phú Quốc recorded the words of four Vietnamese who had left the estate on July 17, 1924, under armed guard to report their abuse. These four workers, all in their twenties, had arrived in late 1923 as part of a group of forty contracted workers from Bình Định province in Annam. A worker noted that over the past eight months, one member of his cohort had died and eleven others had already fled. Collectively, their stories illuminate the ways in which plantation owners flouted the articles of the 1918 labor law. All four workers had been given an advance of ten piasters and a salary of approximately forty cents (cent=1/100 of a piaster) as promised by their European recruiters. Yet, none of these workers were given medication, and all of them had to buy their rice at ten cents per quart (approximately 740–800 grams). Low salaries, physical abuse, extremely hard work, relentless fines, illegal imprisonment, and refusal to sell rice to sick workers all contributed to their desire to quit the plantation.[43]

The stories of the workers show the ways in which some plantation owners openly disregarded the law. For example, every time that the twenty-four-year-old Lê Thật was sick, the boss denied leave, refused to sell him rice, and regularly fined him eighty cents or one piaster. The only medicine he received was fish liver oil. Caporal Tiến was officially ordered to fine and beat him. When Nguyễn Văn Hảo fell sick, he was given five days off, but the director refused to sell him rice past the first day of his leave, and his medications did not cure him. After three months on the job, Trần Kính became sick. He worked through the first three days on the mountain worksite, but on the fourth day, he returned to the estate. The next day, he requested time off and the director refused. He hit Kính with a rattan baton and docked him three days of salary because he was "lazy." Two months later, Kính fell sick again. The director not only refused him time off but also bound his feet and threw him to the base of a coconut tree. He was exposed to the weather for three days and denied food. On the third day, Mr. and Mrs. Thiên secretly gave Kính a bowl of rice, but he later had a relapse. The boss refused to give him medication or sell him his daily rice. Likewise, when Võ Hạnh had an attack of fever, the director refused him time off and levied a sixty-cent fine. As with Kính, when Hạnh had a relapse, the boss bound his feet, hit him with a rattan baton, and left him at the foot of a coconut tree for days. The director forbade other workers to give him food, threatening them with a one-piaster fine. During the night, Võ Hạnh's coworkers took pity and fed him.[44]

Finding such direct evidence of abuse in the colonial archives before the 1927 labor law reforms is extremely difficult, and the experience of these four Vietnamese from Annam probably only remained because of a disagreement between a planter and the administrator of Hà Tiên province. By 1925, however, such incidents had motivated some administrators to call for greater power among the labor inspectorate to sanction planters for unhealthy conditions. While few were ready to advocate for French labor laws, a reform to the 1918 law required plantations to give an account of their workers twice a year.

Sapped Strength

During the forest-clearing work necessary to create Michelin's Phú Riềng plantation, Trần Tử Bình, a plantation worker and communist organizer, observed: "One's strength today was never what it had been the day before. Every day one was worn down a bit more, cheeks sunken, teeth gone crooked, eyes hollow with dark circles around them, clothes hanging from collarbones. Everyone appeared almost dead, and in fact in the end about all did die."[45] Although work inspector statistics from the time suggest that not all workers died, Bình's statement points to the withering effect that plantation labor had on workers. Contrary to French claims that plantation work would strengthen Vietnamese bodies, the punishing climate, long workdays, poor diet, and lack of medical care meant that almost every disease became life-threatening.

Most planters did very little to prevent malaria until the outbreaks sweeping the region in the mid-1920s threatened the existence of rubber production. On the extreme end, the Bù Đốp plantation in Thủ Dầu Một reported that 237 of its 1050 workers, or 23 percent (a 45 percent annual death rate), had died in the first six months of 1927, numbers that were partly responsible for bad press attention. Like de la Souchère, the plantation director tried to blame both the deforestation work and the poor quality of the recruits, arguing that up to 30 percent were already opium addicts, homeless, and even

Malaria deaths among rubber tappers, 1922–1935

	Average number of contract workers			Total deaths (adult workers) and % of workforce	
Year	Cochinchina	Cambodia	Total		%
1922	6,700	n/a	6,700	501	7.5
1924	9,300	n/a	9,300	158	1.7
1926	29,200	n/a	29,200	162	0.6
1927	29,800	4,600	34,400	788	2.3
1928	35,000	5,900	40,900	1,323	3.2
1929	28,500	8,900	37,400	848	2.3
1930	22,200	10,200	32,400	752 (est.)	2.3
1931	14,800	7,150	21,950	n/a	n/a
1932	7,800	4,400	11,985	212	1.8
1933	6,850	4,400	9,345	169	1.8
1934	8,650	6,250	10,566	206	1.9
1935	9,450	6,950	11,789	176	1.5

Sources: Service de la statistique générale, *Résumé statistique relatif*, 2; numbers for 1922 through 1929 from Delamarre, *L'émigration et l'immigration*; numbers for 1927 through 1929 from Delamarre, *L'émigration et l'immigration*; numbers from Morin, "Le paludisme et sa prophylaxie en Indochine," 411. See also IMTSSA Box 191, Cochinchina annual health report, 1933, 196, 198, Cochinchina annual health report, 1934; Box 182, Indochina annual health report, 1935, 300.

blind upon arrival.[46] Plantations such as Bù Đốp, deep in the red earth region, experienced the worst outbreaks, but nearly all plantations were affected. The table above shows the effects of malaria on the plantation workforce.

To form an idea of the relative magnitude of these death rates, the overall death rate in Cochinchina for 1921 was given as 2.41 percent of the population per year, and from 1911 to 1920, the rate was around 2.4 percent of all men, women, and children, young and old. The working-age, mostly male population on the plantations should have had a lower death rate than the general population. If the 1920 numbers remained constant, then in the 1930s, the plantation rate did drop below the general population rate (though plantation numbers did not include the death rate for children, sizable in malaria-stricken areas).[47]

The plantation workforce also influenced overall trends of morbidity and mortality due to malaria in Indochina. The first trend is the link between the number of workers brought down from the north and the growing numbers

Morbidity and mortality among natives due to malaria relative to all hospitalizations
in Indochina, 1915–1934

Year	Hospitalizations and deaths of natives in Indochina		Hospitalizations due to malaria and % of total		Deaths due to malaria and % of total	
1915	78,200	5,400	10,000	12.8	700	13.0
1922	133,412	8,718	16,703	12.5	1,451	16.6
1929	208,600	13,600	21,366	10.2	1,400	10.3
1930	231,602	13,848	21,170	9.1	1,424	10.3
1931	241,518	14,961	18,898	7.8	1,275	8.5
1932	217,030	13,517	19,238	8.9	862	6.4
1933	213,640	13,436	17,329	8.1	793	5.9
1934	243,109	14,441	20,058	8.3	864	6.0

Sources: Service de la statistique générale, *Résumé statistique relatif*, 3, for years 1913, 1914, and
1935–40. IMTSSA Box 182, 1932 annual report, 174, and 1934 annual report, 114. Numbers in italics
adapted from Monnais, who notes that these numbers were not drawn from official publications
but compiled from archival sources. Monnais, *Médecine et colonisation*, 294, Tableau 9,
Le paludisme dans la morbidité et la mortalité indigènes en Indochine (1915–1929).

of malaria patients treated in AMI hospitals. The growth of malaria patients
did not correspond exactly to the number of workers in the rubber industry;
for example, the great increase in the plantation labor force occurring after
1925 was met with a 23 percent fall in the number of malaria patients. This
decrease might show a growing reluctance of planters to send patients to government hospitals, which would increase the plantation's costs. Since building of even rudimentary plantation clinics did not start until the late 1920s
and hospitals were not constructed until the late 1930s, this dip did not indicate a preference of either plantation for on-site treatment but reluctance to
offer any treatment at all. During the peak years of plantation growth after
1922, the numbers of malaria patients grew and remained high. Looking at
numbers like these, the Pasturian Henry Morin emphasized the importance
of malaria in his 1932 lecture to students at the Indochinese School.[48]

Second, the plantations in Cochinchina and Cambodia played an important
role in the overall mortality in Indochina. This role was even more important
in eastern Cochinchina, where the number of hospitalizations was only in the
1,000s for all four provinces combined. Although the world economic crisis
and technological improvements decreased the number of plantation workers in Cochinchina after 1930, this population was subject to a higher level of

medical attention than other Vietnamese, whose use of the AMI remained around 3 percent. Malaria infection continued in other regions of Indochina as workers returned from plantations and, as in Cambodia, the process of forest clearing began to take place in earnest and the relative percentage of hospitalizations in Indochina due to malaria remained high. The worst of the malaria-related mortality had, however, ended.

Third, the growing number of patients who were treated, and died, in hospitals reflected a combination of a growing number of hospital beds and the effects of various epidemics, such as the influenza pandemic of 1918–19 that resulted in a reported total of 34,295 cases and 14,820 deaths in Indochina. This fact helps explain why, despite the steadily growing number of malaria patients during the 1920s, the percentage that these patients constituted in the hospital total remained constant, even decreasing slightly. Of course, as Morin noted, the number of malaria patients treated in the hospitals was only a tiny percentage of the overall levels of malaria morbidity and mortality, as plantations provided little malaria treatment for workers and those living in the countryside had little access to hospitals.[49]

By far the biggest reason for the malaria epidemics was the environmental transformations wrought by the deforestation work that was necessary to create room for the new *hevea* trees. A more detailed understanding of mosquito ecology may help explain what happened on the plantations during the 1920s. Recent investigations into malaria in Việt Nam have focused on the role in transmission of two species of mosquito, *Anopheles minimus* A, or *minimus*, and *Anopheles dirus* A, or *dirus*. These species can act in concert, and in central Việt Nam researchers have recorded *dirus* and *minimus* living in sympatry. Mosquitoes from the *dirus* complex reproduce in water that is shaded in a way available in forests. These mosquitoes are anthropophilic (i.e., aggressive human biters) and inhabited the houses of the Stieng and Phnong, maintaining a cycle of infection. The *dirus* complex was most likely responsible for the southern uplands' general reputation for being unhealthy.[50]

Records show, however, that the most virulent outbreaks of malaria occurred on recently cleared forestlands, and were intimately connected with the road and bridge networks being constructed to bring Vietnamese labor in and take latex and rubber out of eastern Cochinchina. Though it is not clear whether the habitat of *dirus* was destroyed or simply declined in relative importance, mosquitoes from the *minimus* complex came to dominate. The *minimus* prefer clear running streams open to sunlight, an environment that was common to the freshly cut forests. Although not as strongly anthropophilic as *dirus*, *minimus* are efficient malaria transmitters. In fact, the *minimus* mosquito was identified during the colonial period as the main vector of malaria on the plantations, and the role of *dirus* as a primer was not yet understood. As *hevea* grew, the streams became shaded, again creating a habitat for dirus breeding. By that point, a combination of better housing, a quinine regime, and the management of water sources reduced *minimus* and *dirus* numbers, and malaria became less of a problem on the more established plantations.[51]

In response to malaria outbreaks and bad press, quinine was regularly given to workers by the end of the 1920s. At morning roll call, workers were often physically forced to swallow bitter-tasting pills of quinine. This routine led to a famous incident of plantation violence, the killing of Monteil on the Phú Riềng plantation recounted by Trần Tử Bình. According to the official report, the workers were lined up for the morning quinine regime when one worker refused to take the pills. Monteil struck the worker with his cane, after which the workers, who had been living under terrible conditions, reacted by attacking Monteil with their plantation tools.[52]

Most inspection reports from the 1920s give various quinine routines, with less given in healthier regions and during healthier seasons and more otherwise. The Phú Riềng regime that Monteil was administering when he was killed in 1928 amounted to about 180 kilograms of quinine per year.[53] On the Bù Đốp plantation, the daily routine of taking 50 centigrams of quinine was said to take an hour, and because of cost and time it seems likely that many managers were under pressure to cut the quinine regime short. A former worker spoke of having to take medicine to receive his bimonthly salary. This was hardly effective as prevention.[54]

The biological causes, experiences, meanings, and preventions of malaria on the newly created rubber plantations were different from those involved in previous outbreaks. A policy of agricultural development led to the "industrialization" of malaria and other diseases with which the health system was ill equipped to deal. Although industrial diseases are usually thought of as the result of toxic exposure to metals, chemicals, and inorganic materials, this was not the case in colonial Indochina. Industrialization was understood in relation to agriculture, and disease outbreaks on rubber plantations can be interpreted as "ecosystem accidents" like those associated with inorganic industrialization. In other words, malaria outbreaks occurred, in the words of Charles Perrow, because of the "interaction of systems that were thought to be independent but are not because of the larger ecology." Plantation owners were confronted with the problem of designing a system to produce latex for the lowest cost. They likely did not know, or care, about the existence of another system that caused malaria epidemics, and that malaria was an unintended consequence of rubber production. While malaria was the most prominent disease of rubber workers, it was not the only ailment, and colonial labor codes slowly incorporated recent medical findings on a gamut of causes of illness and death.[55]

Malnutrition and Exploitation

Sino-Vietnamese traditions had long recognized the link between food and health, and in the nineteenth century Western medicine started to emphasize a proper balance of ingested elements. A French naval doctor in Asia wrote in the 1890s that "the issue of diet in hot countries is less to give to a body its ration of nitrogen and carbon than to give it these things in the most easily assimilated form, the least harmful." Such interest later led researchers to describe "nutrient deficiency" diseases. Rice-related deficiency diseases became prominent after the introduction of machines that husked and polished rice, a process that drained it of key nutrients. People confined to institutional settings such as militaries, prisons, and, later, plantations were particularly vulnerable to these diseases.[56]

The presence of beriberi reflected the dietary issues on plantations. In the late 1800s, a Japanese naval doctor was the first researcher to argue for a link between beriberi and a rice diet. The existence of vitamins was not known at the time, and medical doctors continued to debate whether diet deficiencies or an infective agent was the primary cause of beriberi. In France, research during the 1910s contributed to the growing knowledge about vitamins, their importance in human diets, and their role in deficiency diseases. Yet the vitamin necessary to prevent beriberi, now known as vitamin B1 or thiamin, was not isolated until the 1920s. In 1921, a French-trained Vietnamese medical doctor, Nguyễn Văn Thịnh, concluded that a deficiency in "factor B" was the most likely cause of beriberi, but did not rule out the possibility that a microbial infection played a role.[57] As medical doctors came to agree on the causes of beriberi, they worked to introduce their understanding into agricultural labor laws. In May 1923, the Goucoch issued a circular setting limits on the husking of the "Java" white rice that was destined for workers. That same year, beriberi was an important topic of a Far Eastern Association of Tropical Medicine (FEATM) meeting in Singapore. Noël Bernard, a Pasteurian working in Indochina, returned from the meeting enthusiastic about setting up a beriberi commission. In addition to improving worker health, Bernard's recommendations aimed at staving off higher rice standards that would have devastated Indochinese exports.[58]

A key task during the creation of plantations was feeding the migrant workers. Since little rice was grown in eastern Cochinchina, plantation owners turned to imports from the Mekong Delta. The poor state of roads throughout much of the region made establishing food supply chains difficult, especially for plantations that were located farther from population centers, so directors sought local sources of food. Workers foraged in the forests surrounding plantations and were encouraged to tend their own gardens after work. Planters depicted these gardens as freeing workers from the small shops that sprung up around plantations, whose prices were twice as high as or higher than the prices of goods found in the nearest population center.[59]

The second rubber boom of the middle of the 1920s stretched these food supplies to their limits and contributed to the spread of beriberi on plantations. According to reports from employers in Cochinchina, Cambodia, and Southern Annam to the labor inspectorate, the number of workers who died from all causes averaged about 140 per year from 1919 to 1926. In 1927, this number jumped to 788; in 1928, it increased to 1,323; and in 1929, it decreased to 848 workers.[60] These deaths forced business interests to recognize the need to address malnutrition diseases. In the spring of 1926, de la Souchère was assigned to the commission to study worker diet (*ration alimentaire*) in Cochinchina. This commission included medical doctors such as Jean Guillerm, an expert on fish sauce. Commenting on advances in medical care in Indochina late in 1926, Henri Cucherousset, editor of *L'Eveil économique de l'Indochine*, noted that targeting the native diet (*ration alimentaire des indigènes*) was a priority. With the study of the nutritional "value of native food almost complete," it was time, Cucherousset stated, to begin the difficult task of "changing the traditional behavior, the immemorial and defective customs."[61]

Worker desertion and death also made it impossible for the government to pretend that labor protection laws were adequately enforced. Workers began to desert plantations en masse, and, after more than 300 laborers left Mimot early in 1927, the French colonial government sent Émile Delamarre, its inspector of political and administrative affairs, to investigate plantation conditions. Delamarre wrote a scathing one-hundred-page report strongly criticizing both the individual brutality of a young Belgian assistant and the orders for physical punishment given by an older Belgian manager. Although the cruelty of these two could not be excused, Delamarre argued, the ownership of Mimot was at fault, as it was unwilling to pay for any improvements in living conditions. The local health director remarked that the Vietnamese workers' housing resembled chicken cages and that European housing was not much better. He complained further that the scarce resources of his service were being diverted to care for the plantation's sick.[62]

Delamarre's report generated a strong reaction in the French and Vietnamese language press. An author in a 1927 issue of *Presse indochinoise* wrote of the workers' lives sacrificed to the "Moloch of high finance." The author argued that his motivation for speaking out came not from a "puerile" sense of humanitarianism but from a concern about worker unrest. For the French public, deaths on plantations became a concern mostly as a threat to social order. Meanwhile, French officials and planters in Tonkin who were unhappy to see "their" labor go south spoke against plantation conditions, which provided material for exposés such as *Les jauniers*, Paul Monet's indictment of labor recruitment practices.[63]

Because of the pressure generated by this report and concurrent press campaigns, the GGI issued a revised labor code on October 25, 1927, establishing uniform laws for all of French Indochina. In addition to strengthening provisions for health, these laws set up a workers' savings fund and created an Indochinese labor inspectorate. As in 1918, the labor code was intended to simultaneously protect workers' and business rights. Most of the articles in these *arrêtés* that dealt with health sought to improve the quality of recruits and improve workers' efficiency on the plantation itself. The newly created labor inspectorate had the right to enter plantations and to suggest improvements to the planters. If these suggestions were not followed, the inspectorate had the right to assess fines on the plantations, but could do little else. While on paper these laws appeared to check the worst of owner abuses, in practice they did not change behavior on the plantations. If labor was plentiful, most owners viewed it as an expendable resource.[64]

While the experience of workers on plantations showed the urgency of the questions of malaria and beriberi, medical science attempted to provide the answers. The government tried to position itself between the interests of the planters and those of Indochinese labor by speaking the language of science. Based on medical articles calling for a daily ration of 750–800 grams of rice, labor laws required that at least 700 grams be provided to every adult worker as a supplement to their salary. This amount of rice was greater than the 555 grams that the average Red River Delta peasant consumed. Some companies went beyond this minimum, with Michelin providing 850 grams for workers and additional rice for their dependents.[65]

Despite these reforms, beriberi and other diseases continued to cause sickness and death, partly because the provision of rice required by the 1927 labor codes did not apply to workers who were already under contract when the law went into effect. Some plantations did not provide this ration, while others gave poor-quality rice. As in prisons, many plantations contracted out the provisioning of food. Often the contractors, both Chinese and Vietnamese, cut corners to increase their profits. But many workers contracted beriberi even when the full portion of Java number two white rice was given. The Agricultural Company of Thành Tuy Hạ, which was established around 1912, received a bad evaluation at the end of 1928 in part because of several cases of beriberi. Forced by economic pressures to respond, this plantation installed its own polishers, which retained roughly 20 percent of the husk. Some plantations took to mixing red-grain and white rice to increase its nutritional value, and others provided supplements of bananas, potatoes, and yams.[66]

Most planters continued to hold workers morally responsible for their own malnourishment. As historian Webby Kalikiti has pointed out, the editors of *L'Indochine: Revue économique d'Extrême-Orient*, a pro-planter journal, held that if workers "appeared poorly fed and sickly, it was because they often lost their food rations in gambling and did not take their doses of quinine as required." The editors went on to suggest that the new labor laws were driven by "philanthropic sentimentalism" and the best solution "was the application of very strict, quasi-military discipline in work places."[67]

Deflected Responsibility

By the late 1920s, working conditions on rubber plantations had become an embarrassment for colonial officials, as continual worker protests and journalistic exposés drew attention to poor conditions. Though plantations did record evacuating workers to local clinics and even hospitals in Biên Hòa or Sài Gòn, the first resort was the plantation clinic. By the late 1920s, plantations had begun to pool their resources and create a few centralized health-care centers. While not yet the model hospitals of Michelin of the late 1930s, these health centers (such as Suzannah's) accepted workers from surrounding plantations. Far from curing, these places, and the nurses who ran them during the 1920s, often added to the problem. In 1928, Bùi Bằng Đoàn, a mandarin from the north with a special interest in the fate of northerners working in the south, noted the two complaints of brutality lodged against the nurse of BIF's Cây Gáo plantation, with four marks from a rattan cane still visible on the sufferers' bodies. Even as late as the early 1950s the burden of proof of illness remained on the workers and their bodies. Doctors did not ask workers whether they were sick or healthy but merely checked temperatures and felt heads and then gave a "yes" or "no," regardless of how the worker felt. Treatment was even worse immediately after World War II, when no doctors were stationed on plantations. During those times, plantation managers often simply looked at the workers to determine fitness for the day. Rarely did they find the workers too sick to go into the fields.[68]

Though Vietnamese women received some protections, they too faced problems. For instance, women were supposed to have two paid months off for childbirth, more than a woman in France was legally entitled to. The tasks, too, that they were assigned were often lighter. But there were numerous disadvantages, such as being given less food, paid lower salaries, and given little physical protection from the men around them. Though plantations were supposed to recruit 30 percent women, males mostly dominated the populations. For example, on BIF's Cây Gáo plantation, contracted workers included 866 males, 86 females, and 8 children, along with 20 "free" workers (i.e., day laborers) from Annam.[69]

For their part, workers attempted to mitigate the effects of plantation conditions. In the 1910s, labor protests on plantations over issues such as nonpayment of salary were carried out in older forms, with workers signing their names in Chinese characters in a circle. During the 1920s, Vietnamese laborers learned how to draw on resources provided by medical science to claim rights under the colonial government. These resources included medical reports, and some government physicians were aware that the establishment of plantations had largely brought about malaria epidemics. Furthermore, a former manager spoke of workers heating rocks and applying them to their bodies to give the impression of a fever. This strategy could work since medical doctors were often pressed for time during roll call inspections. Workers also used Vietnamese medicine, or *thuốc nam*, and gathered mushrooms in the forests. In the 1930s, some plantations gave workers space to grow their own vegetables, which also helped the plantation defray its costs. Not all workers were tappers from the north, and a small but important percentage came as skilled workers. Plantations provided other employment chances, and a tiny number of Vietnamese professionals, such as drivers, metal workers, overseers, carpenters, secretaries, nurses, and mechanics, worked for the plantations. The 200 "specialized" Vietnamese and Chinese workers on the plantation of the Société Agricole et Industrielle de Cẩm Tiên could earn up to three piasters a day.[70]

One of the ironies of the colonial period was that a government labor inspector, Delamarre, was one of the first to bring to light the horrendous death rates on plantations under his responsibility. The labor inspectorate often noted in its reports that the management was trying hard to improve

Protest note. Complaint in June 1912 by workers against BIF. It was withdrawn with the new contract. Signing in a circle was an old technique meant to prevent government officials, or management, from easily identifying leaders. Source: NAVN2 IIB.54/112.

conditions and agreed that many workers were opium addicts. Furthermore, the labor inspectorate argued, after the initial clearing work, mortality numbers had dropped down to those experienced locally by the native militia and secretaries, and even the European assistants. This extreme case of negligence shows the amount of slack given to companies that made even minimal efforts to improve health conditions. The scandals created by the press over the miserable conditions on some plantations pushed the government to pass a sweeping reform of the labor law in October 1927, which was more favorable to the health of workers.[71]

Yet the plantation industry avoided meaningful reform, causing critics to charge the government with exercising insufficient control over transnational capital. Trần Tử Bình recalled in his memoir that "the owners had nothing to fear from the labor inspectors. The government was their government, and everything was fine so long as they could keep matters quiet." Some labor inspectors agreed that little had changed, though they placed the blame on planters. During Bùi Bằng Đoàn's high-profile tour of the south in 1928, he stopped at Michelin's Phú Riềng plantation, where a French assistant had been killed in September the previous year. Đoàn asked tough questions and recorded some damning claims by workers; however, no immediate changes resulted from his visit. While Bình may not have perfectly understood the relationship between the government inspectors and the plantations, he correctly observed that nothing happened after inspection tours.[72]

Business saw matters otherwise. The 1918 regulations contained strict guidelines concerning housing, but the planters argued that such guidelines were not in line with reality. In February 1927, the financier Octave Homberg, speaking before the pro-business lobby Comité de l'Indochine, argued that while regulations regarding buildings on plantations were admirable in theory, strict application gave results contrary to what was desired. He cited the example of laws requiring housing to be built on cement, meaning that houses could not be raised, which supposedly protected from reptiles and wild animals. Homberg also argued that building *paillotes*, or straw huts, meant that they could be burned every two to three years—as the "moïs" of the region did—but "expensive" housing costing thirty to forty piasters per head could not be replaced. Finally, he argued that "coolies" often preferred beds of bamboo to beds of wooden planks.[73]

In 1929, the Comité d'Indochine blasted the 1927 labor codes as overly harsh on employers. This pro-planter organization framed its critique of the law largely as an attack of the government's own record of health protection. It pointed to a lack of a colonial health-care system and the miserable health of those working for the public works as undermining the government's credibility. The committee cited figures comparing the 300 doctors in Indochina for a population of 20 million with the 20,000 medical doctors in France for 40 million people, and pointed out the lack of hospitals in the provinces. It further charged that the administration had failed to deal with the problem of starvation in the Red River Delta, which meant the Vietnamese were undernourished and weak on arrival at the plantations. Industrialists and agriculturalists also called for government aid. The Comité urged "that the Administration give real support to the plantations of southern Indochina in combating, by all the means in its power, the propaganda that is carried out in the anti-French native press not only against these plantations but also against all of French colonization." The Comité argued that the civilizing mission could be carried out through private enterprises for the good of all and held that a comparison of health among Vietnamese working on plantations, on public works projects, and living in the Delta, would show that plantation conditions were not different from those of other places.[74]

Part of the government's reluctance to criticize plantation practices arose from a concern that the "violent campaign against the hiring of contracted labor for southern Indochina . . . [was] anti-French." This campaign included the 1929 assassination in Hà Nội of a French director of labor recruitment. Instead, government and capital used science to reach a compromise over labor. Explanations of death rates focused on beriberi, malaria, and other diseases partly as a convenient metric, and partly as an attempt to naturalize the epidemics.[75] Delamarre, whose 1927 report had caused a stir, wrote in 1931: "It is, in effect, currently unanimously known by the doctors and the functionaries that have dealt with labor employed on the plantations of southern Indochina that the unavailability, the hospitalizations, the deaths, the repatriations, the infant mortality and the abortions recorded among the agricultural working population, above all on the concessions located on *terre rouge* and, particularly at the beginning of new exploitations or in a period of extension of old [exploitations,] should be attributed, for the most part, to malaria."[76]

JOURNALISTS REMAINED MUCH MORE inclined to criticize plantation conditions. During the 1930s, journals such as *Sài Gòn* often published stories of workers who were hurt on plantations. For example, a June 1935 issue reported on twenty-eight-year-old "coolie" Nguyễn Văn Lạc, who was born in a village outside Thừa Thiên near Huế and worked for Quản Lợi. While climbing up a *hevea* tree to cut off a branch, he slipped and fell, breaking a leg. He was taken to the Chợ Rẫy hospital for treatment, but it was not clear what would happen thereafter. What was, the journalist wondered, the responsibility of the plantation toward workers who were the victims of such accidents? Would Lạc be taken care of or would he "have to accept unemployment like 100s, 1000s of other people?"[77] The specter of communism overrode these concerns in the minds of many officials. In a letter evaluating the causes of worker protest on Michelin's Dầu Tiếng plantation, the administrator of Thủ Dầu Một province wrote in closing of "the events of 1931." He likely had in mind the Nghệ Tĩnh Soviets in northern Annam, during which hungry peasants carried out mass demonstrations. These demonstrations were brutally repressed by the colonial government, which dropped aerial bombs on the protestors. The administrator was worried about the influx of workers from Annam, including those who may have participated in the revolts and wished "to continue their subversive propaganda" in the south.[78]

The Indochinese labor inspectorate that was set up in 1927 went through many changes, with some officials calling for biannual visits and the granting of police powers to inspectors. The depressed economic conditions of the 1930s dampened such efforts, and in 1932 the labor inspectorate was placed under the Department of Economic and Administrative Affairs. The rise of the Popular Front government in France in 1936 brought some momentum back to labor reforms in the colonies. The 1927 law already covered contract workers, so the reforms of the 1930s focused on "free" labor, which had lacked any form of protection. The early regulations dealt with working conditions for women and children, the length of the working day, workers' hygiene, and the role of the labor inspectorate. Many unanswered issues remained, some of which were addressed by a labor code passed on December 30, 1936, and then promulgated by an *arrêté* of the GGI on January 27, 1937. According to the labor inspector at the time, this law formed the basis of "a veritable labor charter clear and complete." He noted that some "extremists" had criticized this law for not going far enough in protecting workers' rights. Yet, this law, he maintained, made genuine changes while "not leading to disruptions too abrupt in the economic and social structure of the land."[79]

Continuing Profits, Continuing Protests

By the time that the socialist and former minister of health Justin Godart arrived in Indochina in the spring of 1937, initially limited technical solutions to the problem of worker health had morphed into more comprehensive developmental visions. Godart was sent by the Popular Front government to inspect conditions in France's colonies, a response to the frequent worker strikes and unrest among colonial populations, and his tour included two days in the rubber-growing region of Indochina and on French- and Vietnamese-owned plantations. During his brief visit Godart noted what he called the brutality and stupidity of French colonial society, and he brought the insight of a well-informed, well-placed, and critical outsider to bear on the colonial situation. In his report, he remarked on the beneficial changes rubber production had brought to the region. He wrote about the "remarkable administration . . . [and] important clean-up work" that had been carried out on the plantations he visited, which included those of Michelin. Yet, he also perceived the "dreariness of the people living in the companies' houses" and likened the plantations to "a forest without birds." This image of a bird-less forest captured the uncanny feeling that Godart had as he walked around the facsimile of nature found on the working landscapes of plantations. And while he approved of the plantation managements' efforts to improve worker health, he noted the biological and social poverty of plantation landscapes.[80]

Some planters understood the need to adapt to the changing colonial politics. In response to a questionnaire sent out by the Guernut Commission, the fact-finding mission headed by Godart from January to March 1937, the SPCI emphasized material development, arguing that the steps taken on the plantations should be a model for the countryside. At the same time, the organization argued that the rural population was not ready (*mûrir*) for political involvement. Yet, the reform efforts of these planters did not impress all the administrators. In response to worker beatings administered by French personnel, the Goucoch wrote of the "regrettable feudal spirit" of plantation owners. He also pointed out that projects aimed at ameliorating plantation conditions remained largely in the planning stage, as demonstrated by photographs of actual villages of the Red Earth Plantation Company (SPTR). While the houses in the images are indeed separated, the dirt paths turned to mud in the rain, and the construction materials consisted of wood and thatch.[81]

In practice, plantation owners continued to do little to improve working conditions. The results of the Guernut Commission were indicative of the failure of the French colonial empire to respond to frequent plantation worker strikes and general unrest among the populations of Indochina. Before Godart's visit, Vietnamese intellectuals had suggested a range of reforms. In the end, his visit generated more hope than fulfillment and there was much disappointment. The failure of the French government to institute meaningful reforms caused many intellectuals to turn against colonial rule and helped lead the way for revolutionary nationalism to take hold. As a satirical cartoon

about Godart's visit implied, many activists came to realize that "the liberation of the colonies will be the work of the colonized themselves."[82]

Workers continued to push for changes in labor conditions on plantations.
One example of such activism was the Cochinchinese Agricultural Mutual
Aid Association. With the government's approval, this association held its
first meeting on May 18, 1935, at the EPA in Bên Cát, Thủ Dầu Một. Its goal
was to help workers who had lost their jobs and assist families of members in
need. The association fees were an initial three piasters followed by fifty xu
(cents) each month. It is difficult to evaluate the effectiveness of this organization, but it did not receive much attention in the late 1930s.[83]

Over a decade of worker activism had placed Michelin's three plantations,
in particular Dầu Tiếng and Phú Riềng, among the targets of reform. Worker
unrest caused the labor inspectorate to submit an eighty-five-page report to
the GGI on June 19, 1937, about workers' lives on the Dầu Tiếng plantation.
As Webby Kalikiti notes, this report casts Dầu Tiếng in a very unfavorable
light, especially when compared with neighboring plantations in the Hớn
Quản region such as Quản Lợi, which was owned by the SPTR. The inspector criticized the mentality typical of what he called "a great industrial and financial power" when the Dầu Tiếng plantation sought to impose industrial
principles and Taylorization that were used in factories in France. According
to the inspector, these measures were not appropriate and showed a lack of
psychological understanding of the labor force. Unlike peasants from Tonkin
and Annam, workers in France had been accustomed to factory discipline for
two or three generations. The peasants had only known work on the land,
which varied according to season and day, unlike work in an indoor factory.[84]

The inspector also criticized the choice of personnel, who generally fell into
one of two categories. Many French managers had come directly from France,
with no knowledge about life in Indochina. Michelin policy was to hire men
who did not have preconceived ideas about how to run a plantation on the theory that they were open to innovation. This had the unfortunate effect of
sending employees who were unprepared for the job, knew little about their
workforce, and committed serious errors that threatened both Michelin's plantations and the stability of the colony. But the inspector reserved his sharpest
criticism for those who directly managed the tappers and multipurpose
workers, noting that some of these employees were Eurasians, whom he
viewed as "low-quality" humans when compared with other personnel in
the region. He surmised that perhaps these Eurasians had too much of the
"Annamese mentality"; in other words, they were too given to gambling, to

making loans, and to expecting "thanks" for services rendered. This "flexible" attitude may have exposed workers to additional abuse.[85]

The inspector's report offers detailed calculations showing quantitatively the poor conditions for workers on Dầu Tiếng as compared with those on the Quản Lợi plantation. The inspector divided the tasks of tappers into "leg" and "arm" work. He argued that the former predominated on plantations such as Dầu Tiếng, which had young trees, and that the latter predominated on plantations such as Quản Lợi, which had more mature trees. Tapping on Dầu Tiếng was much more labor intensive and placed great strain on the workers' legs. On Dầu Tiếng, each tapper was responsible for a parcel that had 82 trees in 10 rows, or 820 trees. Owing to variations in land and tree, the number of trees tapped in practice ranged from 375 to 435. This did not necessarily decrease the distance traveled by the worker, however, as he still had to travel the full row. Dầu Tiếng adopted a 6×6 planting scheme, which meant that each tree was separated by six meters on all sides from the others. Multiplying the number of trees in a row by the space in between each tree meant that workers had to travel 500 meters for each row. With ten rows, one round of tapping required five kilometers of travel, and one round of collecting latex required another five kilometers. In addition, workers had to move from village to plot and back, a roundtrip distance on Dầu Tiếng of three to four kilometers, which meant a morning trek of thirteen to fourteen kilometers.

On Quản Lợi, by contrast, tapping and collecting patterns for its more mature trees required less movement. Tappers were responsible for plots consisting of 2 rows of 145 trees each, for 290 trees, placed in a pattern of 6.5 × 6.5 meters, which would require a walking distance of four kilometers. Furthermore, on Quản Lợi, villages were located at the center of tapping plots and the maximum distance traveled by any one worker to and from the village was three kilometers. The total travel came to seven kilometers—about half the distance required on Dầu Tiếng. The Quản Lợi terrain was hillier than the flat Dầu Tiếng ground, but Quản Lợi had less ground cover, which made travel easier. It appeared that Dầu Tiếng, with its average of 400 to 420 trees per tapper, required more effort from the workers' arms. But Dầu Tiếng trees averaged about a half meter in circumference and were tapped with a simple spiral. Quản Lợi trees ranged from 1.35 to 1.60 meters and were tapped with a double spiral in the Sounderland system. The arm work on both plantations was roughly even, and the author concluded that tappers on Dầu Tiếng had a much more physically demanding task.[86]

Vietnamese intellectuals, including physicians and journalists, also advocated for improvements in plantation conditions by urging widespread improvement of rural health. Several leagues to combat specific diseases such as tuberculosis were formed, and the Ánh Sáng league organized efforts to build clean housing for the poor, while the Lạc Thiện and the Hợp Thiện societies were formed to provide public assistance in Huế and Tonkin, respectively. The press had long been a driving force for change, and journalists continued to publish articles about investigations into rural conditions. *L'Effort indochinois*, whose director, Vũ Đình Dy, had become interested in plantation conditions during his reporting in the late 1920s on plantations in Cochinchina and Cambodia, published several articles about the plight of peasants. These articles included a three-part editorial called "Roman de Nguyen le Nha-Que, La Justice Sociale en Indochine ou les 90 millions de caoutchouc" by Pierre Toussaint. This series, which appeared in November 1937, questioned the political priorities of a government that approved the ninety million piaster loan for plantations while allotting very little for disaster relief. Even during the Popular Front period, however, direct press attacks on plantation conditions remained risky. On December 13, 1938, Nguyễn Văn Mai, the manager of the journal *Dân Mới*, was sentenced by Sài Gòn's Tribunal Court to three months in prison, a 1,000 franc fine, and 100 piasters in damages to be paid to the Indochinese Company of Hevea Plantations (SIPH) for publishing a negative article on the enterprise. In 1939, at the end of the Popular Front, a newly restrictive censorship regime made open discussion of worker reform much more difficult.[87]

Soon thereafter, more radical forms of dissent emerged. From 1936 to 1939 Nguyễn Văn Bát organized a women's Association at Suối Tre in Biên Hòa. Bát used the phrase "family is far, neighbors are close" to convince the workers to organize and join his association. The association was based on regional ties and tried to instill a love of *quê hương*, or homeland. Bát also discussed communist doctrine and universal harmony with the father and uncles of Lê Sắc Nghi, later a communist revolutionary fighter. According to Nghi, after this compatriots' association was formed, religious, economic, and social differences no longer divided the workers. Although the communist party attempted to take sole responsibility for worker agitation, something the French Colonial Government was happy to assent to, workers acting on their own or in groups were often key drivers of change in the region. Innumerable worker protests that pressured owners to ameliorate plantation conditions were a force for reform. Most of these protests focused on higher wages, which had been reduced to bare minimums during the economic crisis of the 1930s, but plantation owners often responded to worker action, at least rhetorically, by highlighting projects for churches, schools, soccer fields, and hospitals.[88]

THINKING ABOUT THE LIVES of tappers on plantations builds on studies of biopolitics by moving beyond a singular focus on bodily processes, and prompts one to consider how the interactions between bodies and their environments should be considered in light of colonial situations. In the case of French Indochina, plantations, like botanical gardens, agricultural experimental stations, hospitals, leper colonies, and laboratories, were designed to act on both the environment and the body. These spaces produced knowledge about plants and pathogens on and in working landscapes, and served as demonstrations about what could be done with nature based on science. In other words, plantations acted on the environment and the self, for a profit.[89]

Discussions of the relationship among colonial labor requirements, the health of those living in French Indochina, and the emerging French medical conception of the populations of Indochina as a "problem" to be solved have been largely ignored, and one purpose of this chapter has been to place human conflicts generated by malaria and other diseases within a literature of environment and health. As discussed in the previous chapter, ecology in French Indochina was envisioned as useful for profitably exploiting the environment, and medicine was envisioned as useful for maximizing the output of the human body. The French physicians working on the plantations brought this perspective to their work, but this understanding clashed with the other actors involved. After the 1930s, the environmental context of malaria, while not completely discarded, was being pushed to the background by a more intense focus on the human carriers of the disease, as medical authorities came to see upland ethnicities as particularly dangerous reservoirs of disease.[90] The next chapter considers the efforts to map and naturalize new spaces of malaria and race in rural Southeast Asia.

CHAPTER FOUR

Turning Tropical

Everything proves that if they [humans] are willing to submit to the necessary sacrifices, all human races may live and prosper in almost every climate.
—Armand de Quatrefages

The uncultivated state of the land explains the great unhealthiness.
—Mathis and Léger, "Le paludisme au Tonkin," 1911

This disease [malaria] appeared more and more like the greatest natural barrier to colonization, that is to say settling on the soil sedentary, agricultural, peaceful, and hardworking and therefore tending towards population growth and progress.
—Henry Morin, *Feuillets d'hygiène indochinoise*, 1935

Pierre Gourou (1900–99), one of the founders of French tropical geography, spent many years in Indochina. Gourou was based in Tonkin, his full-time teaching job leaving him little opportunity to visit the southern region. He also preferred to study traditional Vietnamese society, which he viewed as the rice-growing Kinh villages of the Red River Delta. Like many intellectuals of his day, Gourou felt uneasy about cultural mixing, and while he did not openly criticize the colonial project, he tended to ignore some of its most important consequences. Thus, it is little surprise that Gourou did not write much about the rubber plantations of Cochinchina.[1]

When Gourou first arrived in Indochina, the French empire was transitioning from the colonial sciences and their racial categories, to the tropical sciences, with the environment, including climate and disease, as organizing principles. While the idea of the "tropics," a place vaguely defined by high rainfall, warm temperatures, and human diseases, had long informed Western culture, the tropical sciences gave this concept a new intellectual precision.[2] A major contributor to this formulation was Gourou. In his writing about the tropics, he initially differentiated "techniques of production" and "techniques of territorial control" but by the 1970s, he had brought these two sets of techniques together through the term *encadrement*. As the geographer Michel Bruneau has made clear, "what mattered particularly to Gourou were the framings and frameworks pertaining to civil society (family, language, property regimes, 'mentalities', religion, and so on) as well as political society (the village, tribal structures and state frameworks) that were implicated in landscape development." In this way, Gourou helped fashion an expansive vision of tropical geography that informed first the colonial and then the developmental sciences. Furthermore, his concept of *encadrement* served to underline the point that "'the cultural' and 'the political' were strongly linked," both to each other and, in complicated ways, to the environment. Even before the crises of World War II and the two Indochinese Wars, Gourou and other experts promoted concepts and tools that were influential during the process of decolonization.[3]

Gourou's tropicality provided a unifying framework for the colonial and postcolonial problems of agriculture, health, and governance. As the geographers Gavin Bowd and Daniel Clayton have argued, "there were close, if never always direct or entirely transparent, connections between the production of academic knowledge (often in the form of fieldwork monographs) and the creation and maintenance of colonial forms of governmentality." In French Indochina, malaria outbreaks on rubber plantations helped to cement knowledge-producing networks that connected planters, government officials, and medical doctors. In 1928, for instance, Michelin partially sponsored the Pasteur Institute's efforts to combat malaria. A year later, Henry Morin convinced the GGI to adopt the institute's antimalaria measures for public works projects. By the 1930s, such networks stretched across imperial borders as American, British, Dutch, French, and Japanese colonial experts met and discussed problems they organized around the theme of the tropics.[4]

Likewise, plantation agriculture in general and rubber in particular provided the means and the motivations for the introduction of the science of ecology to Indochina. This science was neither antithetical to the practice-oriented ways of plantation agriculture nor subversive to the colonial mission. Colonial experts and their associated institutions helped introduce ecological concepts, which then moved through scientific, government, business, and eventually popular circles. Officials, experts, and planters found ecology to be a useful language with which to communicate as they attempted to frame ecology as an "apolitical" science in the highly politicized space of late colonial empires. Rather than subversion per se, the sciences of European empire planted the seeds of developmentalism.[5] By untangling the mutual influences and growing divergence of geography, medicine, and ecology during the late colonial era, this chapter helps historicize these sciences in twentieth-century Southeast Asia.

Race for Plantations

Questions of race were intimately tied to notions of the tropics. Race thinking in Indochina drew from nineteenth-century scientific racism, or the reliance on biological characteristics to understand and define human difference, and it was used to try to make sense of a complex society. Yet, scientific racism repeatedly failed to apprehend the quotidian reality faced by those in the colonies and French colonialists often turned to older distinctions between "civilized" and "barbarian" current in Vietnamese, Chinese, and other Southeast Asian cultures.[6]

At the turn of the century, race thinking intersected with colonialism through the promotion of projects such as human and animal acclimatization, which drew from a Lamarckian tradition that held that environmental conditions affected heredity. This tradition coincidentally helped soften French medical views of heredity vis-à-vis Anglo-Saxon theories of heredity, which emphasized the role of genetics in passing on human traits. In the early twentieth century, thinking about race was achieving new scientific precision as anthropologists studied phenomena such as *métissage*, or race mixing. Never a numerically sizable group, métis nevertheless loomed large in scientific racial thought, and the colonies provided a veritable goldmine of data as theorists headed into the field to study the consequences of colonial empires for human biology. As Ann Stoler has shown, colonial administrators fretted over the question of métis as they threatened to muddy clear legal categories. Yet, in Indochina, their hands were often tied as French legal codes did not allow for de jure racial discrimination. Furthermore, officials in the French empire paid little attention to the more theoretical writings about race; as Emmanuelle Saada has shown, they essentially ignored the results of a 1908 survey on métis carried out by the Paris Society of Anthropologists.[7]

Race played a prominent role among plantation managers and planters' groups, both as an organizing category and as a practical tool for plantation management. A 1923 government report on the working aptitudes of the different races of Cambodia revealed the ways in which racial divisions on plantations drew from and fed into other imagined assemblages of labor, land, and people. This report highlighted the independent spirit of the rural Cambodians, arguing that they did not like to work for others. This trait made Cambodians appear lazy in the eyes of plantation managers, and for French administrators, the "very independent" Khmer-speakers resisted improving their lives through hard work. Conversely, Chinese labor had become "the essential factor of economic prosperity" and "indispensable intermediaries." Managers and administrators did not understand that, as residents, Khmer-speakers had family ties and other social networks, and preferred to farm their own land rather than work for a plantation, while ready access to capital and social networks allowed the Chinese to fill their role as financial intermediaries.[8]

Plantation practices also led to the creation of racial categories. In 1927, Raoul Chollet, the manager of the Snoul plantation in Cambodia, wrote about the aptitudes of the "races" for the various plantation tasks and assigned different jobs to different "races." With respect to tree tapping, Chollet preferred the "Tonkinese," or Vietnamese from the north. This group also built roads and did other heavy tasks. The second-best group, according to Chollet, was the "Annamites," or Vietnamese from the center (not to be confused with the use of "Annamite" to refer to all Vietnamese during the colonial era). Europeans also employed Chinese, Javanese, and Tamils for tapping, while Cambodians and Malays, by which Chollet probably meant Cham Muslims and possibly Stieng, helped clear the land. Women, children, and the ill generally carried out the lighter tasks, such as tending to saplings and weeding. During the colonial period, most European planters expressed similar views on race and gender.[9]

Such colonial racial views support Pierre Brocheux's argument that the settlement areas, or inner core, of plantations re-created "the image of the global colonial society." In this microcosm, living and working conditions were spatially arranged according to job function and race. Yet, in practice, plantation managers exhibited some adaptability in matters of the division of labor duties, and practical considerations and the availability of labor occasioned a slippage between racial discourse and plantation practices. In 1928, the Agricultural and Industrial Society of Cẩm Tiên listed among its 600 free, or noncontracted, laborers many "moïs" and Cham. These non-Kinh laborers carried out many kinds of tasks, including tapping, which was often attributed to the Tonkinese. Photographs and memoirs show that both Kinh and Montagnards performed clearing work, even though Kinh often suffered more injuries and fatalities from this activity due to their inexperience with cutting down and burning trees. Although plantation discourse divided the races of Indochina into groups with different abilities, individual planters were often pragmatic, drawing on preexisting structures of human difference and re-forming them into a racial vision of society.[10]

Perceptions of race could mask potential fault lines, and hardened racial categories caused management to misunderstand breakdowns in the relationships among those who worked for them as part of the growing anticolonial activism. When tensions arose between Tonkinese and Annamites in

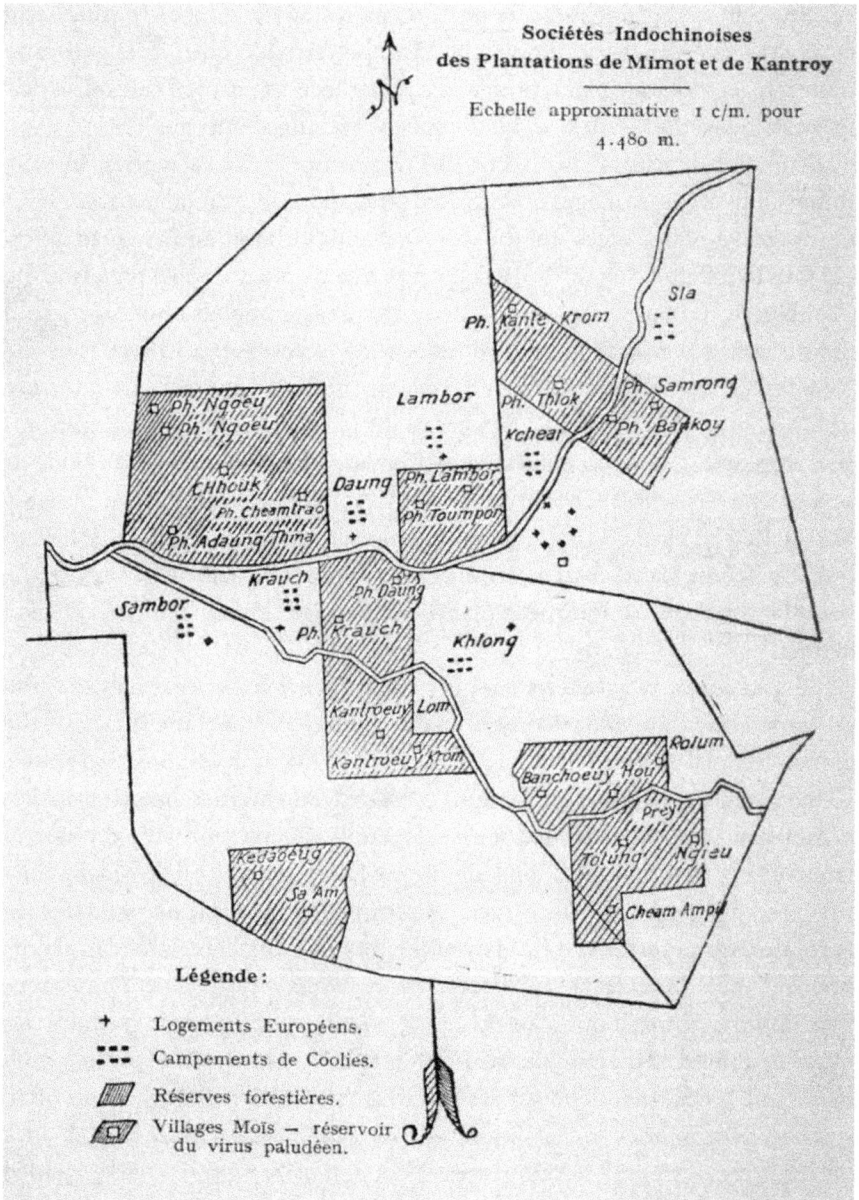

Sociétés Indochinoises
des Plantations de Mimot et de Kantroy

Echelle approximative 1 c/m. pour
4.480 m.

Légende :

+   Logements Européens.

⋮⋮⋮   Campements de Coolies.

▨   Réserves forestières.

▱   Villages Moïs — réservoir
     du virus paludéen.

Organization of Mimot and Kantroy plantations. This sketch shows how French medical doctors and plantation managers viewed the relationship among workers, Montagnards, and malaria. Medical doctors argued for separating workers ("coolies") and Montagnards ("moïs"). Source: *Archives des Instituts Pasteur d'Indochine.*

May 1930 on the plantation of Stung Trang, which was owned by the Société des Caoutchoucs du Mekong, management blamed racial difference. At the heart of this conflict was a Kinh from Tonkin named Dang Van Kien. Dang and some "Annamites" were reported to have been in the room of one of the *cai*, or overseers, when a scuffle broke out in which Dang was injured. He left, but later returned with several others from Tonkin, and in the ensuing brawl Hoang Van Dek from Annam was killed. Such conflicts often arose for several reasons, not simply racial or geographical difference. Planters exacerbated tensions by assigning "Cochinchinese," or those from the south, and métis to low-level managing positions, such as *cai*, while charging those from Tonkin and Annam with tapping duties. Though the details of this incident remained murky to administrators, they appeared satisfied with the explanation that the fight stemmed at least in part from "a certain discord that always had existed between Annamites from northern Annam and Annamites from Tonkin." Yet, the mysterious role of the *médecin indochinois*, or Indochinese doctor, which was noted by the labor inspector, and the fact that he had originally requested three *gardes indigènes* (the native militia) suggest the cause might have been political in nature. The 1930s were a time of growing communist networks, and plantations were favored sites of activity. Nurses worked to spread communist teachings, and the fact that Dang, the single Vietnamese from Tonkin, was in the housing of those from Annam suggests a possible effort to recruit for a strike.[11]

Racial Hygiene

Writing immediately after World War II, Pierre Gourou considered the problem of *mise en valeur* and health. "The primary problem of the *mise en valeur* of hot and humid regions," Gourou concluded, "is a problem of healthiness." Gourou's emphasis on health was a constant, from colonial to tropical to later developmental projects. Nearly fifty years of French attempts to introduce economic projects to Indochina had inspired much more detailed knowledge about the region's environment, health, and peoples, that nuanced earlier theories connecting climate and health. Gourou's statement suggests that tropical lands remained uncultivated because of the insalubrious state of the region, which could not be attributed to a lack of French effort. Industrial agriculture played a key role in beginning to connect environment and health. In this way, plantation experience also shaped colonial subjectivities as the presence of malarial protozoa in blood reinforced preexisting race thinking. The interests of industrial agriculture in the region played a key role in promoting medical science, and environmental knowledge was used to define difference and to locate the proper place and time for bodies. The Pasteur Institute came to blame the upland peoples for malaria outbreaks that killed plantation workers, and recommended a policy of spatially and temporally separating these peoples from plantation labor.[12]

Drawing on knowledge about malaria in Europe, a 1911 paper by Mathis and Léger about the disease, which is quoted in this chapter's epigraph, attributed it to the lack of cultivation of the soil. Medical doctors viewed passive Vietnamese farmers who worked the land as having created the conditions for malaria. Moreover, contemporary thinking about *mise en valeur* was largely hierarchical and envisioned whites as the "brains" and natives as the "arms." A medical doctor participating in the "Reform of Hygiene and the Native Medical Assistance of Indo-China" section at the 1918 Congress of Colonial Agriculture expressed this racist logic when he wrote: "It's an axiom that the white under the tropics cannot engage in manual work without danger. Every European who digs the earth, digs his or her grave, people have rightly said. Since the white cannot carry out hard work of public or private use (land, rail, or water routes, agricultural or industrial exploitations), strength is to turn to native labor. The major part of the future of the colony depends on this labor. They are the *arms* that must obediently carry out what the white *brain* has designed." This statement underscores the connection between health and agriculture in French medical theory. Despite the introduction of the germ theory of disease, which ostensibly broke the link among diseases, bodies, and the environment, the tropics were still considered dangerous for white manual labor.[13]

In 2008, an ex-plantation worker remembered the climate of eastern Cochinchina as mild and healthy. Yet, as plantations moved to less populated areas, malaria among "native labor" became so serious that individual planters began to take measures. One of the first recorded attempts was initiated by one of the more forward-looking planters, Emile Girard, who was the director of the Suzannah red earth plantation and vice president of the SPCI. Between 1917 and 1919, Girard, under the direction of Noël Bernard of the Pasteur Institute, undertook a series of experiments that aimed to break "the intimate relationship between man and mosquito" and reduce the 100 percent malaria infection rate on his plantation. First, Girard tried quinine at forty-centigram doses that took between thirty-six and forty-eight hours to eliminate from the body. When this effort failed, Girard began to destroy the breeding location of the mosquitoes by clearing vegetation from stream banks. The Forest Service director objected, warning of increased erosion and desiccation, or drying up of water sources, that could result from widespread bank clearing. The director also questioned the efficacy of Girard's strategy, correctly pointing out that forest-clearing measures often triggered an increase, not a decrease, of malaria, as *Anopheles* need clear, flowing water to breed. Ultimately, he lost the argument, and by 1919, Bernard stated that the steps taken by Girard had reduced the spleen index by more than 50 percent.[14]

Although some planters were actively involved in antimalaria research, such measures were hardly universal on plantations, and colonial doctors attempted to extend their control over labor and plantation space based on the disease. In a January 1919 letter to the Goucoch, Laurent Gaide, then local head of the health services, discussed Bernard's recent publication on malaria. This publication, Gaide argued, should be circulated among both local authorities and planters. Furthermore, since the question of malaria "is directly bound up with the question of native labor," prophylactic measures should be "obligatory" and carried out "methodically and progressively on all of the plantations, under the technical control of the director of the Pasteur Institute." Gaide recommended that he and Bernard visit the large plantations to organize a medical service in the region. Clearly planters, doctors, and various colonial administrators had a stake in the success of plantations. Motivated by the explosion of malaria rates in Indochina, the colonial government and plantation management moved to enroll (and bankroll) the Pasteur Institutes in Sài Gòn, Nha Trang, Hà Nội, and later Đà Lạt to carry out research on plantations. The Pasteur Institute's researchers derived monetary and political benefits from their relationship with the government and the planters, along with an opportunity to contribute to the growing regional science of malaria and to advance their careers.[15]

Partly through their role on plantations, the Montagnards, or uplanders, also found themselves at the intersection of thinking about race, disease, and governance. Rubber planters regarded the Montagnards as practical labor for the clearing of forests, but their suspected role in the spread of malaria meant that planters sought to segregate them from Vietnamese labor. Montagnard bodies were thus subjected to growing medical intervention, especially on strategic reservations that focused research on Montagnard populations. Moreover, by making visible a common bodily suffering, scientific studies contributed to Montagnard political projects, as some leaders of originally isolated groups of Phnong and Stieng attempted to unite behind a shared medicalized identity.[16]

Although antimalaria measures that targeted environmental vectors were emphasized, these solutions were expensive and prevention efforts often turned to Kinh and Montagnard bodies. Race thinking influenced one of the first examinations of malaria among the Montagnards living near plantations. In a 1919 publication researchers repeatedly noted that groups such as the Stieng appeared to be unaffected by malaria while they still carried plasmodia, a state now known as a healthy carrier. Researchers employed a contrast between exterior health and interior disease, and as roads and plantations brought the Montagnards into closer contact with the Vietnamese, planters came to view the Montagnards as dangerous sources of malaria. Some medical doctors suggested avoiding Montagnard labor, but plantations valued it too highly and claimed that the Vietnamese could not perform clearing work well. Ideally, planters would be able to isolate the Montagnards from Vietnamese workers, for both health and political reasons, but in practice the different groups often lived cheek by jowl.[17]

Early medical studies of malaria reinforced a desire to save the Montagnards that existed among some anthropologists and officials. Based partly on the activism of provincial administrators, the colonial state created Montagnard reservations and focused on malaria prevention and malnutrition. When discussing a 1923 memo that spoke of such reservations, the head administrator of Annam wrote that there was a "Montagnard nation" and he would pursue a "racial policy" to defend it. Like Henri Maître, many administrators opposed mixing Vietnamese and Montagnards, as it would damage the Montagnard "race," an antimixing sentiment that mirrored concerns about *metissage* elsewhere in colonial society. This 1930 map from the Pasteur Institute hints at concerns about the effects of colonization of provinces such as Biên Hòa on the Montagnards. Government-created reservations would function like Native American reservations in the United States and were meant to both contain and protect Montagnards. These proposed administrative solutions for the supposedly precarious existence of the Montagnards aimed at promoting a unified Montagnard cultural and biological identity that could withstand integration into the world economy. Such efforts intensified in the late 1930s, as two 1937 reports blamed Vietnamese infiltration and European-controlled capitalist agriculture for damaging upland society and advocated a program aimed at creating a collective identity.[18]

Medical doctors also showed growing concern for the Montagnards. Drawing on the Italian experience, Prosper Joseph Emile Borel called for a "grande bonification," arguing that only regional economic development could bring about long-lasting changes in health. In 1937, two Pasteurians argued that the Montagnards had shown no "special immunity against malaria, whether hereditary or racial." The authors of this study cited the policy of creating "Inter-

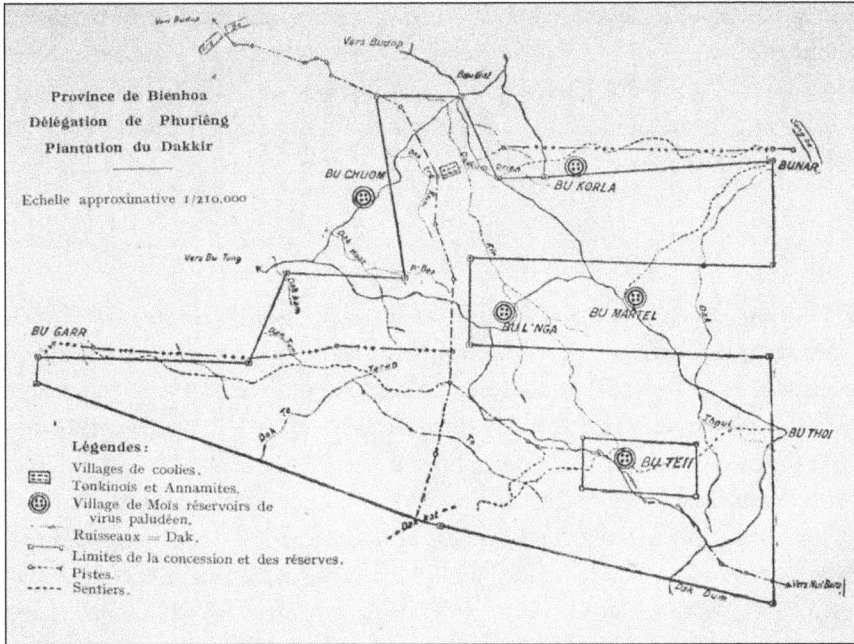

Montagnard reservations in Biên Hòa. This sketch shows how plantations and Montagnard villages competed for the same space. Source: *Archives des Instituts Pasteur d'Indochine.*

nats Moïs," or boarding schools, at the population centers of Ban Mê Thuật and Hớn Quản as evidence that the active introduction of hygienic practices was improving the health of populations. The administration, the authors argued, needed to focus attention on "régions déshéritées," or nonprivileged regions, and while it was too early to speak of eradication, improved treatment would have a beneficial effect. By the mid-1930s, many medical researchers begin to abandon race as a useful concept of heredity and turned their attention to populations and ecology, writing about the demographic threats that faced these "races." In their studies, they attributed low population numbers to malaria and agricultural practices.[19]

Children were a particular focus of study. The 1937 annual report of the Pasteur Institute argued that malaria was devastating the "moï race," especially the children. "If one wants to save the race," concluded the report, "it is necessary to do with the 'moï' as one wanted to do with the black." By "black," the report was referring to French policy in Africa, where the French, according to the author, had worked to improve the health of Africans. A 1943 report by P. Delbove, a Pasteurian, and G. Wormser, a plantation director, employed a map of "moïs" reservation zones similar to those discussed in its argument about the danger of Montagnard children as reservoirs of plasmodia. Such views of race were not limited to medical doctors and officials, and popular views of the Montagnards often combined the sense of a healthy exterior and diseased interior as well.[20]

Seed and Soil

Plantations created malarial conditions, and much of the Pasteur Institute's research was aimed at using empirical studies to address theoretical questions regarding prevention. Its research built on the idea that understanding local conditions and regional differences was a key aspect of preventing malaria. Warwick Anderson has shown how medical discourse in the U.S. colonial Philippines broadly shifted from racial to environmental interventions aimed at controlling the disease during the 1920s. In Indochina, the movement from quinine to environmental control took place, albeit with two important differences. First, both racial and ecological perspectives continued on plantations as industrial hygiene provided the underlying logic of all malaria prevention efforts. Second, theories of heredity were never as hard in the Francophone medical world as they were among Anglo-Saxon medical doctors. Even as older theories of climatic determinism achieved new precision and life through increased collection of environmental and health data, French physicians continued to view races as malleable biological entities. Vietnamese medical doctors increasingly promoted the use of quinine as they adopted racial thinking, and sought to redirect concerns over malaria toward the nationalist goal of strengthening the individual bodies of an imagined Vietnamese community.[21]

Researchers concerned with malaria on plantations made an important contribution to the shift from colonial ecology to tropical ecology. This community of medical doctors and engineers working on antimalaria measures traced moral economies of prevention, or what Lorraine Daston has described as "a web of affect-saturated values that stand and function in well-defined relationship to one another." These actors emphasized the values of quantification, empiricism, and objectivity in their advocacy for antimalaria measures, and connected with international research communities that sought to quantify the mosquito-human-plasmodia relationships. The moral economies of science encouraged researchers to push for small-scale, invasive, and sustainable measures that differed from the large projects highlighted by colonialism. The effects of these measures on colonial society depended, however, on broader moral economies involving peasants, colonial experts and officials, and plantation owners and workers.[22]

Viewing malaria in twentieth-century Việt Nam, and the Pasteur Institute's work in light of race thinking, helps explain several phenomena. First, Pasteur Institute research demonstrates how malaria was understood through a combination of new science and preexisting cultural forms. A potent combination of microbiology, epidemiology, and entomology helped draw malaria into scientific worlds of meaning. Yet, even in this process of uptake, malaria maintained past cultural resonances, both French and Vietnamese. Second, the money invested in finding prevention and treatment methods flowed from capital that could be tapped because it was threatened by malaria. In other words, practical steps were taken to reduce the effects of this disease only where modernization projects were in danger. When engineers and physicians tried to move malaria treatment and prevention off colonial enclaves during the 1930s, these efforts failed, mostly due to lack of funds and government interest.

Throughout the 1920s and 1930s, medical doctors at the Pasteur Institute attempted to make sense of the interactions of race, disease, and labor on colonial projects ranging from military campaigns to public works. Two concepts provided an umbrella for such thinking: *mise en valeur* and the germ theory of disease. Unlike *mise en valeur*, which highlighted the ties between diseases and their environments, the germ theory of disease encouraged doctors to view the two as isolated from each other. Before the advent of the germ theory in the nineteenth century, physicians viewed water as one of the most important media through and over which goods, bodies, and diseases moved. Malaria remained linked to water by running streams, stagnant pools, and mosquito breeding sites, but the germ theory helped displace malaria from the environment to Vietnamese and Montagnard bodies. This theory also helped shape attempts to mitigate the effects of malaria on European and Vietnamese labor by placing a premium on the physical separation of healthy and diseased populations. The germ theory was adopted slowly among most French medical personnel in Indochina, and as late as 1908, a guide to hygiene in Indochina warned of the possibility of contracting malaria either from mosquito bites or from drinking infected water. Even as belief in miasmatic theories slowly faded among medical doctors, the soil was still viewed as a potential source of disease and remained a useful organizing concept for French medical research.[23]

The growing acceptance of the germ theory fostered the use of the "soil and seed" metaphor for a body and its germs, an association more visible in the French word *paludisme*, which comes from the Latin word for "swamp," than in the Italian word *mal'aria*, which refers to the air. In this way, water and soil remained useful organizing concepts for French medical research. In 1926, Prosper Joseph Emile Borel, who had recently begun work at the Pasteur Institute in Sài Gòn, published an article considering the possible mechanisms connecting soil composition and malaria in Cochinchina. Borel justified his inquiry by pointing to the close match between newly published maps of soil type and maps of malaria. The alluvial soils, of both ancient and more recent origin, were comparatively free of the most pernicious forms of malaria, while the soils derived from the breakdown of volcanic rock, which formed the so-called red earth region, sustained terrible malaria epidemics. The idea that the chemical composition of the soil might play a role in the rate of malaria through its effects on mosquito breeding was considered and rejected. Instead, Borel identified a mechanism linking the physical structure of soil, in particular its ability to drain water, to malaria. The alluvial soils were composed of a high percentage of sand and thus drained quickly after the rainy season, while the higher clay content of the volcanic soils ensured year-round breeding sites for malaria-carrying mosquitoes.[24]

Although Borel's focus on the role of soil in the prevalence of malaria seemed to be more in line with older miasmatic theories that linked land and human health, his conclusions were more equivocal. The research of Borel and others helped replace explanations of malaria based solely on geographic distribution with mechanisms of propagation. Borel's larger research program focused on identifying potential mosquito vectors of malaria, and his findings connected malaria not to soil but rather to the types of agriculture that took place on it. He introduced a historical element into his argument by reasoning that the poor drainage of the volcanic soils meant that they had never been subject to intensive cultivation. A vicious cycle was thereby set up between the sparse populations of upland peoples, who had immigrated a few hundred years earlier but were able to work the region only lightly, and malaria, which prevented the population from expanding. Explanations pointing to human agricultural activity as a key determinant of malaria rates had already formed the basis of antimalaria efforts in Italy, and the link between the intensity of land cultivation and the subsequent changes in disease environments had become the basis for understanding malaria by the 1920s.[25]

Borel was not the only researcher at the Pasteur Institute attempting to untangle assemblages of bodies, environments, and diseases in Indochina. Henry Gabriel Sully Morin (1889–1969), who authored many articles on malaria, served as head of the Pasteur Institute's malaria division and had a model career in colonial medicine. He was born to a missionary doctor during a time when religion and medicine came together to heal bodies and souls. Young Morin faithfully followed in his father's footsteps, receiving his medical degree from Bordeaux in 1913. His early career reflected the military's role as the source of training and research; most colonial medical doctors and many of the Pasteur Institute workers had served in the military. Morin's dissertation on amoebas was no doubt influenced by colonial medicine's focus on the water and the digestive tract. After his military service abroad, Morin returned to France, took a course at the Pasteur Institute, then departed for Indochina, arriving in 1925. By 1931, Morin had become director of the malaria prevention program at the Pasteur Institute.[26]

Morin focused on the nexus between industrial agriculture and medicine, and many of his publications dealt explicitly with plantations. As the head of the malaria laboratory in Sài Gòn, he carried out *recherche appliquée*, or applied research, on industrial hygiene. Morin's research was also directed toward scientific audiences, which entailed drawing on empirical studies to address theoretical questions about prevention. Though the Pasteur Institute possessed both political and financial backing, it faced the danger of becoming a mere laboratory for testing blood samples and a supervisor of ditch digging if it became too closely tied to economic projects. The Pasteur Institute in Sài Gòn witnessed a dramatic increase in blood tests, from 700 in 1918 to almost 6,000 in 1928. The research conducted under Morin on malaria effectively became a malarial survey of French Indochina, and was communicated to the public through a series of publications, both monographs and articles in journals such as *Archives des Instituts Pasteur d'Indochine*, *Bulletin de la Société de pathologie exotique*, and *Bulletin de la Société médico-chirurgicale de l'Indochine*. This chance to publish offset some of the tedium, and distraction, of thousands of routine blood sample examinations. Morin enabled researchers to contribute to "pure" science and advance their careers in part by creating a hierarchy of institutions concerned with malaria. According to Morin, the colonial government's Health Service would oversee identifying malaria outbreaks, and the Pasteur Institute would conduct a general investigation. Then local antimalaria engineers and authorities would carry out the "applied" work, thus relieving the institute of dealing with details, while still retaining its key position in antimalaria action.[27]

Morin believed that basic research into local mosquito environments had great practical value, and his early research explored regional differences in the uplands, including studies in Kon Tum in the central highlands and in Lạng Sơn and Cao Bằng in the north. He also undertook entomological studies of mosquito species to determine which ones were potential plasmodia trans-mitters and thus should be the focus of prevention efforts. Entomological studies, such as a 1928 compendium of mosquito biology in Cochinchina, which went into painstaking detail about breeding sites and mosquito parts, were justified because almost all *Anopheles* species were potential plasmodia carriers. As in France, studies of insect biology were not cited in debates about Mendelian and Darwinian versus neo-Lamarck models of inheritance. Instead, the focus and dialogue were with other researchers working on the question of sanitation and hygiene, which led to rote antimalaria efforts such as swamp draining, while little attention was paid to the details of species distribution.[28]

Morin's discussions about hygiene were infected by lingering notions of race, and Morin argued unsuccessfully for the complete segregation of the Vietnamese population at the hill station of Đà Lạt. Morin also believed that cultural understandings of malaria expressed the experiences of people on the Indochinese peninsula. For example, coastal dwellers believed that those in the uplands had sorcerers who thwarted attempted invasions with fevers, while the upland peoples saw malaria as a sign of evil water spirits that sucked the blood of their victims. In either case, the disease and its representation had kept the peoples of the uplands and the lowlands separate throughout the centuries. In 1935, Morin published a summary of malarial knowledge, which he opened with a chapter on the characteristics of the disease. After discussing the relevant geographical, climatological, and population factors, Morin focused on legends and the disease's role in "Sino-Annamese" military and political history. In recounting the decisive thirteenth-century battle be-tween Đại Việt and Mongol forces, Morin noted that in the "Annals," malaria, and not the heroics of Đại Việt troops, was credited as the principal cause of Mongol defeat. Just as nationalists in Southeast Asia attempted to extend their imagined communities back in time, physicians needed to give modern diseases their histories.[29]

Under Morin's leadership, the Pasteur Institute targeted malaria on planta-tions through both the environment and the body. Much of the antimalaria effort on plantations focused on mosquitoes, and in 1933 Morin coauthored a pamphlet offering a range of solutions. These solutions included the use of chemicals such as Paris Green and oil, which Morin argued was an econom-ical method to control the size of adult mosquito populations. The *Anopheles* species breeds in flowing water, and Pasteur Institute researchers devised new application systems such as a nail and box contraption for treating streams. Other strategies focused on the management of water flows, and the centerpieces of such measures were the semipermanent and permanent vector-based approaches that involved the placement and design of villages and the engineering of the surrounding landscapes. Researchers also urged planters to build living settlements one kilometer away from any potential source of *Anopheles* mosquitoes, while maintaining access to water for daily use, which would ideally reduce the exposure to *Anopheles* at night. Researchers advised plantation owners to dry out land around the village with either underground or open-air drains, to eliminate certain types of vegetation on the banks of streams, and to build small dams to create mini-floods to wash larvae away. All of these methods were based on the idea that environmental modifications, including the alteration of water flows, could limit *Anopheles* populations by killing their larvae.[30]

Human bodies (and races), like mosquito species, were also subject to medical investigations, and bodily suffering was extended to advance knowledge of malaria on plantations. Morin repeatedly referred to the usefulness of Vietnamese children and immigrants from Annam and Tonkin as virgin populations that could demonstrate levels of endemicity, or rate of disease in a place, and Vietnamese bodily suffering became a source for refined visual representations of malaria. Only a few species were dangerous to humans, and scientific authorities boiled down complex ecological information into easy-to-use diagnostic tools. For example, a poster submitted to planters and local officials communicates a clear contrast between the small, silent, painless, nocturnal, yet deadly *Anopheles* and the large, loud, painful, diurnal, yet ultimately harmless *Culex*. Theirs was a simple message: while there was a large biological diversity of mosquitoes, only the plasmodia transmitter mattered.[31]

The Pasteur Institute under Morin's direction strongly supported antimalaria measures that targeted the environment of mosquito vectors, which rendered peasants as passive bystanders. However, the institute also pushed for solutions that directly treated human bodies and attacked plasmodia. Quinine had long been a staple, if problematic, tool in the kit of antimalaria measures, and during the 1930s the introduction of synthetic substitutes such as quinacrine and paludrine continued this trend. While these drugs were more powerful than quinine, they remained a source of tension; planters enforced drug compliance from workers but were willing to purchase only enough drugs to keep workers on their feet. Seeking to extend worker time in the field for as long as possible, planters limited the distribution of this drug to seasons with already high rates of malaria infection. Moreover, quinine or quinacrine was often forced into the mouths of workers. French efforts to popularize knowledge about malaria and quinine reinforced the message of individual responsibility, blaming workers for their own ill health while still assuming an essentially passive worker-subject.[32]

Colonial Complicity

In 1937, Jean Adam, a professor at the French National Institute of Colonial Agronomy, published a twelve-page article calling for a "colonial ecology." Ecology, Adam argued, studies "the connections existing between living organism and the surrounding environment." Agricultural ecology, he continued, "has as its purpose the adaptation of the use of plants and of animals to the agricultural environment, with the goal of obtaining, under the best conditions, the highest possible yield of products helpful to mankind." By focusing too narrowly on material nonhuman nature, agricultural ecology had limited its scope of action and failed to consider the souls of humans. To redress this imbalance, Adam sketched out his vision of a colonial ecology. He stated that such a science could offer solutions to the interrelated problems of both the natural and human worlds. It would "study the colonial environment [*milieu*], natural and human,—study the tools of production, plant and animal, to which it would be good to add the underground resources, that equally have something to contribute,—study man, European and Indigenous, in his action in the colonial environment." It would serve as "the summary freeing the mastered ideas from the chapters where the details are given, to scan the horizon and grasp in a synthetic view the generality of the ideas to be acquired." Such an endeavor would also help the French empire catch up to the British and the Dutch.[33] Published under the Popular Front government in France concerned with the conditions of the colonized, Adam's article emphasized "the well-being [*bien-être*] of populations" and the "betterment [*mieux-être*], material and moral" of the indigenous peoples. Even though ecology remained oriented toward race and the colonial order, this science was constructing knowledge that served later tropical and developmental projects.[34]

At the beginning of the twentieth century, the use of ecology remained idiosyncratic, and it was absent from sources such as the multivolume treatise on the flora of Indochina. Although these texts are taxonomic rather than predictive in their orientation, it is not unreasonable to expect to see the term in an era when plant ecology had become an established discipline at many universities in the United States and Europe. Agronomist and meteorologist Paul Carton helped to spearhead the effort to spread ecology in Indochina. After graduating from the National Institute of Colonial Agriculture in France in 1913, Carton joined the editorial staff of the *International Review of the Sci-*

ence and Practice of Agriculture, which was published by the Bureau of Agricultural Intelligence and Plant Diseases of the International Institute of Agriculture in Rome. He subsequently joined the Meteorological Service in French Indochina, eventually becoming the head of the Agriculture, Breeding, and Forestry department. As a researcher, Carton did not make much of an impact on the international scientific community with his investigations into rainfall and sunlight distribution patterns and their effects on plant life. As an educator and popularizer, however, Carton had a much greater influence as he adapted the new subdiscipline of *écologie agricole* (agricultural ecology) to French Indochina.[35]

When the ESASI in Hà Nội was reorganized in 1925, Carton included agricultural ecology, a term coined by one of his former colleagues in Rome, in the school's curriculum. Reflecting Carton's influence, this section on ecology was included under the heading agricultural meteorology (*météorologie agricole*) and consisted of thirty lessons of theoretical instruction and a ten-lesson practicum during the general curriculum of the first year. In these lessons, students examined the influence of weather and climate on plants and animals and considered questions of plant varieties, acclimatization, and genetics, all of which focused on increasing yield.[36]

The cultivation of rubber, too, contributed to the production and dissemination of ecological knowledge. Carton pushed for laboratories and agricultural stations to act synergistically, as "the [agricultural] station," he wrote, was "the natural extension of the laboratory." By the late 1920s, rubber planters, from small and medium owners to the largest multinational companies, such as Michelin, were taking note of the climate knowledge produced by Carton and others in the laboratory and on experimental fields. Such knowledge had previously been considered esoteric, but an economic journal from the late 1920s noted that "agriculturalists have begun to notice that the study of atmospheric phenomena, which influence the development of plants, can be of interest to them and that it is not a priori a waste of time to follow such work."[37]

While Carton, like most contemporary agricultural researchers, published little on "native" agricultural practices, by the late 1930s the attention paid to the environment promoted by, and useful to, rubber plantations engendered critiques of inappropriate agricultural practices. As experts gained a better appreciation of local conditions, some even valorized "indigenous" farming techniques, a stance that could be construed as anticolonial, especially in British colonial Africa, where Helen Tilley sees "the roots of . . . a 'farmer first' mindset." But even as researchers began to consider native knowledge, in

Southeast Asia they retained their top-down, Eurocentric approach to knowledge, a dynamic that continued within French tropical geography until well into the postcolonial period. In the late 1930s, Carton reaffirmed the economic usefulness of agricultural ecology, offering an extended argument in his 136-page textbook for a first-year ecology course at the ESASI. Published in 1938, this textbook highlighted both the usefulness of ecology for increasing plant and animal yields and the potential of ecology to help humans adapt to new climates and resist diseases. Carton again defined the purpose of agricultural ecology as discovering the optimal conditions for crop growth. "Humans," he wrote, "try to get a maximum yield, in both quality and quantity, of products necessary for their needs from the plants that they exploit or cultivate."[38]

Studies of ecology also addressed the relationship between human health and climate. In 1934, Carton teamed up with Henry Morin to publish an article discussing the relationship between climate and malaria. This article looked at three ways in which "climatic factors" acted "upon microbial diseases transmitted to humans by intermediary hosts." These were "(1) on man, by putting his body in a more or less receptive, or resistant, state with respect to the microbe; (2) on the different species of transmitting insects that, according to their effect, could live or not, more or less abundant and with variations in their biological cycle; (3) on the microbe itself." The article attempted to explain convergences and divergences between entomological studies of mosquitoes and epidemiological studies of malaria. Carton and Morin contended that because of climate, the "European organism" was weaker in certain places and thus attacked more easily by plasmodia. The authors concluded by emphasizing that the "resistance of the human organism" was the most important factor in fighting malaria, as plasmodia strains were essentially constant and local adult *Anopheles* species were sufficient in numbers to effectively transmit plasmodia.[39]

Such research was generalized by Carton in a series of articles that articulated a neo-geoclimatic determinism reformulated in the language of statistics. Even as Carton presented a series of "climatograms" that plotted yearly cycles of temperature and humidity, he drew on Lamarckian-style heredity to explain disease environments. In a 1935 book chapter entitled "Climate and Man (Climatic factors in human ecology)," Carton quoted a 1907 work on hygiene in West French Africa that argued that "the climate has a natural influence on the cause of all diseases, either by its effect on the organism, or by its influence on the bacteria, or, in the tropics in particular, by its action on intermediary hosts of known and unknown parasites, which play a considerable role in this region." From this link between climate and disease, which was taken as axiomatic, Carton expanded his thinking to address human races. "Thus one sees," Carton concluded, "that the effect of climatic factors on men of the different races and their influence in the etiology of diseases constitutes a very great, still unexplored, field of study. Programs of observation, statistics, laboratory experimentation, etc. have been undertaken and followed with the object of resolving the numerous, very complex problems that arise."[40]

Research data generated by the Pasteur Institute's attempts to manage malaria on plantations also encouraged scientists to put forward more general theories of disease environments that mixed older medical authorities with recent advances in research. This combination created environmentally informed views of race. As the historian Mark Bradley has shown, ideas about the enervating effects of tropical climates on the "annamite" race continued to structure much of the perception of the native inhabitants of French Indochina among observers in the United States who knew next to nothing about French Indochina. Inside Indochina itself, however, theories of geoclimatic determinism were generating increasingly sophisticated data and nuanced theories that eschewed broader conclusions about the relationship between civilization and climate. As discussed earlier in this chapter, French medical personnel began to view Stieng and Phnong as reservoirs of plasmodia despite their healthy appearance. Doctors labeled the Montagnards "healthy carriers" and highlighted the threat they posed to French economic projects. At the same time, the Montagnards were depicted as a vulnerable population at risk from the disease, and some French worried about the possible disappearance of these "races." In fact, blood and the malaria-causing plasmodia traveling in it marked Montagnard bodies more strongly as racial others and contributed to new forms of biological identity, in the same way that terms such as "moï," "phnong," and "kha" were applied by lowlanders to upland groups in earlier centuries.[41]

Meanwhile, thinking about race and environment had evolved among some of the human sciences in colonial Indochina. Susan Bayly argues that race science and cultural thinking could run in parallel, but in the 1930s, some forward-looking thinkers separated biology from culture and medicine from anthropology. Bayly focuses on the role of Durkheimian sociology in the thinking of Paul Mus, who analyzed cultural phenomena symbolically without reference to biology. Mus, Bayly argues, played a key role among anthropologists in moving racial thought away from hard biology and toward soft culture. His writings "reflect a remarkably sophisticated recognition of the fluidity and political awareness of 'traditional' Asian thought." Bayly focuses on the changes in thinking about the Cham of those working within the

Durkheimian tradition to show the various intellectual currents at play in Indochina. These writings moved from Antoine Cabaton's linking of a low birthrate to degeneracy in the progressive *Revue du Monde Musulman* in 1907 to the writings of the Vietnamese demographer Nguyễn Thiệu Lâu, a student of Mus, who argued in 1943 that while Cham culture was being swamped by the more "energetic" Vietnamese and Montagnard cultures, their population numbers were not in decline.[42]

Bayly's argument for a global shift from biology to culture in the human sciences is supported by publications such as *We Europeans*, which aimed to dismiss the "vast pseudo-science of 'racial biology' . . . which serves to justify political ambitions, economic ends, social grudges, class prejudices." Likewise, by the 1930s some anthropologists were proposing cultural theories of humanity to replace racial ones. Yet thinkers such as Mus, though influenced by Gourou, retained concepts of tropicality that linked environment (if not race) and culture. Many doctors in Indochina remained tied to racial concepts, as shown by the work of Pierre Huard (1901–83). Moreover, in the sociologist Emmanuelle Saada's work on race-mixing in the French empire, she spells out the limits of an intellectual approach to race thinking without reference to social practices. By the 1930s, the French colonial administration had come to view race as a politically important category, and the 1937 Guernut commission inquiry includes many questions about métis.[43]

These typologies drew from efforts to make Asian and Western bodies distinct biological objects. Vũ Văn Quang's thesis "The Eurasian Problem in Indochina" for the Ecole de Médecine et de Pharmacie in 1939, for example, reproduces questions that were common in Paris at the turn of the century. The work of Vũ's colleagues at the Institute for Social Studies (Institut indochinois de l'étude de l'homme) further reflects the limits of the diffusion of Durkheimian sociology and the continued acceptance of the reality of biological races. According to John Kleinen, this institute, established in 1938 in Hà Nội by Pierre Huard, head of the Indochinese Medical University, and Paul Lévy, a researcher at the French School of the Far East (EFEO), "marked a shift in orientation in colonial knowledge production towards more specialised rural fieldwork." The Institute for Social Studies also published articles that analyzed racial characteristics such as the "Giao chi" toe, a marker that supposedly distinguished Vietnamese from Chinese. Such racial discourse continued after the end of World War II with a 1948 article by Huard on the discovery of the "yellow" body in the Far East. Yet race thinking among scholars was not always exclusionary; in the French scientific world, the idea of positive racism aimed at improving the human stock. In the hands of Vietnamese medical doctors, who also subscribed to racial typologies, race thinking could serve Vietnamese nationalism.[44]

Tropical Medicine on Plantations

Rubber planters initially eschewed the science of ecology but grew to appreciate its insights, making the plantation a privileged site to examine the shift from colonial to tropical ecology. Rubber plantations were laboratories in which the control of human and plant diseases was tested and applied to the Indochinese terrain. Race was a key concept in exerting such control, and in the 1910s emerging biomedical understanding of malaria combined with the growth of industrial agriculture to form new ideas about race. Just as agricultural techniques defined natives and Europeans, the presence or absence of the plasmodia that resulted in malaria was used to ascribe identities to the Kinh, Khmer, and Montagnards. Certain bodies and populations became the target of government programs; plantation labor received most of the modest resources available to combat the effects of malaria, but the clear majority of the rural population had no access to health care. In the 1930s, the writings of scientists such as Paul Carton, economic logic, minority politics, and military security combined to racialize malaria prevention and treatment. Thus, even as medical sciences released malaria from its miasmatic ties to the land, the disease was reinserted into already racialized bodies and their tropical environments.[45]

One framework for incorporating French Indochina into regional and global networks was the 1934 International Rubber Regulation Agreement, which was signed in London by the major rubber-producing countries. As discussed in chapter 2, the London agreement formalized knowledge production and circulation networks not only among Southeast Asian colonies but also between the colony and the metropole. Simultaneously, diseases associated with rubber plantations such as beriberi and malaria precipitated new regimes of health and strengthened growing medical networks across rural Southeast Asia. By comparing health conditions on "colonial enclaves" in different countries of the region, researchers and plantations contributed to a "tropical governance" that was made possible and was embodied by medical thinking about rural hygiene in the late 1930s. Although medical doctors failed "to translate science into social action" regarding the problem of beriberi, their research strengthened growing medical networks. In

French Indochina, the colonial government successfully resisted efforts to alter rice standards to address vitamin deficiencies, even as nutritional science developed at the FEATM meetings informed reforms of its labor laws.[46]

Henry Morin served as a French liaison for these tropical projects. As director of the Pasteur Institute between 1931 and 1940, Morin presented the institute's findings at several international conferences and taught malariology in Singapore between 1932 and 1937. In 1934, the League of Nations organized a conference in Singapore to celebrate the first international course on malariology to be held in the Far East. Morin attended the conference, and after summarizing the state of knowledge about the disease and its prevention, he addressed the explosion in malaria that had taken place in Cochinchina. He accepted the centrality of the plantations in these outbreaks during the 1920s and 1930s. Yet he was still willing to defend the colonial project and plantation owners' goodwill. "When one has seen with what passion all of the plantation directors and all of the assistants were attached to carrying out this thankless work, over a long-term, with what worry they followed their health statistics, with what joy they announced their victories over the diseases and over the deaths of their coolies, one asks where one can find strict relations of exploiter and exploited that certain self-interested voices vainly search to substantiate the legends present in the ill-informed opinion of the Metropole."[47]

ALTHOUGH MORIN SPOKE ABOUT the role of planters in reducing diseases, he was also increasingly interested in native participation. In 1937, the League's Health Organization coordinated the "Intergovernmental Conference of Far-Eastern Countries on Rural Hygiene" in Java, which some scholars view as a seminal twentieth-century moment in thinking about global health. A key theme that the French delegation for Indochina submitted for this meeting was the importance of self-development among peasants, and their willingness to self-regulate and participate in economic development. "More may be expected," the report read, "from measures which appeal directly to the individual." The report continued, "Moreover, when improvements have thus been made, there is no need for constant external assistance and supervision; they become part of the life of the rural population and have a deeper and more lasting effect." Old power networks were often reinforced in the process of encouraging villager involvement and "personal influence," the report argued, "whether that of the mandarin, the doctor, the teacher, the plantation inspector or the sick-attendant, is always the most effective." This emphasis on "personal influence" encouraged health care on French plantations to remain a paternalistic affair.[48]

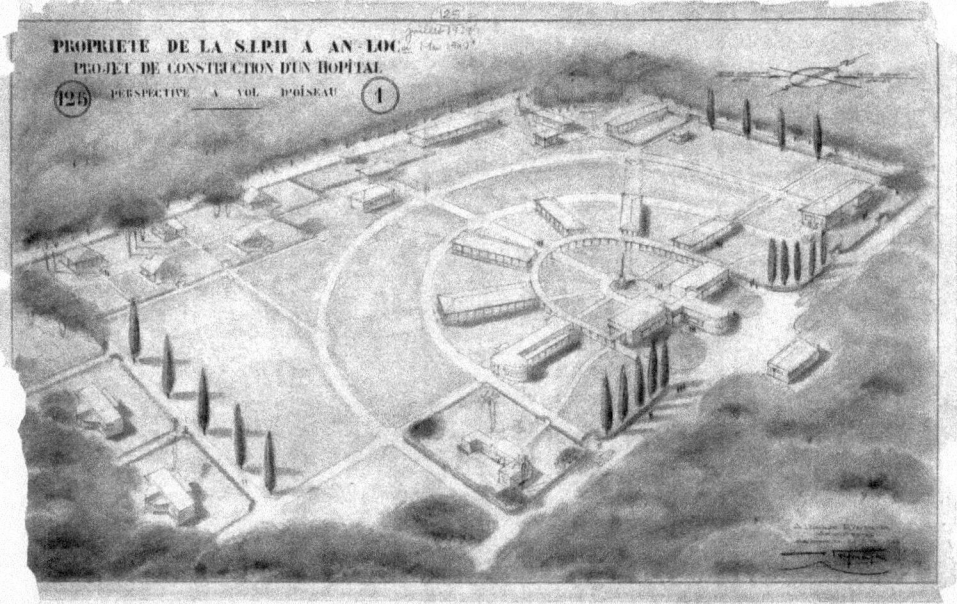

SIPH hospital, 1939. This design for a SIPH hospital resembles other hospitals that were constructed on plantations in Indochina. Note the straight lines and visibility that gave both a modernist look and ability to keep patients under surveillance. Source: Jacques Veysseyre, 1939.07, Hôpital SIPH à An Lộc.

This intercolonial conversation about the governance of health in the tropics fed back into practices in Indochina. In 1935 Jean Canet, who trained at the Pasteur Institute in Paris, wrote about the challenges he faced while implementing plantation health care. Canet had recently been hired by the SPTR, one of the largest rubber producers in Indochina, to head a reformed medical system responsible for the health of 10,000 workers who tapped 35,000 hectares of rubber, on lands extending over 250 square kilometers. By the late 1930s, a few large plantations built hospitals similar to the one planned for SIPH at An Lộc, where sick workers, like latex, would be processed in an orderly, efficient manner. But most workers had limited access to medical infrastructure and Canet had only one Vietnamese medical doctor and twenty-five nurses working under him. With such an inadequate staff, Canet could only write wistfully that he would use both "medical prophylaxis" and "drainage" to combat malaria, which continued to be a significant cause of illness.[49]

Like other managers and medical doctors in the plantation industry during the 1930s, Canet published articles and pamphlets on the topic of plantation health care. For the 1937 International Exposition in Paris, Canet authored "Consideration on the organization of preventive and curative medicine in the large rubber plantations in Indochina," which highlighted some of the changes in the previous twenty years of plantation medical work. He argued that private industry and the colonial government had been cooperating to create healthful conditions on the plantations for Vietnamese workers. "This regulation," Canet continued, "permeated with the local habits and the psychology of the natives, is established in a way that simultaneously considers the interests of the employees and the employers and notably, in Cochinchina, tries to protect the labor coming from the deltas of Northern Indochina. The action of this health policy, governmental and private, was translated in some years by a general improvement often remarkable: death rates, quite high ten years ago, have currently reached a general level of less than 1 percent."[50]

Canet highlighted legibility, sanitation, and assumptions about "native psychology," all of which were common themes in writing about tropical governance. He believed that ignoring the rules of hygiene in a hyperendemic zone of malaria "would create a cemetery where one wanted to establish a center of human life." To keep workers alive, healthy, and productive, Canet prescribed a modern, clean village consisting of blocks of houses separated by wide roads, drawing from hygienic rules that were being implemented in rural places across the globe.[51] The implementation of such projects encouraged medical doctors to view workers as complicit in their own ill health. The Pasteur Institute published a poster portraying three workers, two of whom have taken quinine and have a good appetite, and a third who is hunched over in pain or fatigue. The language of the poster conveys a moral tone, using the term *đứa* (a person of lower position) for the sick man and *người* (a neutral term) for the two healthy individuals.[52]

The houses in these villages also reveal the assumptions about "native psychology" that informed the practices of French medicine in Indochina. Housing on the plantations was divided into "European" bungalows, used by the managers and owners, and barracks, used by the workers. Worker housing had evolved from the hastily constructed shacks of the 1910s and the 1920s to newer models for individual families offered in the late 1930s. Owners realized that to retain "free" labor (i.e., workers without long-term contracts), they needed to provide better living conditions. House models were to resemble "native" housing but be built from durable construction materials. Both aspects intended to address "native psychology" while following hygienic principles. This model used bricks, with either beaten-earth or cement-lined floors. The roof was thatch, which both lowered costs and addressed worker complaints that the corrugated iron roofs of earlier models created oven-like conditions. These houses were also designed to keep out the *Anopheles minimus* mosquitoes that fed in the evening. Most importantly, the newer models had spaces for gardens so that workers could grow vegetables and raise animals.[53]

Like other European employers throughout Southeast Asia, French planters and medical doctors viewed native laborers as unreliable patients and incapable of protecting their own health, even within highly regulated plantation spaces. In the 1920s and 1930s, this attitude led to the adoption of environmental control measures aimed at breaking the cycle of plasmodia transmission between humans and mosquitoes. While expensive, these measures did not depend on active native aid or even passive compliance. In Indochina, some doctors and officials sought to apply to the countryside the often-successful methods of malaria prophylaxis that were developed in cooperation between the Pasteur Institute and the rubber plantations. The solutions developed for colonial enclaves, however, did not spread evenly to rural settings, where preventing malaria had its own moral calculus. Doctors in the Vietnamese countryside were unable to log even the sometimes perfunctory medical statistics that were recorded on plantations, and colonial administrators balked at the attempt to extend a heavy state presence among the rural population. In his role as administrator, Henry Morin stated that the inability of communities to pay for antimalaria measures restricted their implementation. In the late 1940s Morin asked rhetorically: "Without financial resources, without competent personnel and without suitable equipment, how can one apply, in a village, preventative methods which necessitate at the least, three operations per month, all very carefully conducted and supervised?" A rational organization, he stated, would consider "the needs, the resources, and the customs of the country." Even as most colonial administrators doubted native capacity for self-regulation, they preferred lower-cost alternatives like education and quinine distribution.[54]

Increased external and internal threats to French rule, however, encouraged colonial administrators to assert political, military, and medical control over rural Indochina. The central highlands seemed to offer a vital perch from which to fend off potential adversaries, and a growing emphasis on native participation in rural health programs, promoted by international organizations such as the League of Nations, encouraged greater investment in rural health measures. Several local military outposts (*postes des guards indigènes*) were created to project a state presence into the countryside. As Mathieu Guérin and Annick Guénel have shown, malaria was a key problem in places like Haut-Chlong in Cambodia, since the disease greatly reduced its effective manpower and exposed it to devastating attacks from local groups like the Stieng. The use of isolationist logic based on malaria is revealed in designs for rural native guard posts, where villages can be read as a symbol for Kinh and forests for Montagnards. Although these posts were tightly controlled, the prophylactic measures proposed by doctors and engineers shed light on the problems of endemic or hyperendemic malaria prevention, which were comparable to those of the surrounding countryside.[55]

After noticing many deficiencies in malaria prevention measures during his 1937 tour of the central highlands, the head of the Protectorate of Annam (Resident Superior d'Annam [RSA]) commissioned the head of the Pasteur Institute laboratory in Hue, along with the head antimalaria engineer, to draft clear, simple prevention instructions for those in command of the local military outposts. Both pamphlets focused on environmental modifications aimed at protecting the outposts rather than nearby villages. The laboratory director divided the antimalaria efforts into two types, one focused on humans and the other focused on *Anopheles*. He guessed that some French officers might be put off by what they had previously seen or heard about mosquito control and he assured them that the measures advocated in Annam would be much less drastic than those shown in films about mosquito control efforts in Italy and along the Panama Canal. While the officers may have been assuaged, applying the outpost solutions to local populations remained difficult. This laboratory director and the sanitary engineer were encouraged by the moral economy of their scientific disciplines and tried to extend prevention methods to all rural settlements. The laboratory director had taken a trip in 1934 to study malaria prevention efforts in Java, Madagascar, and South Africa and came away impressed by the methods of Ross Park, a medical doctor who had brought a malaria outbreak under control. These methods, including the spraying of individual houses, were quite invasive. Park's measures were taken only during outbreaks, but those at the Pasteur Institute considered extending antimalaria measures, including the possibility of spraying rural housing on a more regular basis. At the end of 1935, the antimalaria engineer was sent on a study mission to Cochinchina to prepare for cleanup work planned for Pleiku in the central highlands. He expressed admiration for the work accomplished on the posts and plantations of the south, and was sanguine about the possibilities of repeating this success in Pleiku.[56]

The RSA and others in the colonial administration, however, argued that the focus of attention should be on the military outposts, as the complete cleaning up of rural Annam was seen as unachievable. The RSA argued that eradicating malaria was unrealistic because "it is currently impossible to hope for the necessary cooperation from the local populations." Considering malaria control on plantations a success, administrators viewed rural health as an oil-spot project that focused on small areas that could be closely managed. Promising to reduce malaria throughout the countryside would only result in disappointment for the rural residents. A more realistic goal was to ameliorate hyperepidemic conditions and return them to mere endemic levels. Administrators pushed for the use of quinine, which was relatively inexpensive, and endorsed the distribution of educational pamphlets. These approaches again illustrate the moral economy of malaria prevention: the solutions were cheap and placed the responsibilities of care on the villagers instead of the state.[57]

Applying solutions developed for military outposts to local populations presented several challenges. The first was the level of knowledge required to successfully carry out the measures. For example, identifying a location was, even for a small camp, no easy task, as many variables had to be considered. Like the commanding officer, the village head had to establish control of the countryside in a 1,000-meter-radius circle (or three square kilometers). This meant identifying three potential sources of *Anopheles*: natural sources of water, such as springs; managed sources of water, such as irrigation streams; and brush in the countryside. For natural sources of water (i.e., those not being used for agricultural and drinking purposes), solutions were less politically challenging; one could simply remove the water or change its rate of flow. While the actual building of the pipes was straightforward, designing and constructing the drainage system required a high level of hydrology knowledge. Dealing with water sources used by the local population was more difficult, and a neglected water control system was useless at best. The state was reluctant, and the village could not afford, to spend resources on the constant surveillance required by such interventions.[58]

Villagers also had many reasons to prefer quinine to vector-based prevention. Antimalaria campaigns throughout Indochina had long been associated with military campaigns, and vector-based prevention also interfered with agricultural tasks. Furthermore, villagers had a range of medical options to choose from, including *thuốc bắc*, or Sino-Vietnamese medicine that, despite protests from biomedical doctors, was efficacious. The Chinese herb *Artemisia annua* is especially effective against plasmodia *falciparum* and currently forms the basis for some of the most powerful antimalarials. *Thuốc nam*, a southern version of *thuốc bắc*, was another low-cost treatment. Yet the emphasis on individual responsibility was tempered by growing pressure on colonial governments to live up to their promises to care for their subjects and by a fundamental distrust of the ability of natives to care for themselves. The head of the Pasteur Institute laboratory in Huế acknowledged the argument that malaria was best cured through the cooking pot (*marmite*) and the rice bowl (*cái bát*), meaning that only greater prosperity would eliminate the disease. The laboratory head conceded, however, that this was a very general argument, and since Annam would likely remain impoverished for quite some time, quinine would be much more useful than rice.[59]

Patriotic Hygiene

The 1930s saw important changes occurring in imperial thinking about rural society, and the formation of regional and global organizations for rural health conditions such as FEATM and the Health Organization of the League of Nations affected health on and off plantations in southern Indochina. Thanks largely to the efforts of American doctors in the Philippines, FEATM was formed in 1908, and among the medical problems this regional organization first tackled was beriberi, an important disease of workers across Southeast and East Asia that came from a diet consisting of white rice and little else. In 1938, Hà Nội hosted the last FEATM meeting, where malaria was one of the chief topics of discussion. While the work of these organizations on diseases and their encouragement of the circulation of medical doctors influenced plantation societies, political struggles among planters, government officials, and scientists over plantation conditions, along with World War II and the First Indochina War, also played important roles in restructuring industrial agriculture relations. Finally, the activism of Vietnamese workers created pressure for reforms and influenced changing notions of health on plantations and in French Indochina more generally.[60]

Ironically, even as a few medical doctors, agronomists, and anthropologists were challenging the basic tenets of scientific racism, many Vietnamese nationalists were repurposing race thinking for their own purposes. Colonialists and nationalists converged around the discursive field of eugenics as race thinking and environment became linked on plantations through women's bodies. The 1927 reforms in Indochina's labor laws discussed in chapter 3 were intended to simultaneously protect the rights of women as workers and reproducers of colonial labor. Among other things, these laws emphasized mothers as "producers and reproducers," provided time for the family, and protected the right of families to work together, which had been a major problem and bitter experience for many. A pronatalist movement in France and its colonies helped encourage the protection of mothers as producers. Businesses too started to find it in their interest to provide more health care, and Octave Homberg noted how "a maternity, a vaccination center established today in the brush will give us each day a work force of a powerful factory, a vast agricultural exploitation."[61]

Concerns about mothers and maternal medicine were often tied to displaced eugenics fears. France's 1870 loss of Alsace-Lorraine to the Prussians provoked much concern over a supposedly diminishing and aging national population. Fears of degeneration haunted France, and many thinkers therefore emphasized the need for puericulture, a sort of positive eugenics that encouraged the breeding of "desirable" individuals and promoted an overall increase in reproductive output. In contrast to the hard eugenics, this notion of "soft" heredity also encouraged the improvement of social and economic conditions. Support for eugenics in France led to the 1912 formation of the French Eugenics Society. Created partly in response to the Universal Race Congress that took place in Britain the previous year, the society drew strength from the writings of provincial intellectuals such as Georges Vacher de Lapouge, a librarian who advocated artificial insemination as a means of reproducing human populations. William Schneider argues, however, that propositions such as the "1926 legislative proposal for a premarital examination law was probably most significant for bringing the tactics of French eugenics in line with the activities of eugenics movements in other parts of the world."[62]

Moving beyond concerns about plantation labor, by the 1930s the logic of positive eugenics had encouraged many Vietnamese to try to improve the racial strength of Vietnamese vis-à-vis other "races" on the Indochinese peninsula. In 1937, Phạm Văn Quang, a member of the House of Representatives in Annam, articulated the stakes of rural health to the Guernut investigative commission. Of all the works of the protectorate in Annam, Quang found that only the AMI had done much good. But Quang argued that it offered medical service only to the rich and concentrated its services in the urban centers, barely touching the countryside. For Quang and other nationalists, malaria represented a threat to the Vietnamese race, and growing nationalism persuaded Vietnamese physicians to adopt moral economies of race and malaria prevention. In the mid-1930s, Đặng Văn Dư, the chief of the Hà Tĩnh provincial hospital, created two pamphlets written in Vietnamese to advertise the values of quinine. Fifteen thousand copies of these pamphlets had already been distributed throughout the province when they came to the attention of the RSA in 1938 and were eventually chosen for wider distribution to rural audiences. In one pamphlet, Dư examined some of the general aspects of malaria. He argued that of all the diseases in Indochina, malaria was the most harmful for "our race" (*dân tộc mình*). While other diseases such as cholera, plague, and smallpox killed many, and killed quickly, they did not strike often, and they did not have a significant effect on population numbers. Malaria, on the other hand, prevented children from growing up healthy, women from giving birth, and men from working hard. The effectiveness of these arguments in the countryside is difficult to determine, but among middle- and upper-class Vietnamese, these views of gender and age roles were very common.[63]

Another effect of malaria pointed out by Dư was that it prevented Vietnamese from settling in the uplands. A strong theme during the colonial period was the idea of *nam tiến*, or southern expansion, which involved the spreading out of Vietnamese society from its cradle in the Red River Delta by expelling, killing off, or absorbing the peoples and cultures of the central and south. This process was complicated, episodic, and uneven, but by the beginning of the twentieth century, Vietnamese speakers inhabited the lands from north to south in current Việt Nam. Yet *nước độc*, or poisoned waters, in the upland region had prevented migration to the highlands, according to Dư. This was one reason for overcrowding in the deltas and on the coastal plains, as Vietnamese eked out a perilous living that made them vulnerable to natural disasters and famines.

Dư noted three ways to prevent malaria: use a mosquito net, carry out works around the village to reduce the numbers of mosquitoes, and take quinine. He claimed that nets were useful, if properly employed, but could not prevent all bites since people could not stay under nets all the time. Environmental modifications, as carried out on plantations, were time-consuming and too costly for most villages. This left quinine as Dư's ideal solution and he was a strong proponent of the state quinine service, recommending four tablets per day for adults, with lower dosages for children and half of a tablet for newborns. He argued that quinine was ideal because it was relatively cheap and subsidized by the state, and thus spending larger amounts of money on effective quinine was more economic than wasting smaller sums of money on ineffective *thuốc bắc*.

In the third version of the handout, Dư went into more detail about what Vietnamese had commonly held to be the causes of malaria. The term used by doctors during the colonial period was *bệnh sốt rét*, which translates as "the disease of fevers and chills," with the words *ui, ngược*, meaning "remittent" or "recurring," sometimes added at the end. This more technical term replaced older terms, some of which referred to malaria and other malaria-like diseases.

Dư needed to convince the population that Western-trained doctors knew what malaria was and that they knew how to cure it. He had earlier explained why *nước độc* did not cause malaria, calling the suggestion "completely wrong." He argued that a better way to understand malaria was in terms of "poison land." He went into detail about the link among malaria, mosquitoes, and plasmodia. In one version of the handout, he argued that Western medicine had demonstrated that malaria germs (*vi trùng*; i.e., plasmodia) could be found in blood samples of all those with malaria and that if the blood sample did not show these plasmodia, then the illness, no matter how much it resembled malaria, was not malaria. His conclusion was that for these diseases, quinine would not work, and *thuốc bắc* might be more useful.[64]

Dư's approach to malaria prevention in the Vietnamese countryside drew on the rhetoric and practices developed for malaria prevention on plantations. Yet differences remained in the models of power that operated in the discourse of the colonizer and the colonized. Colonizers tended to express a biopolitical concept of power, emphasizing quality of family life and the productive aspects of colonial rule. Hospitals, schools, and improved housing were major themes of administrative and planter rhetoric, as officials sought to enact needed reforms. On the other hand, writings by the colonized, even those with a relatively high position in colonial society, like Dư, often focused on territorial conceptions of power. Even when not explicitly anticolonial, these writings underscored the idea of a Vietnamese nation with territorial sovereignty. However, biopolitical and territorial conceptions of power were not mutually exclusive; the two were present in both discourses, with varying degrees to which one or the other concept was emphasized.[65]

Ecology and Nation

On the eve of World War II, some Vietnamese medical doctors and agronomists began employing the concepts of tropical medicine to advance nationalism. Railroads, rubber plantations, and other imperial economic ventures throughout Southeast Asia had gained a deservedly poor reputation for work conditions. Noting the reluctance of Vietnamese to immigrate to the south in 1939, Bùi Kiến Tín, who went into private practice after World War II, wrote in his medical thesis about "those who want to emigrate don't dare to repeat the experience that cost the lives of their loved ones, during the construction of the first railroads or the establishment of the rubber plantations." Tín stated that until the late 1930s, the Pasteur Institute had been solely responsible for hygiene in the colony, and he argued for the creation of a centralized institute of hygiene. The Pasteur Institute, Tín continued, was oriented toward research rather than practical ends, and while strong scientifically, the institute's efforts took a long time to yield results. He quotes Morin to argue that antimalaria measures remained at the experimental level. Tín also inserted quotes from other French physicians to assert that these measures had had minimal impact outside of plantations.[66]

Working conditions on the plantations during the 1940s, notwithstanding the optimistic predictions of medical doctors in the late 1930s, remained poor. In noting sickness among workers in 1943, Jean Canet argued that one means of increasing health was to decrease the cost of living and the price of food. He then listed all the inconveniences of such a policy, including the necessity of having a special organization on large plantations dedicated to controling the price of food and monitoring its distribution to prevent resale off the plantation. It seemed much simpler to pay workers a higher salary. Yet this policy was not adopted, according to Canet, because of "moral failings" among the workers, specifically their passion for gambling. Canet's article noted: "Among the Annamites, in general, and among the Tonkinese in particular, a passion for gambling for itself is so violent that 90 percent of the individuals only have but one preoccupation when they receive their salaries or when they have sold whatever product; it is to play. An Annamite cannot have a cent in his pocket without being tempted by gambling."[67] Like the Pasteur Institute poster that urged workers to take their quinine, this problematic view of Vietnamese culture continued to place responsibility for illness squarely on the shoulders of the "native."

Another challenge to reform was the European control over the rubber industry, which created a clear racial divide between European owners and managers and Vietnamese, Cambodian, Chinese, and Montagnard workers. This control was theorized as necessary by leading French scholars of Indochina. In the summer of 1939, Charles Robequain (1897–1963), who had recently returned from Hà Nội to become a professor of geography at the Sorbonne, published the results of his study on the effects of over fifty years of French administration on the economy of Indochina.[68] Building on his doctoral thesis, which was completed in 1929, and contemporaneous work at the EFEO, Robequain used the emerging framework of tropical geography to address the question of agricultural colonization in Indochina. "It is generally recognized today," he wrote "that the European cannot engage in the same kind of agricultural tasks as the Indo-Chinese without being degraded to the status of the 'poor white,' struggling vainly against the competition of the colored man and forced, in order to survive, to reduce his needs to a minimum, lowering his standard of living to the point of destitution. . . . He does not engage in actual farm work, but instead manages and superintends the cultivation of a relatively large holding, whose yield is high in comparison with native plantations."[69] Robequain continued his analysis of European involvement in colonial enterprises by raising questions about the size of holdings, the proper balance between company-run estates and smaller plantations, the nature of the relationship between Europeans and natives "for the best legitimate interests of both," and the status of land held by Europeans. Robequain also noted a hierarchy of access to knowledge within Indochina, and he believed that the French should control the large agricultural estates that required the application of scientific knowledge.

Anticolonial activists denounced living and working conditions on rubber plantations and the concepts employed by tropical science. In response to bad press and the worker strikes that were buffeting the plantations, owners expressed their fears of labor shortages. Some owners attempted to create better living and working conditions, or at least a better image of these conditions, and they began to adopt the idea that medicine, along with schools, religious institutions, and housing, could serve as a symbol of French goodwill in the colonies. The colonial state aided these efforts by advertising the social works being carried out by the owners. The Indochinese Communist Party (ICP), which was formed in 1930, controlled few of the protests of the 1920s and 1930s and sought instead to harness labor activism under its banner of anticolonial politics. Many planters found evidence of communist organizing, such as a banderole pronouncing "the formation of an Indochinese Communist Government in union with the Chinese Communist Government for the world revolution" along with other revolutionary materials. Such activities led to the founding of organizations for rubber workers, and in September 1944, the communist-led "Rubber Organization of the East" assigned comrade Lê Đình Cự to organize the Liên đoàn Cao Su (Union of Rubber Workers) at Long Thành. Meanwhile, laborers were also organizing, and at the end of 1944, 500 workers from the nearby An Viễng and Bình Sơn plantations, which were both part of the SPTR group, marched to protest their unpaid wages. According to official Vietnamese history, Delamarre (Đờ-la-mông), the work inspector, refused to concede and employed a Vietnamese corporal (*cai*) named Minh to carry out his bidding. The protestors subsequently captured *cai* Minh, and Đờ-la-mông was forced to capitulate to demands. While official histories credit the Communist Party with helping to organize such protests, they also show that workers acted with a high degree of autonomy and that the Bình Sơn party cell was not organized until the end of the year.[70]

After World War II, several agricultural engineers, including Nghiêm Xuân Yêm, decided to side with the Việt Nam Độc Lập Đồng Minh Hội, or Việt Minh for short. In May 1941, Hồ Chí Minh and the ICP leadership formed the Việt Minh in anticipation of a protracted struggle against fascism. Yêm, who later served in the government of the DRV, contributed several articles to the journal *Thanh Nghị* between 1941 and 1945. In his role as agricultural engineer, Yêm had a chance to talk with many farmers, and his analysis of conditions in the countryside reflected his awareness of rural conditions. It was the Vietnamese agricultural agents and engineers who made use of the subversive potential of ecological knowledge by pointing to the failures of the colonial state to adequately manage agricultural production or care about the deaths of ordinary Vietnamese.[71] In 1944, for example, Yêm published the findings of a "small investigation" on the policy of "voluntary peasant migration." This policy was based on granting concessions to smallholders set up in the north by the law of November 23, 1925. He argued that this policy had been a total failure, pointing to the tiny numbers of concessions granted in the past twenty years. Yêm pointed out that most peasants were too poor to make the move, they had no capital saved, and they had no means of getting started. Yêm argued that this policy worked only for those farmers who had some capital and a "new mind," one that valued profit. But those farmers tended to stay put in their homelands. The psychology of the *dân quê Viet-nam* (sic), or Vietnamese peasant, meant that these farmers neither dared nor wanted to move if they could help it. But the problem did not merely concern a frame of mind, and Yêm discussed the difficulties farmers faced in new environments. They did not know the environments they were moving to, and some plants, such as *tranh* grass, were difficult to deal with. Farmers also rightly feared malaria, and even though techniques existed to manage this disease, the farmers could not afford quinine.[72]

Furthermore, Yêm discussed the relationship between young intellectuals and Vietnamese farmers. Like other authors in *Thanh Nghị*, Yêm held that science and action should go hand in hand. Yêm focused his attention on the plight of the peasants and in his articles proposed concrete social and economic plans dealing with, among other topics, rice, raising cattle, and the land system. He placed the blame for the 1945 famine, which resulted in the deaths of between one and two million people, squarely on the shoulders of the French government, arguing that the government should have moved rice from the south to the north, even during the war. He wrote about the northern uplands, but his analysis applied to the southern uplands as well. Some young intellectuals heeded Yêm's call to fight for the people, and for many this meant joining the Việt Minh. For instance, Đặng Văn Vinh and Nguyễn Khoa Chi from the graduating class of ESASI in 1941 settled in the north and took on responsibility for setting up rubber production and processing for the DRV.[73]

To defend their interests during World War II, those in the French-controlled sector of the rubber industry collectively undertook an intense propaganda campaign proclaiming the healthiness of the plantations. The onset of the war in Europe, and then in Asia, did not initially bring many changes to plantation conditions in Indochina. The Japanese empire imported its rubber from Java and Malaya, and Japanese planners essentially left Indochina's plantations alone. As the war continued into 1944, however, worker transport from north to south began to be seriously hampered by the Allied destruction of French and Japanese transport boats. Because of the lack of transport between north and south, maintaining the health of workers who were already in Cochinchina became more crucial. With planters desperate for regular workers, the rubber industry intensified its efforts to improve its image. Publications advertising soccer fields, schools, and churches were common methods of promoting awareness of plantations among a French readership, as was encouraging the visits of political figures such as the Cambodian king. When famine struck in the summer of 1945, sentiments turned further against the Asian "liberators." "The Japanese," one former rubber worker told me, "were very cruel. Very cruel!" In comparison, "the French were a bit more civilized."[74]

THEORETICAL DEVELOPMENTS IN tropical agriculture and medicine encouraged the Pasteur Institute to shift from interventions aimed at race (and human bodies) to those aimed at ecology (and environments) as the basis for prevention and treatment. Yet, during the 1930s race continued to appear in discussions about the causes and cures for malaria. Medical researchers continued to link "race" with the nation and when Vietnamese medical doctors justified the expenses of applying malaria prevention efforts to the countryside in the 1930s, race was implicit in much of their nationalist discourse. Furthermore, by linking measurements of blood and spleen to the distribution of malaria among populations, medical doctors encouraged demographic manipulation. In other words, malaria science served to re-territorialize this disease in already racialized bodies that were located in environments, and in nations. Medical research both increased concern about the fate of Vietnamese and Montagnard populations and made the integration of supposedly unhealthy bodies into colonial society problematic. As medical expertise advanced in the rubber region, reservations were created for locals to "protect" them and their land, and efforts were made to separate plasmodia-carrying locals from plantation labor.

Government officials, planters, and scientists sought to govern human diseases and human bodies through "internal frontiers" of race and place, on and off the plantations, which did not simply draw on preexisting environmental and social conditions but reworked them as well. Plantation managers created new racial and ethnic divisions to exploit social tensions, and biomedical doctors helped inscribe race into colonial disease environments through their application of industrial hygiene on plantations and the surrounding countryside. Like their counterparts in Southeast Asia, French officials used bodily secretions to judge colonial subjectivity and viewed agricultural practices as the point of connection between bodies and their environments. Racial categories were the common currency of the colonial experience and they were woven into the colonial sciences as well. Many scholars have made clear that "postcolonial" can simply refer to a time after colonialism and does not necessarily indicate an end to the ideas and prejudices prevalent during colonialism. One of the organizing concepts linking humans to their biophysical environments that survived the transition from colonial to postcolonial sciences was the idea of race. As chapter 5 shows, concerns about race morphed into concerns about modernization.

Part III
Rubber Wars

CHAPTER FIVE

Maintaining Modernity

This is my family's plantation. It has been such for seventy years. And it will be such until we are all dead. We want to stay here because it's ours, it belongs to us.

—Hubert de Marais, *Apocalypse Now Redux*

What gave us a start, and put us on the defensive, was the people who came out to meet us.

—Trần Tử Bình, *The Red Earth*

At the end of World War II, the rubber industry faced several formidable challenges, including the reversion of plantation land to scrub forests, the destruction and loss of equipment, and, most importantly, the dispersal of its workforce. Before the war, the industry had yielded over 60,000 metric tons of rubber annually, but by the end of the war, production had ceased almost completely. At the end of 1945, rubber harvesting in Cochinchina and Cambodia resumed, and in 1954 output exceeded prewar levels. This achievement was made possible by higher productivity per worker, which resulted from improved techniques, utilizing trees that had not been tapped during the war years, and unsustainable tapping efforts that maximized short-term profits. By 1953, Vietnamese and Cambodian plantations were producing close to 75,000 tons combined.[1] This seemingly straightforward narrative of recovery, however, masks the many battles that took place in and around plantation landscapes. These landscapes became a means for those in both the communist and the capitalist worlds to imagine a "new life" that would leave feudalism and colonialism behind and lead to a modern society and a tropical modernity. For communists, it meant an end to exploitation and foreign domination. For capitalists, it meant a modern Vietnamese society full of satisfied workers and consuming individuals. Workers, too, were searching for a brighter future and possibly the *đoàn thể*, or organization, that seemed to have eluded Vietnamese society during the colonial period.[2]

This chapter explores the symbolic and material struggles over nature, labor, and plantations in two ways. It considers how the First Indochina War (1946–54) shaped the environment on and around plantations and how the plantation environment in turn shaped the war. Techniques of *encadrement*, or "landscape-moulding techniques and systems of spatial organisation," were simultaneously appropriated and challenged by those fighting the war. Plantations were in a region that served as a gateway to Sài Gòn and the Delta to the south, and the mixed environments of rubber trees, forests, and rice paddies helped structure the battles over how plantation lives would be lived. This chapter also examines how anticolonial forces and rubber workers challenged the colonial order embedded in plantation landscapes. This was not an easy task. Both land and society resisted easy reformation, and racial attitudes toward the proper relationship between nature and culture persisted, even in those who ostensibly supported Vietnamese independence. Activists, soldiers, workers, planters, experts, and officials made history, but not on the landscapes of their choosing.[3]

This chapter begins by tracing the efforts of Vietnamese laborers to improve plantation conditions, including hours worked and benefits received. A legacy of colonial labor abuses prompted many Vietnamese to vigorously oppose plantation agriculture and initially encouraged the anticolonial Việt Minh to destroy the plantations as part of its "scorched earth" tactics. The First Indochina War made rubber plantations strategically invaluable for anticolonial forces, prompting the French military to try to either eliminate these sanctuaries or co-opt their spaces. The chapter ends by examining how stories of hardship and survival later led the Việt Minh and rubber workers to reenvision rubber plantations not as hated colonial institutions but as beloved centers of nationalist production. The First Indochina War paved the way for the incorporation of rubber and other economic activities into the Vietnamese nation, symbolically refashioning plantations into revolutionary landscapes in the minds of both southerners and northerners.[4]

Scorched Earth

The year 1945 was a momentous one for the Vietnamese and the rubber industry. First, a March 9 coup d'état resulted in the Japanese military rounding up and imprisoning French officials, soldiers, and civilians. This coup effectively ended French colonial rule, left the rubber industry in disarray, and promoted nationalism among Vietnamese workers. One plantation manager recalls a Japanese official telling Vietnamese workers that they were now responsible for their own actions, a statement that aligned with the pan-Asian vision of race that was promoted through the Greater Asian Co-Prosperity Sphere. That year, a famine that left over a million dead in Tonkin and Annam was also peaking. On August 14, after the dropping of the second atomic bomb, Japan surrendered to the Allied forces, leaving a military and political power vacuum throughout Southeast Asia. On August 19, the Việt Minh seized power in Hà Nội and multiple provinces in rapid succession. On September 2, Hồ Chí Minh proclaimed Vietnamese independence and announced the Democratic Republic of Vietnam (DRV). Less than two weeks later, nationalist Chinese troops occupied northern Indochina and British troops landed in the south. On September 23, French forces that had been released from Japanese camps and rearmed by British troops seized power in Sài Gòn and slowly worked their way into the countryside, including the plantations. They proved unable to do the same in the north, which was being occupied by Chinese troops. By the time that the French Expeditionary Corps had landed on October 3, a complex negotiation for power among several political and military forces had begun. Finally, the day after Christmas, the French signaled their intentions to fund a war by revaluing the piaster from ten to seventeen francs, which gave the piaster more purchasing power while creating hardships for people living in Indochina.[5]

The French, the successive governments of South Vietnam and Cambodia, the Việt Minh, and the Cao Đài all viewed the rubber industry, including plantations and laborers, as a significant asset in the coming military and political struggles. The Việt Minh destroyed some of the rubber produced during World War II, but the French sold off most of their stock after the war. Memories of the colonial rubber industry informed the attitudes toward plantations of many of these actors. From the French perspective, the rubber industry provided a means to restart the Indochinese economy, and plantations were key sites in efforts to modernize agriculture through technical, social, political, and fiscal reforms. Old colonial hands viewed plantations as organizations that could solve complex problems of colonization involving labor and bring about simultaneous improvements in rural living conditions and increases in agricultural exports.[6]

One of the most urgent postwar questions facing the rubber industry was how to rebuild its workforce. In 1946, according to the Rubber Planters' Union (Union des planteurs de caoutchouc), only 15,000 workers were active in Cochinchina and Cambodia, while the rubber industry had the capacity to employ up to 60,000 workers. The French government expressed support for planter recruitment, and at a June conference in Đà Lạt the French consul dedicated one point in his seven-point plan to increasing the number of planters. Many proposals to address the issue of worker scarcity were put forth, including further mechanization of the plantations and incentives to help retain workers in the south. When these initiatives did not work, planters considered forced labor measures that had been approved by the Labor

Inspectorate, including recapturing former workers and extending existing contracts. Planters were also interested in using Vietnamese returning from abroad, but acknowledged that these workers would seek higher salaries and would rather use the more advanced skills that they learned during the war than work on plantations. Finally, planters discussed reviving colonial-era schemes for foreign labor recruitment and looked to Tamils, Montagnards, and even Javanese laborers left behind in Siam on the receding tide of Japanese empire.[7]

The ICP was also busy preparing for war and in a strategic gamble Hồ Chí Minh issued a resolution on November 11, 1945, to dissolve the party. Hồ wanted to convince a skeptical public including intellectuals, Catholics, and others wary of ICP rule that the Việt Minh was not simply a communist tool. While Hồ never intended his controversial resolution to disband the ICP, "many party members in central and southern Vietnam were left confused and demoralized for months."[8]

Local cadre did not remain still, and they evoked memories of plantation landscapes to frame the battle against the return of the French as a war against exploitation. They also coined several anti-plantation slogans and poems, many of which emphasized the connections between workers' bodies and rubber trees. A verse in one poem stated that "rubber grows well in this place; Every tree is fertilized with the body of a worker." The equation of one worker for one tree highlights the Việt Minh's recognition that the plantation structure pitted workers against trees. Such slogans also appealed to notions of barbarism, accusing French imperialist capitalism of valuing trees more highly than human beings. This idea was captured by the phrase "workers weren't worth one rubber tree." Socialism, the Việt Minh claimed, would bring civility to plantations by reversing this inequality and subjugating rubber trees to humans.[9]

Initially, Việt Minh publications called for the destruction of rubber trees and plantations. A saying from the time encouraged workers to "transform the rubber plantation into a battlefield to kill the enemy." A combination of practical and ideological motivations encouraged the Việt Minh to launch the Rubber War Front (Mặt trận cao su chiến), which sabotaged trees and rubber-producing equipment. Whereas "bandits" and other nonstate actors had previously relied on the unruly forests and a complex environment to delay would-be pursuers, the Việt Minh and the French navigated their way through the straight lines and monocultures of the plantations.[10]

Economic and strategic motivations also drove attacks on plantations. Việt Minh leaders argued that the French were drawn to Indochina purely for profit, and that the rubber industry contributed directly to the French war effort. A North Vietnamese journalist, Điệp Liên Anh, observed that "one rubber tree equals one enemy. To destroy one rubber tree is to kill one invader." Not surprisingly, direct assaults on plantations were part of the Việt Minh's broader strategy to wreak as much havoc in the colonial economy as possible. Anh noted that "the rubber plantations became the main target in the attack plans of the Việt Minh in the eastern region [southeast Vietnam] to weaken the French economy and weaken the vitality of the French Expedition." Historian Pierre Brocheux has pointed out that Việt Minh documents were full of the terms "scorched earth, blockade, and *guerre économique*" as "the economy . . . became an issue, a battlefield, and a weapon." The goals of "destroying the economy of the enemy" and "contributing to building the economic basis of the war of resistance" were often in tension, as places such as plantations could serve both objectives.[11]

These arguments convinced many rubber workers who were seeking better conditions for themselves to join the Việt Minh. Some participated in the general uprisings that had taken place in August, and later many workers filled the ranks of armed units such as Battalion 1 (Chi đội 1), which formed in 1945, and Battalion 10, which formed in 1946. Consistent with slogans depicting trees as fertilized with the bodies of workers, most early campaigns attempted to destroy both the equipment and the trees. According to one history of the rubber workers' movement, "all of the rubber workers struggles during the period of resistance were aimed at the sole purpose of 'sabotage.' " These early efforts of resisting the French reinvasion, however, met with considerable difficulties, and by February 1946, the French and British forces had reoccupied most of the plantation region. Faced with these setbacks, the Việt Minh attempted to set up a parallel economy, and in the spring of 1946, they issued the đồng to compete with the Indochinese piaster.[12]

The colonial government grew increasingly concerned about the vulnerability of workers to Việt Minh recruitment, with good reason; one Vietnamese history states that 34,000 out of the 55,000 workers took part in the revolution. In the aftermath of the August Revolution, leaders became aware of the possibility of violent confrontation. Mark Lawrence has discussed efforts to avoid conflict, including the Franco–Việt Minh accords, which were signed on March 6 by representatives of the French government and the Việt Minh. Some accused Hồ Chí Minh, a key broker in the deal, of selling out the anticolonial cause, but the accord created an opportunity for the Việt Minh to prepare for prolonged resistance. It published several leaflets and periodicals, which circulated among rubber workers and provided an alternative to French news outlets. The materials gave general guidelines for action and encouraged resistance via a steady stream of heroic stories and poems.[13]

One of the Việt Minh's publications was *Công Đoàn* (Trade Union), later renamed *Cảm Tử* (Suicide, literally "to volunteer for death"), which boasted a circulation of up to 4,000 copies. *Công Đoàn* was meant for distribution throughout the southern region, but after the Việt Minh established province-level organizations, they produced their own newspapers. In mid-1946, the Biên Hòa information service began printing *Đồng Nai*, which was named after the province where it was published. This paper was printed on only two pages, with 300–500 sheet runs. Battalion 10 members later took responsibility for writing it. Run numbers were limited compared with *Công Đoàn*, and by 1947, *Đồng Nai* was suffering from a critical lack of materials, which meant that some of these journals were printed with seaweed ink. The first Rubber Union, which was started in Biên Hòa with 4,000 members in September 1946, produced its own journal, *Sinh Lực* (Vitality, alternatively Sap, an apt name for a journal directed at rubber tappers).[14]

While fighting had been taking place sporadically in the south since the French reoccupation, the violence of December 19 and 20, 1946, that rocked Hà Nội and other Vietnamese cities above the sixteenth parallel marked the official beginning of the First Indochina War. The beginning of the war pushed southern cadres to further centralize, and on December 20, the Tây Ninh Rubber Union was formed with 2,533 members. Ten days later, the Thủ Dầu Một Rubber Union was formed, with 1,635 members out of a total of 12,000 workers. Both unions published their own newspapers, *Cao Su* (Rubber) and *Cần Lao* (Workers). No organization existed for Bà Rịa, an area with a relatively high number of Vietnamese and Chinese owners, likely because rubber production had not yet restarted and few workers were available to organize.[15]

Bleeding the Colonial Economy

During the 1940s and 1950s, plantations existed at the edge of the mobility frontier and they were among the first places to build airstrips. Airstrips helped rubber plantations stay in business throughout the Indochina Wars, and military and political leaders commandeered these airstrips for their use. Anticolonial forces also sought to turn these landscapes to their advantage. The mixed plantation-forest environment offered several advantages for the Việt Minh. The neat rows of trees on plantations both increased the mobility of the nationalist forces, which traveled largely on foot, unlike the mechanized

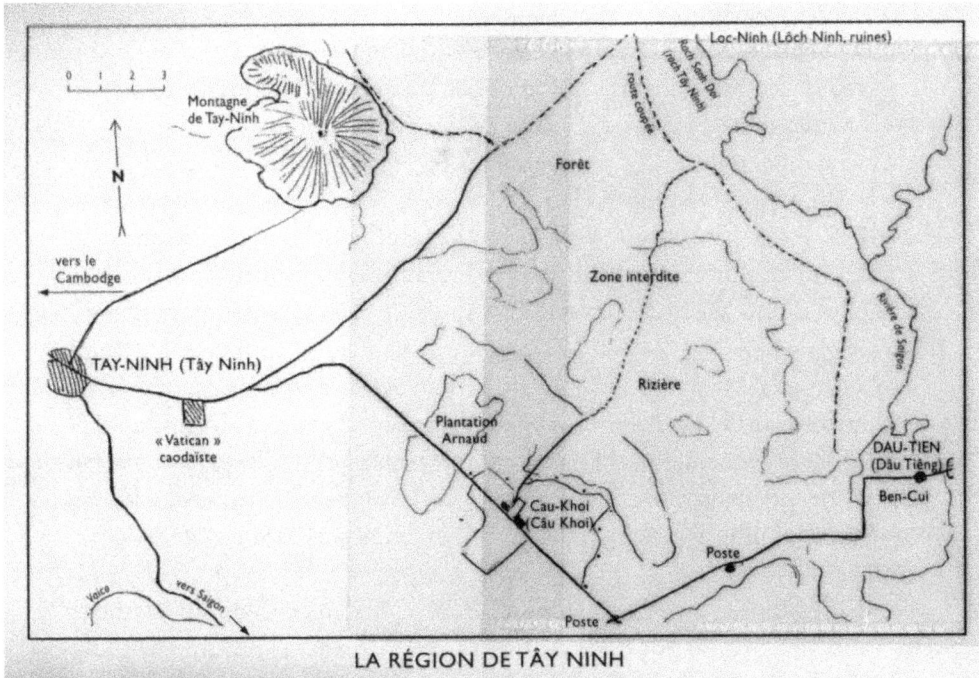

Mixed environment, Tây Ninh. This map by Motte shows the mixture of forest, rice paddy, and plantations that was very useful to the Việt Minh during the First Indochina War. Source: Motte, *De l'autre côté de l'eau.*

French army, and neutralized French airpower. Plantations provided the guerrillas with places to sleep, ready-made paths, wood and water, and places to hold meetings and to train. The nearby forests provided a space that was virtually impenetrable for French-led operations and into which the Việt Minh forces could disappear after engagement. Both forests and plantations provided a convenient place for the Việt Minh to evade detection, especially from French aircraft, and served as a shield during attack and retreat. Plantation laborers represented an important pool of potential allies, and production could be held hostage for supplies and money.[16]

Geographically, plantations played an important role in control of the border region of southern Indochina and served as the gateway to the Mekong Delta and its rich rice reserves. Since plantations straddled both sides of the Cochinchinese and Cambodian border, rubber tapping served as a useful cover for travel between the two lands. As a countermeasure, the French military and planters attempted to control movement and militarize the plantation perimeter, transforming sites such as Courtney into military camps. Michelin's

Dầu Tiếng plantation eventually held eight posts and 300 soldiers, and the Quản Lợi plantation maintained five posts and 200 soldiers. The French also engaged in the destruction of rubber trees at small and medium-size plantations to minimize their strategic value to the Việt Minh.[17]

Meanwhile, the Việt Minh continued to organize, founding the Rubber Trade Union (Nghiệp Đoàn Cao Su) at Long Khánh in 1947 to coordinate activities among the various rubber unions, provincial organizations, and military units in the region. The welter of organizations provided a measure of resilience for the Việt Minh, as small groups learned to act autonomously. Following general policy, the union's two main goals were to "turn the plantations into a battlefield to kill the enemy" and "destroy the enemy's economy." The rubber company (Đại Đội Cao Su) belonging to Battalion 10 was responsible for destroying rubber and helping secret organizations on the plantations kill the militia members who controlled the plantation space. These organizations continued to publish, with newsletters such as *Mission* (*Sứ Mạng*) and *Voice of the Forest* (*Tiếng Rừng*).[18]

A Việt Minh report titled "French Plantations" recorded four methods for sabotaging plantations: (1) destroy the workforce, (2) destroy the tools, (3) destroy the trees, and (4) destroy the means of transportation. In the section on trees, the report describes two ways to kill trees and several ways to decrease the production of latex, including the introduction of *hevea* diseases. While the Việt Minh bemoaned their lack of a mycologist to carry out biological warfare, they could remove the bark from the trees. When plantation owners quickly devised methods to regenerate the lost bark, the Rubber Union countered with a practice called girdling, which entailed cutting a deep ring around the tree trunk that prevented latex from traveling up from the roots and eventually killed the tree. Simply burning the trees became a more widespread form of resistance.[19] This campaign to sabotage French economic interests succeeded early in the war, and even served as a model for the Communist Party during the Emergency, or communist insurgency, in Malaya. According to *Rubber Workers' Fight*, which was published by the Workers' Publishing House, from 1945 to 1948 rubber workers had carried out 1,600 incidents of rubber sabotage, destroying over seven million high-quality trees, or 10 percent of the total number of *hevea* trees in Indochina, and ravaging 17,000 of the 150,000 plantation hectares. Specific worker action continued throughout the year and often coincided with meaningful days in the Việt Minh calendar. On May 1, International Labor Day, workers at Courtney in Bà Rịa went on strike for higher wages, and observers noted communist symbols everywhere on Dầu Giây. Bastille Day was also marked by worker actions.[20]

It is difficult to verify the numbers involved in these events and the role that the Việt Minh played in organizing them. The author of *Rubber Workers' Fight* claimed that on May 19, 1947, in honor of Hồ Chí Minh's birthday, the workers of Biên Hòa destroyed an astounding 300,000 rubber trees. Yet the "French Plantations" report calls this total into question. The report states that in a single night, one person could girdle about one hundred trees, which meant that for the night of May 19, about 3,000 workers would have had to have been involved. With 4,000 workers in Biên Hòa early in 1948, this level of activity would require an extraordinary amount of participation. The report does list 250,000 trees destroyed on Michelin's Dầu Tiếng plantation on December 8, 1947—the highest single-night total recorded for that year—but in Biên Hòa the earliest sabotage incident is listed as June, and the highest total recorded is 20,000 trees.[21]

While it is possible that the author of *Rubber Workers' Fight* had access to other documents, the strong didactic component to the book suggests inflated numbers, especially for actions that were carried out on Hồ Chí Minh's birthday. The text was published in 1950, during the height of the First Indochina War, at a time when the Việt Minh needed to score political points, and promoting the successful efforts of partisans was a priority. *Rubber Workers' Fight* contains grisly passages describing workers who were decapitated after being caught carrying out acts of sabotage, or simply being in the wrong place at the wrong time. Such descriptions drew from colonial tropes of French soldiers displaying severed heads during the late nineteenth and early twentieth centuries. These images aimed to stir public outcry and to warn partisans about the consequences of being captured. Such publications garnered international support, and on May 19, 300 French intellectuals signed a petition urging negotiations with the Việt Minh. A day later, Hồ Chí Minh and Hoàng Minh Giám rejected a cease-fire agreement with the French, due in part to the failure of the 1946 truce.[22]

In early 1947, most French military leaders were predicting a quick victory over the Việt Minh as the anticolonial consensus that had formed around the August Revolution began to fray, especially in the south. The Hòa Hảo decided to support the French after the Việt Minh assassinated its leader, Huỳnh Phú Sổ, in the spring of 1947. Likewise, many members of the Cao Đài, who had fought alongside the Việt Minh after World War II, shifted their alliances and left the "liberated" Việt Minh zones after their supreme spiritual leader, Phạm Công Tắc, agreed to side with the French military. The Cao Đài's Holy See is in Tây Ninh, which at the time served as a hub for rubber plantations, and this swing signified a major victory for the French.

The United Southern Rubber Trade Union (Liên Hiệp Công Đoàn Nam
Bộ Cao Su), with Vũ Văn Hiên as its chair, continued to try to destroy the
French economy in Indochina. By 1948, however, the Việt Minh admitted in
private that worker actions were starting to lose their effectiveness. The defense of plantations by private militias, which were often composed of Nùng
and other Montagnard groups, curtailed the success of sabotage efforts.
These attacks were further limited by the organizational weakness of the Việt
Minh in the south, "practically a 'non-entity,'" according to the historian
Christopher Goscha. A call to workers on Cầu Khởi to join the revolution
recognized the difficulties they faced, including the threat of job loss. The results of the first full year of the Indochina War for the Việt Minh were mixed.
It had been driven out of Hà Nội and much of Cochinchina and had lost
important allies in the south, including the Cao Đài and the Hòa Hảo, effectively losing its connection to the region. However, the French military,
which had predicted a quick victory, was finding it hard to pin down the Việt
Minh in the north.[23]

Planting a Resistance Economy

As the Indochina War dragged into its second year, neither side was making
clear progress. Rubber workers and the Việt Minh employed a range of tactics
in this complicated political and military landscape, including capturing
radio stations, spilling latex, breaking collection bowls, and leaving the plantation altogether. These activities had a different effect than the simple sabotage
of trees, as they reduced company profits without destroying the potential for
future production. Others did not make such distinctions, and attempted to
disrupt plantations however possible, with protesting workers often combin-
ing multiple strategies. For example, on January 28, 1948, workers at the An
Viễng and Bình Sơn plantations (the SPTR group) set fire to plantation
facilities and fled to the liberated zone.[24]

Việt Minh sabotage did more to sustain the morale of the anticolonial resistance than to slow down French economic activity, and Việt Minh leaders
began to coordinate rubber sabotage with other anticolonial activities. In
1948, Battalion 1 became Regiment 301 (Trung Đoàn) and Battalion 10 became Regiment 310 as Việt Minh leaders attempted to incorporate sporadic,
local activism into larger structures. Plantations fed these military units, with
rubber workers composing much of Regiment 310 and forming significant
portions of Regiments 300, 301, 309, 311, and 312. By drawing workers away
from the industry, the resistance posed a real threat to the plantations, which

faced a severe labor shortage. According to *Rubber Workers' Fight*, 8,000 members of the Trade Union originated from the ranks of rubber workers, a high percentage of the 21,000 rubber workers of the south.[25]

In January 1949, seventeen representatives of the southern forces of the Việt Minh met to review the previous year's activities of the Rubber War Front. The Việt Minh's success varied considerably by province, but the record of the meeting noted many achievements, including the destruction of the Vên Vên factory. Cadre in Thủ Dầu Một, the province closest to Sài Gòn, destroyed the most trees during the first half of 1948 and hit other essential plantation areas, including factories. After cadre in Biên Hòa organized a Rubber Company in May, the province recorded the highest destruction rates. Biên Hòa's Rubber Company had also been particularly adept at coordinating with Regiment 310, the provincial regiment. In Bà Rịa and Tây Ninh, numbers were low throughout 1948 because of sporadic Rubber Front activity. In Bà Rịa, Regiment 307 assisted in rubber-destroying activities, as did the ethnic minorities (*dân tộc thiểu số*), while cadre in Tây Ninh made sacrifices but accomplished little. In Cambodia, according to the report, the Việt Minh had been able to organize very few sabotage activities, and the French had successfully retaken control of the rubber industry.[26]

Although the news that the People's Liberation Army had taken Beijing in January may have heartened the delegates, other news was less optimistic. The delegates admitted that the plantations were still earning great profits for certain French owners, which, given the limits on destruction efforts, did not come as a surprise. Furthermore, the cadre from Thủ Dầu Một had failed to attract the support of ethnic minorities or specialized workers. In 1948, the Việt Minh also noted an increase in alienation among workers and a decrease in sabotage acts. Several hypotheses were put forward: the enemy had prepared effective defenses; organizational problems within the military had disrupted rubber-destroying activities since June; a lack of food had forced soldiers to concentrate on growing crops rather than military activities; and confusion among the soldiers meant that the local cadre did not pay as much attention to sabotage whenever a Rubber Committee was formed. These trade-offs pointed to tensions regarding who had the authority to decide strategy, how to distribute the war loot, and how to assign credit for actions. In some cases, the secret activities of the Việt Minh cadre interfered with those of the Rubber Union and vice versa. Concern over the alienation of workers reached the point that, in June, the Việt Minh announced 2/NB, a measure that acknowledged that, in addition to destroying rubber, the Việt Minh needed to pay attention to the cost of living and the salaries of workers.[27]

The French engaged in vigorous military efforts to complement the political developments and stabilization of Cochinchina. This "pacification" program was led by General Pierre-Georges Boyer de la Tour du Moulin. The "de la Tour" towers represented an attempt to exert control over the southern region, including the villages, rice fields, and forests of the Mekong Delta and plantation region. Việt Minh cadre attempted to neutralize the towers, and attacks on towers feature prominently in Việt Minh mythology about their military campaigns. By the fall of 1949, the French had built around 3,000 fortified posts using 70,000 soldiers to defend them, effectively turning the southeast region into a tightly controlled camp. After a series of failures, the Việt Minh had been forced to change military tactics, and leaders in the north sent instructions for Nguyễn Bình to prepare for more direct military confrontations with the French. On August 18, the Standing Committee of the Central Committee of the Party (Ban Thường vụ Trung ương Đảng) issued instructions to shift fighting from guerrilla warfare (*du kích chiến*) to mobile warfare (*vận động chiến*). In response, the Regional Party Committee of the South (Xứ ủy Nam Bộ) began preparations for a general counterattack (*tổng phản công*).[28]

On October 1, 1949, Mao Zedong announced the People's Republic of China (PRC), and People's Liberation Army troops arrived at the northern border of Tonkin soon after. Heartened by this development, the Southern Command (Bộ Tư lệnh Nam Bộ) sent out instructions on November 18 that created the inter-regiments and substructures. The Việt Minh formed the 301 and 310 inter-regiment (Liên Trung Đoàn 301–310), which concentrated its activities in the plantation-dominated southeastern region and attempted to bring order to a chaotic situation. Similarly, the communist victory in China had reverberations on rubber plantations in Malaya during the Emergency.[29]

Việt Minh plans for 1949 had called for three main activities on plantations: organize, defend, and attack. The Việt Minh also reiterated the need to destroy the trees, and distributed knives for cutting them down. However, on November 17, Lê Duẩn, who was by then the secretary of the southern region (*Bí Thư Xứ Ủy Nam Bộ*), issued new directives for southern action. These directives included an order to destroy the profits of the French but to leave rubber trees in place. This decision may have been made in part to avoid alienating rubber workers from the Việt Minh, but Duẩn also asserted that rubber plantations were part of the future wealth of the country. "Rubber trees," Duẩn stated, "are a large source of profit of our Fatherland, we must take care of [*chăm sóc*], protect [*bảo vệ*], and not destroy [*chặt phá*] them."[30] While getting the message to the cadre on the ground was difficult, and both conservation and destruction activities continued after 1949, Duẩn's call signaled a long-term change in strategy. From that point on, the Việt Minh leadership began to view rubber plantations as an immediate source of money and supplies and as a natural resource that would remain after the departure of the French. As a result, they adopted more finely tuned methods to preserve rubber trees and garner material and people from the industry. In the first six months of 1950, the workers of An Lộc plantation burned 2,100 tons of rubber and destroyed six lorries, causing 2,000 piasters of damage, while carefully avoiding the trees. The workers of Trảng Bom celebrated Hồ Chí Minh's birthday in 1950 by burning 34,000 kilograms of rubber and twenty-eight vehicles with a value of 600,000 piasters, but they did not destroy the trees.[31]

Although fighting was less widespread in Cambodia than in Cochinchina, it remained a site for anticolonial efforts. The year 1949 witnessed both the creation of a Cambodian state and the founding of the Syndicat des ouvriers du caoutchouc du Cambodge (Union of Rubber Workers of Cambodia), which was created by, among others, Thanh Sơn. This organization, created late in 1949, supported "the struggle to improve the living standard of the workers, the sabotage of plantations, and gathering for the BO DOI [Viet Minh] of the region." To conserve limited resources while combating such groups, the French military attempted to create "autoprotection," or self-defense, forces, and some plantation personnel joined the so-called Garde Volontaire de la Libération (Voluntary Guard for Liberation), who acted as local guides. But many European plantation employees expressed little enthusiasm for such schemes, recognizing the realities of their region's remoteness and of the impossibility of patrolling the vast network of plantation roads. Furthermore, the difficulties of recruiting sufficient labor for rubber production left little manpower for military duties. Plantation managers eventually worked out a modus vivendi with anticolonial forces in which the flow of materials to and from the plantation could continue in exchange for a cut of food, cash, and other supplies.[32]

Growing a National Economy

As part of the war effort, the French attempted to establish a viable Vietnamese government that was amenable to French interests. An important step was taken on October 1, 1947, when General Nguyễn Văn Xuân (1892–1989?) became president of the Cochinchinese Republic, while also serving as the defense and interior minister. This South Vietnamese government faced entrenched colonial-era interests and attitudes, particularly regarding the economy. Most French officials and planters, like their British and Dutch counterparts, refused to imagine independent colonies of any political system. The difficulties of breaking from colonial patterns were illustrated late in 1947 when an argument broke out over the composition of a new Rubber Trading Board (Comptoir du caoutchouc). At issue was the lack of any "Cochinchinois" or Vietnamese government representatives on the board. Nguyễn Văn Xuân wrote on November 6 to Nguyễn Khắc Vệ, the industry and commerce minister, inquiring into the trading board, which he had only learned about through the October 1 *arrêté* that was published in the *Journal Officiel Français* two weeks later. A week after Xuân's letter, Vệ wrote back advising him about the composition of the board and about the amount of non-French plantation owners.

When the Vietnamese leaders wrote to the French, their complaints fell on deaf ears. The French military commander, General Pierre-Georges Boyer de la Tour du Moulin, dismissed Vietnamese concerns, arguing that "this text does not make any racial discrimination and that, in these conditions, there is nothing preventing the Cochinchinese [meaning Vietnamese] producers and exporters from forming part of the Board, that is if it is justified by the level of their production and trade; it would be in the interest of Cochinchinese producers and exporters to agree on this subject with the Planters' Union and the Exporters' Union, which are open to them." De la Tour's laissez-faire approach left Vietnamese, Chinese, Khmer, and Lao to fend for themselves against French planters. De la Tour also argued that the Cochinchinese government was making no contribution, via subsidies or tax breaks, to the rubber industry, and therefore had not earned a spot on the board. The board would remain limited to the French empire and its planters.[33]

The colonial government had been willing to bail out French planters on several occasions, so its position struck Vietnamese observers as hypocritical. A handwritten note dated December 12 colorfully stated skepticism toward de la Tour's position. Xuân wrote on December 19 that, while he was happy to have the assurances of de la Tour that there was no racial discrimination, the proof of the pudding would come in the eating. Xuân also stated that, while he appreciated the industry's contribution to the Cochinchinese government, a representative on the board would help the government oversee the well-being of the plantations. Finally, and perhaps most importantly, he wrote "that straightforward reasons drawn from the political developments of the moment militate in favor of the participation of the Government and South Vietnamese interests in the management of the organization in question." There is no record of de la Tour's reply.[34]

While colonial attitudes persisted, French and Vietnamese nationalists sought ways to defeat the anticolonial movement. In 1948, under French pressure, Bảo Đại agreed to establish a Vietnamese government within the framework of the French Union. Bảo Đại placed this provisional government of South Vietnam under the leadership of General Xuân. These measures set the stage for the Halong Bay accords, which were signed on June 5 and began the process of unifying the different governments of Việt Nam, Cambodia, and Laos under one central authority. Labor issues continued to be a sensitive issue for this government. For approximately five years between 1942 and 1946, the labor inspectorate had been unable to carry out its routine inspections of the plantations. When the labor inspectorate did restart its inspections, it documented low salaries and other hardships on plantations. The inspectorate noted in a report from October 1948 that many houses at the An Lộc and Suzannah plantations still had thatched roofs and dirt floors, which contradicted the plantations' claims that they had improved their housing. The report also noted the levels of malaria, which it called "unexceptional."[35]

The length of the workday was also discussed. Legally it was set at nine hours, but the inspectorate argued that workers often stayed on the job for only eight and a half hours, with some "voluntarily" putting in more than nine hours for higher pay. The inspectorate argued that such hours were comparable to those of Vietnamese farmers, which was flawed given the different rhythms of the day. Given the continuing number of strikes, the lack of worker complaints recorded by the inspectorate is highly dubious. It is likewise remarkable that the inspectorate recorded only two workplace accidents for An Lộc, both of which involved workers losing arms in rollers, and one death on the Suzannah plantation. This death occurred on July 2, 1948, when a thirty-year-old worker named Lê Mừng was crushed by a truck. The SIPH paid for the funeral and provided one year of salary to Mừng's widow. Workers were largely left to fend for themselves, and the report notes that workers were growing their own food. Meanwhile, the SIPH directors themselves lodged a long list of complaints, most notably against restrictions on labor recruitment.[36]

Cold War Bodies

In addition to the environment, plantation bodies became sites of Cold War contestation. The establishment of the Associated States of Vietnam (ASV) in June 1949, led by Bảo Đại, resulted in a reorganization of political and legal structures that favored some Vietnamese control of governmental functions, including the management of labor migration patterns. The Ministry of Labor and Social Action's more combative attitude in negotiations over plantation labor recruitment reveals the push for Vietnamese independence. The

Labor Inspectorate was initially reluctant to allow recruiting activities, fearing that such recruitment would offer a way for Việt Minh cadre to travel south and incite worker activism. Other officials were concerned about Vietnamese labor being introduced to Cambodia, which might create tensions with locals. Yet the rubber industry's entrenched influence made it difficult to reform, and Vietnamese officials had a difficult time instituting controls on the industry. Such difficulties encouraged critics to argue that ASV leaders actively supported the rubber industry. In *White Blood, Red Blood,* Anh cites the visit of former agricultural engineer and then head of the ASV Trần Văn Hữu (1896–1984) in 1951, where Hữu supposedly praised the French plantation owners and managers for their bravery in facing, among other threats, the "poison water" of the forests (referring to diseases such as malaria) and Việt Minh ambushes. Anh wondered about the ability of a Vietnamese government official to ignore the suffering of Vietnamese laborers and their vital role in making the plantations. He also noted the response of one of the rubber workers, who said, "That's really a case of the mother singing the praises of her children."[37]

The use of political prisoners and prisoners of war as labor, the place of Vietnamese labor in Cambodia, the need for labor reform, and the role that foreign-owned enterprises would play in an independent Việt Nam all became major points of contention. One of the first issues to arise was the political backlash for drafting onto plantations a broadly defined category of northern "prisoners of war" (POWs). Most of these so-called POWs were simply people in northern and central Vietnam who were displaced by the fighting, many of whom had not shown any sympathy for the Việt Minh. The use of displaced persons housed in camps raised red flags for human rights groups and those who were concerned with the use of forced labor, which had been officially banned in French-controlled territories on February 8, 1947. The Việt Minh criticized such recruitment practices, and members of the southern Vietnamese government had their own concerns as well. Hải Phòng became a focal point of these protests; the city continued to be a departure point for many leaving to the south, and the head of the Labor Bureau in that city delayed the recruitment process. Rather than convince skeptical northern Vietnamese authorities to relocate, planters appealed to sympathetic French and southern Vietnamese officials, and in March 1949, one plantation received authorization to recruit 780 "coolies." With governmental approval, contingents of workers consisting from 80 to over 200 people arrived by airplane into Sài Gòn—an action that was condemned by several articles in the northern press.[38]

With discussion about prisoners from Tonkin under way, workers from Annam continued to arrive in the south. In July 1949, Đinh Văn Ngọc received authorization to recruit eighty workers from Quảng Nam for the Mekong Rubber Company in Cambodia. On September 10, Ngọc arrived at the village of Hòa Vang and asked the district chief for a collective contract for those on a list of forty-two names. After the district chief refused, Ngọc simply absconded with the workers. The governor of Central Vietnam received word of this action and protested to the governor of South Vietnam that any recruitment had to be approved by the local authorities, in this case the provincial authorities of Quảng Nam. On November 18, Trần Văn Huê, the labor and social action councilor for the south, wrote to the governor of North Vietnam about the case. Huê noted that Ngọc had not traveled through Sài Gòn; therefore, he must have gone directly to Cambodia and worked through the Labor Inspectorate there. Huê also noted that Ngọc was at that time in the central prison for fraud and the abuse of confidence, leaving those from Hòa Vang stranded in Cambodia.[39] This case illustrates several important characteristics of labor recruitment and employment in the rubber industry during the First Indochina War. As the scholar-official Paul Mus pointed out, the mechanics of the recruitment process revealed a continued reliance on Vietnamese *cai* to gather recruits. And rubber plantations could circumvent local authority with relative impunity; Ngọc may have been put in jail, but the plantations still received their workers.[40]

Meanwhile, planters continued their desperate attempts to recruit labor. They put up a new demand for 2,500 workers even as previously recruited workers began to arrive. The commissioner of the Republic of Cochinchina put this request on hold until more regular recruiting channels could be set up. As part of the process to determine the best recruiting methods, Prince Bửu Lộc wrote a letter dated November 19, 1949, that requested the opinion of Nguyễn Văn Xuân about the recruitment of northern POWs. Xuân replied nearly a month later that this recruitment was an extremely delicate question—an assertion that was supported by the outcry in the northern press. Politically, using "prisoners" gave a major propaganda victory to the local communists, who used the theme of colonial-era abuses during recruitment and on plantations to critique the French and French-supported government. While recognizing the urgent (*impérieux*) need of plantations for labor, Xuân pointed out that using "people displaced during military operations" could easily give the impression that the French military was rounding up people to be sent to the plantations that were *bagnes*, or penal colonies, in the eyes of many in the north. In addition, the north required much reconstruction work, and losing too many workers to the south would make that work difficult and could upset the social equilibrium. Perhaps the situation would change, Xuân wrote, but he recommended withholding approval for the time being.[41]

The dispute over whether to grant rubber companies the right to recruit workers from the north continued through the spring and early summer of 1950. Reflecting regional tensions as well as political differences, many northern officials, including the representative of the minister of labor and human services, Mai Ngọc Thiệu, opposed the unrestricted recruitment of labor while southern officials, such as Trần Văn Hữu, governor of South Vietnam at the time, were often willing to immediately grant permission to recruit. In a letter dated February 15, Thiệu opposed any movement of northern workers, writing that French recruiters could not be trusted. Like Xuân, Thiệu cited unfavorable public opinion and stated that northerners were afraid to fall into the hands of the capitalists. Thiệu did, however, acknowledge reasons to support such migration to the south, including unemployment, the need to refurbish plantations for the good of the country, and the possible use of foreigners such as Chinese or Malaysians to fill labor needs. Instead of outright rejection, Thiệu argued for serious reforms in labor contracts, insisting that workers have the right to end their contracts if bullying or beating occurred on a plantation. He stated that recruitment could be acceptable if the law of October 25, 1927, was strengthened and its articles enforced. To test the efficacy of enforcement mechanism, and the owners' goodwill, Thiệu argued, recruitment should be limited to fewer than 1,000 workers, with contracts limited to one year. Thiệu anticipated strenuous owner objections, but pointed out that if they treated workers well, plenty of people would sign up for plantation labor.[42]

In response, the governor of North Vietnam Nguyễn Hữu Trí suspended all recruitment while waiting for the decision of the labor secretary. Meanwhile, the powerful Planters' Union, which had the ear of Trần Văn Hữu, the president of the Vietnamese Council of Government, continued its advocacy. Hữu thought that this suspension and the proposed changes to labor laws would unreasonably damage the rubber industry and threaten profitable production. Moreover, he argued that Việt Nam was trying to attract international capital and investment, and that stringent labor conditions would scare away financiers. Similar concerns about protecting foreign direct investment were expressed elsewhere in Southeast Asia. Ironically, this Vietnamese perspective on labor migration reproduced a divide in colonial interests in which southern French officials and planters wanted to access the north's supposedly dense population. The governor of South Vietnam proposed a commission to meet in Sài Gòn that would be composed of representatives of the labor secretary, the governor of North Vietnam, the governor of South Vietnam, and the Planters' Union.[43]

At the meeting, which reflected and reified the close ties between government and business, representatives agreed that the model contract created in 1927 should be redrafted, that the government should have more tools to enforce the provisions of the contract on planters, and that the workers' standard of living should be improved. The representatives agreed to refrain from recruiting POWs or using forced recruiting methods, to shorten the contract time from three years to thirty months, to shorten the work week from sixty hours to forty-eight hours, to increase the amount of paid vacation from eight to eleven days, to give labor inspectors more oversight over the recruiting process, and to increase the material living conditions of the workers and their families. These concessions were minimal, and while northern delegates wanted one-year contracts, they could wrangle only a six-month reduction from the 1927 contract. This change had, moreover, already been instituted during the Popular Front era. The new contract thus represented a victory for business and was either in line with minimal international standards or so vague as to allow delayed implementation.[44]

The compensation that was issued to families when workers died on the job also became a point of contention. In 1950, automobile mechanic Phạm Văn Ban died while repairing a small truck at the Quản Lợi plantation in Thủ Dầu Một. His grieving family was paid 4,529.70 piasters (less than US$300) by the plantation—the equivalent of one year of salary. While this was more than the five piasters that a worker's life was worth in the 1920s, it was still a paltry sum. Furthermore, managers continued to go largely unpunished for violence against male and female workers, though workers did put up some resistance. Điệp Liên Anh noted that between 1945 and 1954 "the rubber workers became a reserve for the 2 sides. . . . The Việt Minh attempted to steer plantation workers into the struggle against the French." Regiment 303 in Thủ Biên province was formed in 1951 from Thủ Dầu Một and Biên Hòa, and 95 percent of its members were rubber workers. Workers also appropriated plantation materials, from cows to chemicals, and poems were written to celebrate female workers who provided supplies to male soldiers.[45]

The Value of Tropical Labor

The year 1950 saw the rapid escalation of the Cold War in Southeast Asia and a shift in the perception of the Indochina War; it had been historically viewed through the lens of colonialism but began to be viewed through the lens of communism. In January, the PRC, and soon thereafter the USSR, recognized the existence of the DRV. Although many in the U.S. and British governments had been skeptical of French efforts in Indochina, they recognized the ASV early in February to counter the growing threat of communism. The Korean War, which broke out a few months later, caused an increase in prices for rubber, which provided a boon to the Malay State and aided the British in their fight against communist insurgents during the Emergency. In a Cold War environment, government officials and rubber planters were forced to deal with the newly emerging legal and political subjectivity of plantation workers. Officials and planters could no longer contemplate subjects without legal or political standing. However, discussions about this standing often focused on socioeconomic questions, including salaries and compensation for death. This commodification of labor reveals changing assumptions about the value of workers' lives and a renegotiation over the role of Vietnamese laborers in society, the status of their health, the worth of their lives, and views of their bodies.[46]

One of the most contentious issues in the rubber industry before World War II had been worker pay, and debates surrounding salaries continued throughout the 1940s. The government began to assert control over the industry in the spring of 1946, creating a committee to set a minimum wage for contracted plantation workers. While the outcome of this commission's work is not clear from the archives, it seems to have had little effect.[47] Most Vietnamese officials accepted the medico-legal framework for labor that had been established during the colonial era. Early in 1949, the minister of labor and social action, Ngô Quốc Côn, wrote a report to the provisionary governor of Việt Nam concerning the question of minimum salary among agricultural workers. In his report, the minister discussed at length the monetary and nonmonetary benefits received by plantation labor. Basing his reasoning on colonial-era science, the minister argued for increased benefits. In a section on the "theory of minimum salary," he stated that part of the minimum salary was derived from the workers' daily bodily needs. This amount was based on caloric and nitrogen requirements and could be rationally determined by "men of science." He argued that, while the monetary value of the maintenance ration could vary in space and time, this ration was a constant, since Vietnamese bodies in the north had the same needs as those in the south. Thus, this ration could serve as "a scientific base for determining a part of the salary," with other human needs such as housing and heat related to these food costs by a fixed ratio. In the final section of the report, the minister applied the theory of minimum salary to agricultural laborers. He noted that, unlike urban businesses, agricultural employers needed to provide housing. Adding up the food, housing, and other benefits stipulated in the 1927 labor laws, the minister concluded that current salaries were inadequate and that workers should be allowed more control over how they received their benefits to adjust to their family situation.[48]

Meanwhile, workers were fighting to increase this salary, and on April 30 and May 1, 1949, more than 200 workers on An Lộc and Courtney struck for higher wages. Adopting pan–East Asian peasant practices, they signed their names in a circle so that no ringleader could be identified, and succeeded in raising salaries from 3.5 piasters to 5.25 piasters per day. Such concessions show that planters were in desperate need for workers, but were willing to raise salaries only when workers actively struggled.[49] Critics of low salaries and harsh recruitment techniques often decried the industry by pointing to the perceived profits made by rubber plantations. For example, a report held in the National Archives of Vietnam (NAVN3) that considered the costs and benefits of rubber production of the Red Rubber Plantation Company (SPTR) in Cambodia. The report stated that as of October 30, 1949, the cost of producing one kilogram of rubber in Cambodia stood at 7.20 piasters while the average price of selling rubber for 1949 was 8 piasters, yielding a 0.80 piaster per kilogram profit. With yearly yields of 1,000 to 2,000 metric tons of rubber for the SPTR, profits were from 800,000 to 1,600,000 piasters (US$50,000 to US$100,000). Việt Minh calculations of profits for all of Indochina in 1948 gave 224,200,000 piasters (around US$12 million). The report took the stock market into consideration and pointed out that the stocks (both the Vietnamese term, *cổ phần*, and the French term, *actions*, are used in the report) for Cambodian plantations had risen while those for southern Vietnam had fallen. The French stock market seemed to be responding to the short-term profitability of Cambodian plantations such as the SPTR and the short-term lack of profitability of Vietnamese plantations. The report included the caveat that the cost of "self-defense" in Cambodia was possibly lower than in southern Vietnam and that management decisions, including higher frequency of tapping, meant that short-term latex yields per tree were higher. The Việt Minh used the increasing selling prices post-1949 as a proxy measure of increasing profit.[50]

In the 1950 debate about labor recruitment, Thiệu and the ASV also raised the question of salaries. The ASV pointed out that unskilled laborers in Hà Nội earned 15 piasters per day, while skilled laborers earned 40 piasters. Considering the provision of rice, some form of housing, clothes, transport, and tools, the equivalent salary on plantations would come to approximately 10 piasters and 25 piasters. At the time, actual plantation salaries were set for un-skilled and skilled labor at 4.25 piasters and 5.25 piasters, respectively. Dele-gates were unable to reach an accord on this issue, and the planters' delegate left the meeting with the promise that he would consult with the union.[51] In an undated note that was likely penned in early 1950, the SPTR made its own calculations about salaries. Not surprisingly, the company found that the cur-rent amounts of 4.25 piasters and 5.25 piasters were justified. It factored in every material benefit that the plantation provided, using the family budgets of unskilled and skilled workers living in Sài Gòn as calculated by the In-stitute of Statistics and Economic Studies (known at the time as both Viện Thống Kê và Khảo Cứu Kinh Tế and Institute de la statistique et des études économique). These family budgets did not give absolute prices; they relied on relative amounts spent on medicines, clothes, food, and entertainment. The use of southern prices, and average worker salaries in Sài Gòn of 12 pias-ters and 14 piasters for unskilled and skilled labor, respectively, represented an important difference in the calculation. Using lower salaries from the south, where peasants from the north and center worked, rather than the higher cost of living and salaries in the north, where peasants would live after-ward and spend their government-mandated savings, or *pécule*, meant that workers lost money.[52]

Planters and government officials eventually agreed on a salary of eight piasters and ten piasters. At first glance, the increase to eight piasters was a sub-stantial improvement over the 1939 salary of forty cents (*centime* or *xu*), but in reality the gains were minimal. As indicated in the *Economic Bulletin of Viet-nam* (*Bulletin économique du Viet-Nam*), the cost of living in the late 1940s was 2,000 percent that of 1939. A simple calculation shows that this percentage matched the increase from forty cents to eight piasters, in effect giving rubber workers a raise that merely met the cost of inflation. It was commonly under-stood that salaries in the late 1930s had been held extremely low from the combined effects of colonial rule and years of depression. Plantations had increasing percentages of high-yielding hybrids, an achievement that was trumpeted by planters, and the fact that workers were tapping only the best trees reflected a substantial increase in productivity, none of which ben-efited the workers.[53]

Amid continuing strikes, Nguyễn Văn Tâm's (1895–1990) government is-sued new labor regulations that allowed for organizing among workers. In response, Trần Quốc Bửu and his French ally Gilbert Jouan formed the Viet-namese Confederation of Labor (CVT), which focused on rural workers and attempted to avoid both communist domination on the left and government control on the right. But the governor of South Vietnam did not abandon the industry, and in March 1953 the government passed a law strengthening the hand of the labor ministry in its dealings with worker unrest. French planters responded by claiming that the cost of production was greater than the price of rubber on the Sài Gòn market. They also argued that paying for protection accounted for 10–15 percent of manufacturing costs. According to the *Economic Bulletin,* in 1953 the average price of producing a kilogram of rubber stood at 17.06 piasters, including 3.29 piasters for agricultural work, 1.75 piasters for factory processing, 0.85 piaster for transportation, and 1.50 piasters for security, among other costs. The historian Marianne Boucheret, on the other hand, cites later figures suggesting that "self-defense" accounted for only 1–2 percent of the cost of making rubber. Meanwhile the selling price stood at 13.96 piasters per kilogram, which resulted in a net loss of 3.10 piasters.[54]

The calculation, of course, held little sway among the Việt Minh. While the average monthly selling price of rubber in Sài Gòn fluctuated greatly during the First Indochina War, overall it remained relatively high during the war years, and was driven higher still by the demands of the Korean War. Selling prices increased from a low of approximately 7.50 piasters per kilogram at the end of 1949 to a high of 28.40 piasters during the first semester of 1951, gently sliding back down to a low of 11.70 piasters per kilogram in August 1952, before rising again thereafter. This movement in the price of natural rubber followed the world market centered in Singapore, which saw the highest prices since the 1920s and peaked in mid-1951 at 660 percent of its 1949 summer value. While this enormous price increase resulted in fabulous wealth in Singapore, Boucheret suggests that its effects in Việt Nam were more muted. In any case, the devaluation of the piaster in the first semester of 1953 meant that Vietnamese and Cambodian rubber exports became more competitive on the world market, increasing the sales volume.[55]

During this series of fluctuations, salaries increased steadily from eight to ten piasters in mid-1950 to twenty-two to forty-nine piasters at the end of 1953. This increase partly resulted from the rising cost of living and labor and the devaluation of the piaster in relation to the franc during early 1953, which was meant to help offset the cost of the war. In June 1954, the *Economic Bulletin* published the results of the first-ever survey of rubber workers' salaries, which was based on data gathered from eighteen plantations that employed a total of 13,483 workers. These reports of increased salaries served as free publicity for the plantations. Compared with the salaries of laborers in Sài Gòn-Chợ Lớn at that time, the plantation salaries held steady, with the biggest discrepancy in pay for skilled workers: unskilled male workers earned

31.75 piasters, unskilled female workers earned 27.65 piasters, and skilled workers earned on average 55.64 piasters. The shrinking difference in male and female pay most likely signaled the desperate need for labor. While the highly chaotic situation of the war makes salary differences over time and space difficult to compare, the limits of planters' goodwill are reflected by their efforts to quietly force the French-backed governments to foot the bill for the educational services that they were publicly trumpeting as evidence of their benevolence. The governments denied their requests, arguing that the local schools were private, as they were opened and run by the plantations. Though it routinely rejected these types of pleas, the government tacitly supported the plantations by exercising little control over their activities.[56]

The closing years of World War II and the first two years of fighting between French and Việt Minh forces had both decimated the medical system and worsened disease conditions. According to an annual government report, a loss of personnel and the destruction of material had impeded health-care efforts. The author emphasized the drastic decrease in European personnel, as there remained many Vietnamese doctors for several regions, especially Sài Gòn. Malaria continued to be an important challenge facing the countryside as treatment for this disease remained inadequate. For example, Đặng Văn Cương, the first Vietnamese minister of health, noted that a lack of medical doctors combined with rural violence had ended inspections of plantations. Both the medical and scientific communities were faced with the realities of decolonization.[57]

Socializing Knowledge

The association of rubber plantations with colonial knowledge production and circulation challenged anticolonial leaders to adopt more nuanced views of science. Drawing from Marxist-Leninist analytical tradition, the Việt Minh argued that the colonizers were using capitalist-controlled science and industry to oppress the colonized. Like their PRC counterparts, DRV leaders asserted "the class character of science" and portrayed themselves as standing at the leading edge of a progressive movement that would overthrow "imperial" sciences and replace them with "socialist" sciences. Yet, DRV leaders could never afford to be as dogmatic in their approach to science due to their difficulties in procuring the goods needed for modern life and war, especially before they began receiving aid from the PRC. With centers of science and technology quickly dispersed or destroyed by the French military, the Việt Minh resorted to low-level production in the countryside.[58]

The fate of the Indochinese Rubber Research Institute (IRCI) illustrates some of the tensions surrounding scientific research. The IRCI was assembled in 1941 to fulfill the terms of the 1934 International Rubber Regulation Agreement. Indochina's debt to other colonial empires was the publicly stated rationale for the creation of the IRCI, but in reality it was driven by French planters' desire to remain competitive, the threat posed by synthetic rubber, and concerns about falling prices. The privately funded institute benefited from a legacy of government research on rubber, and many of the IRCI research sites, such as the Lai Khê agricultural research station, were inherited from the government. New sites were to be built on both gray and red soils in Cochinchina and Cambodia, and private businesses would play a central role. The SPTR, for example, contributed its own machine to the IRCI's Office of Chemistry and Technology to study the properties of latex.[59]

The total dominance of the rubber industry in the management of the research institute was reflected in the composition of the board of directors, which consisted of five industry representatives and only one government representative. Yet this private research organization continued to have close official ties. The first director of the institute was an agricultural engineer who served as the director of the laboratory of the colonial state's agricultural services in Indochina for many years. The fact that industrial concerns controlled the research institute largely determined the agronomic science produced by its research stations. These stations fit into a preexisting division of scientific labor between industry and government and focused on economic botany and latex science, leaving government-run stations to carry out studies that entailed the drawing of soil and the drafting of climate maps. The institute's stations also had a regulatory function, as researchers graded rubber for international sale and tracked production statistics.[60]

In 1941, using clones, or hybrid stocks, that had been developed in the Dutch East Indies and Malaya, the agricultural research station at Lai Khê created its own clones. The economic botany that was performed at the institute's stations continued the scientific programs of the 1930s and aimed to produce locally adapted *hevea* and study the behavior of various plant parts in relation to a range of growth factors. These investigations were in line with industry trends of the late 1940s and early 1950s, as researchers figured out the biological mechanisms of latex production. These experiments boiled down to one fundamental question: how to get the highest yield of rubber per unit of input and bring actual yields of clones closer to their theoretical potentials.[61]

The funding for the institute originally came from Sài Gòn's Rubber Office, which itself was supported by a tax on rubber exports that had been mandated by the 1934 international agreement. From 1941 until the 1945 Japanese coup against the French colonial government, the institute functioned surprisingly well. In fact, plans were drawn up for a new research center at Lai Khê that would stress a modernist vision of rubber, focused on straight lines and organized knowledge production.[62]

In 1949, the establishment of the ASV had brought about a new legal structure for organizations such as the IRCI, in which French planters retained control. Yet the pressures of a nation-centered policy and an increase in violence combined to decrease the circulation of individuals and funds from France, and regional cooperation programs were reestablished in the newly independent states of Malaysia, Indonesia, and Sri Lanka. These regional ties no longer signaled imperial cooperation; instead, they reflected Cold War coalitions based on national identities.[63] The annual and technical reports of the 1950s reveal further uncertainty as to the status of the science carried out by Lai Khê. Between 1952 and 1958, reports categorized experiments as either basic science, which were undertaken to better understand the basic properties of *hevea* and latex, or applied science, which answered specific questions that were confronting the industry. The stations' ability to produce original knowledge was ultimately dictated by industrial demands. After 1958, because of a decrease in experiments, the results of work done on production and consumption of latex were published together and became the organizing categories of research.[64]

The First Indochina War had mixed economic effects on the institute. On the one hand, war brought benefits for rubber science, and by 1952, the Rubber Research Institute's budget had received a total of 2,250,000 piasters in war damage reparations for buildings and infrastructure. This sum represented a considerable percentage of the annual budget, which had risen steadily from 6,358,818 piasters (roughly US$430,000) in 1949 to 11,812,000 piasters in 1952 (roughly US$670,000). Proportionally, the research stations constituted a rather minor expense; Lai Khê's annual operating budget in 1949 was little more than 200,000 piasters, and rubber sales from tapped experimental trees, along with taxes on rubber exports, increasingly contributed to the institute's budget. This growth in funding further calls into question the planters' claims of penury during their negotiations over worker salaries.[65]

On the other hand, the institute's laboratories and research fields suffered from their ties to industry. The French military attempted to organize some planters into "self-defense" or "self-protection" forces, but, as Marianne Boucheret has shown, the military often complained that plantation owners were more concerned with profits than strategic interests. For some planters, self-protection represented an unnecessary expense and threatened their attempts at neutrality. They argued that the military disregarded on-the-ground realities and left the plantations to fend for themselves. Most of the local Việt Minh cadres did not discriminate between the institute and other plantations. Almost every year, the institute's annual report listed a French employee who was killed or seriously wounded in attacks. Most Vietnamese tappers and management personnel faced almost daily threats of kidnapping or execution for collaborating with the French. The relationship between the French military and the plantations drifted further apart during the 1950s, when the military became reluctant to fight the owners' battles against labor.[66]

Meanwhile, French planters seemed unable to adjust to the realities of decolonization. Even when the "first Vietnamese Rubber Company" that exported raw or semiprocessed rubber was created in 1951, plantation owners and managers were unable to break with colonial mind-sets. The president of the Technical Commission, M. Bocquet, expressed these challenges in a talk given in 1950 at the IFC in Paris. Bocquet claimed the authority of twenty-five years of living in Indochina, a Vietnamese wife and friends, and the ability to speak the "language of the country," all of which enabled him to understand the "mysterious Asian soul." Yet, during a meandering speech in which he rehashed colonial-era stereotypes, the president insisted that Vietnamese nationalism could not explain what was going on because the Vietnamese masses were incapable of acting on ideals. Echoing the postwar French programs of colonial technicians, the president argued that Việt Nam suffered from material want due to a lack of a technical elite. He relayed a technocratic vision in which agricultural engineers would bring prosperity to the countryside, thus eliminating material motivations for the revolution.[67]

While battles continued in the south, Việt Minh leaders became interested in starting rubber production in the north. Rubber made sense within a newly envisioned socialist economy, and had international value within the bloc of fellow socialist countries. The nation-states and empires forming this network, including the Soviet Union and the PRC, sought access to rubber, and the Việt Minh, and later the DRV, stood to benefit from their climatic location. Interest in rubber growing was also motivated by experiments that were being conducted in southern China. In 1952, Việt Minh leaders learned of groups of Chinese youths sent to replant the southern countryside with rubber trees. In a letter marked "secret" from August 5, 1952, Nguyễn Văn Lưu reported on rubber growing in Guangxi. Đặng Văn Vinh contributed to this report, writing a summary on the margins of the letter. Vinh noted that, in general, rubber could be grown below 23 degrees latitude, an area that included all but the northernmost points of Việt Nam. The main limits to growth were temperature, rain, and strong winds, which could topple *hevea* trees because of their shallow roots. Thus, despite aid provided by the USSR and Eastern Europe, the DRV researchers were unable to cultivate *hevea* stocks that could withstand northern winters until after 1975.[68]

To further develop rubber growth in the north, DRV leaders sought to appropriate the legacy of French colonial plant research. In a letter dated April 14, 1952, the Government Economic Board asked Nguyễn Xuân Cung, the minister of agriculture, to send information about those who had rubber-growing expertise in northern Việt Nam. Four months later, Cung informed the board about a handful of experts who were active decades earlier, including Nguyễn Duy Cân, who had worked at the Gia Rây experimental station from 1931 to 1937, Đào Thiên Tránh, who had worked at Bolovens, and the better-known agricultural experts Đặng Văn Vinh, Vương Gia Cân, and Nguyễn Khoa Chi. The printed material, according to the minister, was even more out of date; Henri Jumelle's 1903 *Les plantes à Coutcho* [sic] and a handful of other materials at experimental stations in the north were the only documented resources.[69]

The Việt Minh were not alone in debating the meaning of plantation agriculture and medicine during the First Indochina War. The IFC hosted talks given by those with firsthand experience of the plantations, and in 1949, the medical doctor Trần Đình Quế spoke about the history of the health-care system in Việt Nam and tried to evaluate the effects of the French presence by comparing Indochina with other countries in the region, notably Japan, China, Siam (Thailand), and the Dutch East Indies. Que concluded that, except for Japan, Vietnam's medical system under French colonial rule, as measured by number of beds and money spent per inhabitant on medical expenses, compared favorably with other societies in Asia. Que also proposed a program for improving the health of the Vietnamese, noting that the laws protecting workers' rights on plantations and other industrial sites needed to be strengthened. He cited England as a good example of the benefits of the "socialization" of health. Que closed by pronouncing medical doctors the "principal artisans of works of peace."[70] Que's view of medicine as important for "human capital" and his emphasis on the commune as the key to introducing health measures were formed in part during colonial times. Que, however, did not view these themes as mere French creations. He cited the work of S. E. Hoàng Trọng Phú, a mandarin who placed people from several villages in hospitals so that farmers could return home and disseminate hygienic rules. Que further argued that public hygiene campaigns would be most efficient if focused on one disease because peasants would more clearly comprehend what they were combating, the effectiveness of their actions, and the more abstract hygienic rules. Tackling many diseases at once would merely disperse efforts. Adjusting to Vietnamese norms, Que continued, the individual would be responsible for the collective and vice versa, and social discipline would be used to enforce norms.[71]

New Life

The First Indochina War was not simply a battle among armed forces; all sides of the conflict attempted to win the "hearts and minds" of the Vietnamese "peasant," including those who worked on plantations. One of the most influential scholars of Vietnamese peasants during the postcolonial era was Pierre Gourou. Gourou's thoughts on peasant mentality continued to evolve after he returned to France, and he eventually assumed a chair in tropical geography at the College de France. Given Gourou's studied avoidance of colonialism in his writing, it is not surprising that he did not directly address the issue of rubber workers. But Gourou's framework for thinking about the relationship between humans and their milieu shed light on the successes and failures of plantation agriculture.

The limits of Gourou's thought are now apparent; even his later tropical geography was informed by an orientalist approach to Asia that saw Asians as an absolute other for Europeans. His scholarship began with the assumption that peasants in the global south lived in different natural and mental universes than he and other Westerners. As Gavin Bowd and Daniel Clayton write, "Implicit in his work is the idea that the Vietnamese and French were locked in fundamentally different cultural worlds." But his later tropical geography also broke with geographical and environmental determinism and allowed for a dynamic interplay between culture and nature. He also raised the possibility of asking what the peasants thought. And he called into question the effects of improvement projects on "peasants," arguing, in Bowd and Clayton's summary, "that France's questionable colonial policy of *mise-en-valeur* would have a limited effect in the countryside and spell disaster for the peasant." This is a lesson that the developmental sciences and modernization theory of the 1960s and 1970s by and large did not heed. For the most part, those projects in Việt Nam, whether big or small, remained controlled from the top.[72]

During the First Indochina War, the civilizing mission was recast into a mission to modernize. French political and military leaders attempted to win "hearts and minds" and to assuage American and other international observers' fears about their intentions in Indochina. Influenced by conversations with Pierre Gourou, the sociologist Paul Mus analyzed life on rubber plantations through the lens of tropical geography. In 1952, at the height of the First Indochina War, Mus published *Vietnam: Sociology of a War* (*Viet-Nam: Sociologie d'une guerre*), which imagined the contrasting views of various actors in the war. This book represented a change of pace for Mus, who was a student of Buddhism in Asia. Mus was no stranger to Indochina, however, as he had spent his childhood there, later returning in the late 1920s and 1930s to a post at the EFEO. After World War II, Mus directed the Ecole nationale de la France d'outre-mer (formerly the Ecole coloniale), but a series of antiwar newspaper articles led to the nonrenewal of Mus's contract in 1950 and he spent the rest of his career at Yale. Rather than condemn either the French or the Vietnamese, Mus juxtaposed a nuanced picture of both. Mus illustrated his thesis with several examples, including a discussion of rubber plantations in a chapter called "School of a New Life: A Qualified Failure" (*L'école d'une vie nouvelle: un demi-échec*). From the French worldview, plantations were glittering visions of modernity, all straight lines and clean surfaces, with hospitals, schools, and places of worship—signs of a progressive social vision. British observers, likewise, were impressed with the "superb living conditions" of some plantations. Furthermore, plantations could help reduce pressure on the overpopulated rice deltas. Despite the early loss of life due to brush clearing, plantations "appear as a net gain for the country."[73]

In addition to simple material benefits, plantations offered a model for an ideal modern village to the Vietnamese peasant. Yet, by judging plantations as semi-failures, Mus questioned the penetration of the plantation mentality into the peasant cosmology. It is important to note that Mus was not making a moral judgment about the French goal of creating *homo œconomicus*, or individualistic, rational, economic actors. He instead focused his inquiry on finding out why this effort failed. Although Mus was not a communist, he drew from Marxism, arguing that the Vietnamese society of his time did not yet have a proletariat. Even if the farmers of the Red River Delta worked at industrial operations such as cement factories, they often left after a few years and returned to their villages. Likewise, most of those who migrated from the north to southern plantations (approximately 3,000 per year), if they survived their three-year contract, returned to their natal village. Mus held that the village, rather than the individual, remained the basic unit of Vietnamese society. Mus argued that the pressures of recruiting and the collusion between *cai* and village leaders led to an expected result: only "undesirables" were sent away from the village. For Mus, "the urban vagrants, the rural unwanted, and the survivors of bad fortune" that found their way to colonization projects were "a clear sign of a forced attempt, artificial, and not a fundamental change in the society and customs."[74]

Mus also argued for the dominant role that the collective played in society, and that money paid by Europeans, either as advances or as salaries, disappeared into the collective. Rather than being dehumanized through the institutions of industrial capitalism, as Karl Marx predicted, Vietnamese peasants were stripped bare of their material gains by and for the collective. Yet the villagers wanted nothing more than to lose themselves in the collective, for this meant happiness, fulfillment, and a well-lived life. "And it is for this reason that there aren't plantation villages that give a sense of being fully effective. They don't have," Mus concluded, "a grounding anchor in the supernatural that everywhere holds Vietnamese society to its customary lifestyle."[75]

Several scholars have pointed out that Mus's overriding belief in culture and his orientalist approach to Asia led him to overemphasize both the role of the village in Vietnamese society and the inability of Vietnamese peasants to understand and follow rational, individualistic behavior. Whether a farmer acted within and re-created a moral economy or was a "rational peasant," researchers have come to understand that twentieth-century peasant society was stratified and full of individual actors.[76] But Mus's misunderstanding of the role of the village in Vietnamese society and his essentializing of peasant life are less relevant here than his attempt to represent Vietnamese actors and, more importantly, his recognition that "liberated" Vietnamese, or those calling for independence, were writing new histories. As Mus noted: "They will write anew this [plantation] history, in their fashion: neither on that which we have considered fair to say about it and on the statistics and information that we have in all intention of honesty furnished for the world, when there was no demand; nor, in short, on our files but now on the things that they have seen, wrongly or rightly, differently from us, and also from within, on a disconcerting number of facts, real or subjective, of which we have not had, in general, any idea."[77]

Mus continued in this vein, citing the philosopher Trần Đức Thảo, who also argued that the two mental universes—one French, one Vietnamese—could look at one set of statistics and see two completely different stories. The French, for example, could point to the amount of rice, coal, and rubber produced, or even the number of educational certificates awarded, "but at the same moment, based on the same numbers, the Vietnamese reckon in the negative space the distance of these achievements with those that they well think that Vietnam, in other circumstances, could have achieved." Thảo was articulating a postcolonial critique of colonial rule before it had ended. Both sides could agree that modernization had occurred under colonization, but for the Vietnamese, plantation landscapes were still a sign of lack, and represented only a colonial modernity.[78]

Oddly, Mus used Thảo's insight to reinscribe colonial-era attitudes and argue against the adaptability of Marxist-Leninist thought to Vietnamese society. Mus viewed both communism and *mise en valeur* projects such as plantations as projects that did not consider the religious sentiments of peasants. "These coolies, peoples rude in their material and mental endowments, but ingenious, often, in the way they use them, don't encounter any difficulty in ending, without visible motive, a full part of their life, that had all appearances of being enough. This contrast in their attitudes appears to us a logical contradiction and remains inexplicable. It appears, to the contrary, to be agreeable to them, and far from bothering them in any way."[79]

Mus argued that the "coolie" had his or her own logic and viewed life on plantations as temporary. Even if life on plantations provided more comforts, peasants preferred to live in their home village, surrounded by their family, their loved ones, and their traditions. Mus criticized the simplistic view of statistics but was unable to see the evolving, fractured, and interactive parts of Vietnamese society. Even though he placed statistics within a lived experience, his holistic view prevented him from seeing the cultural changes that were taking place. Instead of rubber workers agitating for inclusion as citizens in the postcolonial nation-state, Mus saw "Tonkinese peasants" as trapped in tradition.

This type of perspective was not held solely by foreigners; Marxist historians have also tended to emphasize the collective as dominant over the individual in Vietnamese society. Moreover, the governor of South Vietnam sent a letter on May 17, 1952, marked "secret" to the minister of the interior regarding the pacification of plantations that argued that the "States within the State" had better material conditions than the surrounding rice villages, but that they provided no sense of collectivity. He wrote of the plantation that it "is without deep influence on the individuals and their collective mentality." The Việt Minh offered a structure and government that could fill the void left by "overly professional" plantations. He concluded that the military could not stop all Việt Minh actions, such as recruiting workers to work on swidden agriculture, so the best approach was to change the "mentality of the inhabitants." The South Vietnamese government needed "to forge a soul for these villages of the uprooted through skillful publicity and services provided. The day when," the governor of South Vietnam concluded, "the coolies know that they depend on a Government that knows them, that follows them, and that they can appeal to through intermediaries of true notables who they have confidence in, one will be able to ask of them to react as citizens of a state that will have shown them that it exists." The governor of South Vietnam's words betray his desperation to assert the presence of his government, particularly on plantations. A draft of the letter was sent on May 16 to the director of political affairs. The director expressed his strong concern that plantations did not have "villages"; they had only groupings of housing with workers from the north, none of whom were ready to serve as "notables." During the 1960s, the historian Nguyễn Khắc Viện adopted some of the same assumptions in his work on Confucianism and communism. Although opposed to Mus's analysis in some respects, Viện believed in the compatibility of certain Vietnamese "traditions," Confucianism with communism. For Viện, the collective ethos, discipline, emphasis on morality, and opposition to religion were inherent in both systems.[80]

Planter propaganda focused on both material advantages and social connections. A monthly journal titled *Liên Lạc* (*Connections*) highlighted the "heavenly" aspects of workers' lives. This journal ran for at least thirty issues between January 1952 and July 1954, and was produced by the official government printer, the Imprimerie française d'outre-mer in Sài Gòn. The subhead of the journal was "The information service between the Southern Rubber Plantations and northern Vietnam," and true to its name, each of the journal's issues devoted extensive space to short letters from plantation workers to their relatives in the northern provinces. Many of the letters are quite moving; they reassured fathers and mothers, aunts and uncles that the letter writer was still alive, and inquired after the welfare of the worker's relatives. The voices in the letters range from fairly educated factory workers to barely literate rubber tappers. It is reasonable to expect that the publishers would strive for credibility by seeking out real workers for these stories. Yet none of the letters complained about labor conditions on plantations, even as workers' strikes continued to take place, which suggests that the plantation-controlled journal was heavily censored.

The journal also included consistent attempts by the plantation owners to appeal to workers. In an October 1952 article called "The Price of Rice Has Risen," the piece's author, Đàn Tâm (a literary term for loyalty), argued that while the rising price of rice hurt other consumers, especially the poor, rubber workers were unaffected. Đàn Tâm stated that because of the rice provisioning system, whereby each worker received a set amount of rice regardless of price, the plantations, rather than the workers, absorbed the rising costs. In the highly inflationary environment between 1952 and 1954, food was a major concern, and the French believed that the provision of rice, medicine, and other necessities as part of a worker's salary would prove attractive, as they were unaffected by inflation. The journal also published positive stories about the health conditions of the plantations. In November 1952, Cố-Nh stated that while plantations had previously been unhealthy, this problem had disappeared after the land was cleared. One of *Liên Lạc*'s chief goals was to counteract Việt Minh efforts to dissuade northerners from working for plantations. In a November 1952 article titled "Truly a Heaven?," Trần Quỳnh weighed in on the ongoing propaganda war between planters and the Việt Minh. Quỳnh referenced the media attacks and rumor campaigns that were being waged against the plantations in the north, and argued that plantation living conditions were relatively better than almost anywhere else in Việt Nam.[81]

Tapping Independence

In July 1953, the French announced their intention to give total independence to the ASV, and nearly a year later, in May 1954, French troops capitulated to the Việt Minh at Điện Biên Phủ. Meanwhile, labor struggles continued, informing, and informed by, high politics. As part of its ideological struggles with the Việt Minh, the French-backed ASV government had accepted some degree of labor organizing in the hope that it would siphon off support from the Việt Minh. Việt Minh narratives about rubber remained harshly critical of plantation conditions and were aimed at achieving specific political goals.

Enemy soldiers and nature made the closing years of the First Indochina War difficult for the Việt Minh in the south. Despite scattered victories such as the Battle at La Ngà, when the southern branch of the Việt Minh carried out frontal assaults on French military posts, it incurred heavy casualties under the combined assault of French air and artillery power. Nguyễn Bình, commander of the southern forces, failed to break the de la Tour plan and was killed in 1951 en route to a strategic meeting in northern Việt Nam. Tigers, too, caused problems for plantation workers and Việt Minh. Lê Sắc Nghỉ, a former plantation worker born in Annam, and later general director of the Đồng Nai rubber company, recalled in his memoirs the predations of a man-eating tiger in 1950 that supposedly killed nearly one hundred people and terrorized soldiers in the inter-regiment 301-310. The French military understood the fear that tigers could inspire in the Việt Minh and were willing to exploit those fears by using commandos to simulate tiger attacks. Rumors held that the tiger Ba Móng, or "3-Claws," was the former pet of a plantation owner, and others believed that the tiger was a spirit of nature.[82]

Climate, too, caused significant hardships for the Việt Minh. In 1952, a mid-October typhoon hit the southeast region, flooding the Đồng Nai and Bé river systems. The loss of stored food supplies, combined with a spring French offensive, reduced many Vietnamese in the region to eating forest plants, and they were forced to practice swidden agriculture. Plantation workers were forced to grow their own corn as a supplement to their wages and food rations, but the higher ground of the plantations allowed some Việt Minh to escape this devastating flood in 1952, and plantation workers provided them with food and shelter. The Việt Minh devoted their energies to repairing the flood damage while avoiding French raids. In the north, the French could repel a Việt Minh attack at Nà Sản in December, but the Việt Minh survived and were able to counterattack. Playing off plantation labor's oral tradition, a saying from the time supposedly went: "Going to Battle Zone D is easy, returning is hard. Soldiers who go lose their lives, soldiers who return lose their stripes."[83]

Life was also difficult for soldiers who were fighting for the French. Dominique de la Motte, a young lieutenant who led a group of Chinese, Khmer, and Vietnamese commandos in 1951 and 1952 near plantations in the border region, recalled that at one point or another everyone had malaria. Vietnamese and Khmer soldiers suffered, and Motte stated that "contrary to received wisdom, Asians [les Jaunes] were not better off than Europeans [les Blancs] and they died young." French medicine had come to the consensus that bodies of different "races" suffered from tropical diseases equally, but commonsense notions about differential immunity were hard to overcome. Medicine was hard to come by, and Motte served as a medical doctor despite his lack of medical training. According to Colonel A. Rivoalen of the Far East French Forces (Troupes françaises d'Extrême–Orient), during the first year of combat, malaria infected 37 percent of the troops, with the rubber plantations listed as a principal source of the disease.[84]

French soldiers often received a message that was in many ways less scientific than the one being broadcast by the Việt Minh; a cartoon-filled military manual illustrates the complex interactions of Vietnamese and French understandings of the disease. A green image rising from the rice field recalls both Henry Morin's (the former head of the Pasteur Institute's antimalaria section) description of sorcerers protecting the uplands of Việt Nam from intrusion and Alexandre Yersin's argument that the Vietnamese word *nước độc* (poison water) referred to "terroir" rather than water in a literal sense. Science did not completely disappear from this malarial vision, as the sorcerer's crescent shape mimics the *falciparum* gamete, attacking with its henchman, and mode of transmission, the mosquito. The cartoon's words invoke the pharmaceutical industry, as the soldier is urged to take his prophylactic pills.[85]

This sorcerer may summon for us other metaphorical ghosts, namely the painful lives and deaths of countless rubber plantation workers haunting Việt Nam during thirty years of fighting between 1945 and 1975. At the end of World War II, Henry Morin found himself in charge of the Pasteur Institute in Đà Lạt, where he was wounded in a clash with the Việt Minh. Had he, a professional charged with the duty to heal, been troubled by death rates that in some months, on some plantations, reached as high as 30 percent of the workforce under his watch? Morin returned to France in 1947 and became involved with the World Health Organization (WHO) as a consulting expert. Even as the WHO emphasized DDT in its anti-malaria campaigns, Morin remained an advocate for integrating psychological and moral components to these campaigns. Whether he was "driven by the desire to teach and defend third world countries against the scourge of malaria" or trying to extirpate ghosts is impossible to say, but Morin's life reminds us that malaria and its science had varied meanings.[86]

BOTH THE FRENCH AND Vietnamese governments viewed plantations as key symbolic battlegrounds during the First Indochina War as they sought to transcend the colonial legacies of plantations and remake representations of those landscapes. Yet all sides found the social, cultural, political, and economic relationships inscribed on the land through plantations impossible to eliminate. Many plantation owners and workers had aided the French-led military to fight the Việt Minh, but not all plantations were eager to get involved, as the owners and managers faced the daunting task of restarting rubber production. French plantation owners responded to the violent attacks by Việt Minh partisans, a lack of workers, and growing nationalist scrutiny of plantation conditions by portraying plantations as inviting places to live and work. The ASV attempted to navigate its way through muddy political waters by cautiously encouraging Vietnamese nationalism while allowing for foreign ownership of private businesses such as plantations, while the Việt Minh continued to condemn plantations as "hells on earth."[87]

The turbulent years on the Indochinese peninsula from 1940 to 1954 posed a serious threat to plantations. After a steep decline at the end of World War II, the rubber industry revived, and by the time the Việt Minh defeated the French at Điện Biên Phủ, rubber exports exceeded the previous high of

60,000 tons. Between 1955 and 1965, producers benefited from the relatively peaceful conditions, and Vietnamese began to play a more significant role in the industry. The arrival of the U.S., Korean, Australian, and other anticommunist militaries marked a shift to a more violent period that entailed a rapid decline in output. Despite the use of plantations as battlefields, rubber remained key for nation-building projects. South Vietnamese and American planners spent extensive time considering the rubber plant's economic potential, but by 1975, production and profits had again fallen to zero.

CHAPTER SIX

Decolonizing Plantations

The white left, but the yellow stays.

—Philippe de Marais, *Apocalypse Now Redux*

Around 1956 to 1957 and after, then they . . . at that time the plantations, they also paid a little bit more attention, they cared more, were a little less strict, and there were more medications, more plentiful medications and so the diseases decreased.

— Phạm Ngộc Hồng, retired rubber worker

In 2008, I sat down with Phạm Ngộc Hồng, an eighty-one-year-old Vietnam-ese man from Nam Định, who had started working on a Michelin plantation and later managed a section of a SIPH plantation. As we discussed the history of rubber in Việt Nam, Hồng recalled a recruitment poem called "Cái nhà mộ phu" (Recruitment office), dating from the early 1950s:

> No money but you want to take a plane?
> Come for the advance to go to Nam Kỳ
> Your family can come too
> Day and night come together there is no need to be reluctant.
>
> Old Thành Nam still has the recruiting company
> Thái Bình is the street, the house is number three
> After you sign the contract
> A loan of one hundred piaster to spend
>
> And you can come and go often
> The office can accept all, according to common regulations
> Medicine, rice, cloth for free
> Sick child, wife in child birth will have care
> Thirty months is the end of the contract
> The money left over in the ten thousands you can bring it back
> In South and North they come in crowds.[1]

This poem has a "six-eight" form, in which lines of six syllables alternate with those of eight syllables. Moreover, the ending sound of the six-syllable line rhymes with the sixth syllable of the eight-syllable line, and the ending

sound of the eight-syllable line rhymes with the ending sound of the following six-syllable line, and so on. This rhyming scheme is a useful mnemonic
device found in Vietnamese epic poetry and makes the poem an effective advertising jingle.

The content of the poem was also meant to appeal to workers; the mention
of an airplane is both a marker of modernity and a reference to the fact that
during the First Indochina War, plantations were so desperate for workers that
they flew them down from the north, especially after the trans-Indochinese
railway was cut by Việt Minh activity. The second stanza mentions the work
of the recruiting companies. Though the subject of much controversy during
the colonial period, these agencies continued to operate into the 1950s in
Nam Định and Thái Bình provinces, which were traditional sources of plantation workers. One hundred piaster was a large amount at the time, as the
average monthly salary of a worker was twenty-one piasters. The final stanza
promises benefits that workers valued, including food, medicine, clothing,
and family care. Finally, workers were promised they could return home
after their thirty-month contract with some savings. It is difficult to assess the
success of these efforts; while northern and central Vietnamese were skeptical of such advertisements, wartime conditions encouraged migration.

The DRV depicted rubber plantations as "hells on earth," releasing books
such as *Rubber Hell*, which aimed to counter the allure of plantations and a
working life for farmers in the poverty-stricken north. One of the slogans that
supposedly originated in the colonial period was "cao su đi dễ khó về,"
which translates as "it is easy to go to work for rubber plantations but hard
to return." It is now commonly assumed that "hard to return" meant that
most workers died on the plantations before they could return. This interpretation was indeed part of the understanding of the 1930s and 1940s; this
slogan, however, was also understood to mean that workers often got married
and settled down in the south rather than return to their natal villages in the
center and north. These alternative readings have disappeared from popular
memory, leaving only the memory of the hard life of most rubber workers.[2]

While Vietnamese communist histories claim that these sayings arose spontaneously from the rubber workers' movement, it seems that some might have
appeared only after 1945. Phạm Ngọc Hồng offered this interpretation: "That
sentence was a sentence after this, after the day when Việt Nam signed the
Geneva Accords [in 1954, three years after he started working on a planta-
tion], after that then the sentence existed, at that time the people from
the north came to the south, almost all of them. If they knew that sentence before, no one would have come. That was a very complex [decision];

the environment, living activity, [they] could play sports, and they had literature and arts. [Well] they had sports but only some."[3] The real improvements
in plantation conditions, he stated, happened only after 1956, due in part to
pressure from international opinion. He continued: "Around 1956 to 1957 and
after, then they . . . at that time the plantations, they also paid a little bit more
attention, they cared more, were a little less strict, and there were more medications, more plentiful medications and so the diseases decreased; [regarding]
the outside environment, the roads were extended, more houses were built,
the environment was a little more comfortable. And the problem of people's
livelihoods, of the workers at that time was completely dependent on the
plantation, almost like dependency, they provided you with something and
you only had that to use."[4] While this quote points to the continuing ways in
which workers were dependent on rubber plantations and their environments, it signaled an important shift attendant with the formation of the Republic of Vietnam (RVN) and independence stands out as a convenient
turning point for nationalist histories written from a variety of political view-
points.[5]

   This chapter considers the ways that the rubber industry functioned in the
First Republic of Vietnam, which lasted from 1954 to 1963. Although plantations continued to dominate health and labor, a different vision of rubber had
arisen in the postcolonial state, one that relied on small changes at the local
level, ideally made by residents themselves. This vision of community development promoted individual and family-based delivery of health care and
control of labor, which would incidentally lessen the appeal of communism
among rural Asians. This vision, however, failed due to politics, war, and pre-
vailing environmental conditions, as the often-violent conditions of the Cold
War limited its viability. Despite Ngô Đình Diệm's nationalist credentials and
his vision of modernity that focused on "community development," he continued a colonial-era policy of supporting the plantation over smallholder
rubber. And while a vocal, well-placed minority called for greater smallholder
participation in rubber, the efforts of most agronomists and medical doctors
continued to support plantation agriculture.[6]

   The intersection of community development with plantation rubber raises
the question of agency and intentionality. Local institutions, labor-management
interactions, and *hevea's* ecology challenged global ideologies by defining the
limits of land use and by shaping development outcomes toward local ends.
Initially, planners from the United States and the RVN, or South Vietnam,
looked to rubber as both an important source of income for the Republic and
a way to create a class of Vietnamese smallholders, or individuals and families

who owned modestly sized rubber plots. Instead, measures taken by South Vietnamese leaders ended up sustaining plantation production, and wartime pressures pushed visions of development toward high modernism. Plantation production declined and costs grew as rubber producers were forced to pay for self-defense, ransom, and replacement of stolen or destroyed equipment as well as taxes to the RVN and the National Liberation Front (NLF). Such costs made it difficult for farmers to succeed as smallholders, even as the ideology of modernity continued to value plantation output over smallholder production, and formal science over informal modes of knowledge. In this way, colonial legacies continued to structure thought and behavior regarding rubber and to channel human-environment interactions on the plantations of post-1954 Việt Nam into patterns that resembled those of their colonial predecessors.

New Directions in Plantation Labor

In July 1958, Ngô Đình Diệm, then president of South Vietnam, went on an inspection tour of the plantations of the SPTR, Lai Khê (a rubber research center), and Michelin's Dầu Tiếng plantation. On his tour, Diệm commented positively on the new coagulation techniques and the use of Caterpillar tractors for deforestation. Diệm's tour was meant to publicize a rubber loan program that was ostensibly aimed at encouraging Vietnamese to enter the industry, yet ended up funding large French-owned plantations. Why was Diệm, an ardent nationalist, offering support for French-controlled institutions and businesses? How did the rubber industry fit into Diệm's projects of state and nation making? What strategies did other state officials, planters, workers, and communist insurgents adopt to make the best use of plantations? Only two weeks after Diệm's visit, roughly 400 members of a communist force caused considerable damage during an attack on the Dầu Tiếng plantation. This communist attack indicates the ongoing struggles over plantation space.[7]

During Ngô Đình Diệm's First Republic, rubber plantations remained the dominant, if controversial, models of labor and health in the red earth region. This was partly due to the pivotal role plantations played in resettling thousands of northern migrants. In early July 1954, Diệm formed his government and on July 21 signed the Geneva Accords, creating a divided Việt Nam. As part of the Accords, DRV and South Vietnamese representatives agreed to allow people to move freely between the north and the south for 300 days. During this period of "regrouping" (*di cư và tập kết*), intense efforts were made on both sides to influence population movements. Beginning in

August, thousands of Vietnamese Catholics moved from the north to the south and DRV cadre and troops moved from the south to the north.[8] Plantation holdings confiscated from defeated opponents provided Diệm with a means to deal with the massive influx of refugees into the RVN. For example, an RVN-controlled plantation near Củ Chi, which was formerly owned by Lê Văn Viễn, or Bảy Viễn, temporarily housed 8,000 refugees in 1956.[9]

Diệm was not concerned with labor and health conditions on plantations when he arrived in southern Việt Nam during the middle of 1954; rubber workers, however, advanced their agenda through strikes that began in October of that year. After the First Indochina War, the labor movement split into many unions, with some focused solely on advancing workers' rights while others viewed their activism as part of the socialist revolution. Regardless of their political leanings, rubber workers made several demands on their employers: build schools for their children, reduce the number of trees tapped, distribute white rice instead of brown, decrease the number of working hours, provide promised clothing, and increase bonuses and wages. In a letter dated December 10, 1954, Nguyễn Tăng Nguyên, the minister of labor and youth, criticized the workers who were striking without seeking arbitration, but he reserved his harshest words for the French directors. He stated that salaries were low and that many of the workers' grievances were legitimate, and that since the owners prevented labor organizing, the workers had limited channels for dispute. Nguyên argued that worker activism was not due to international communism but rather to awareness of worker organization in other industries. To better understand the situation, the minister advocated an in-depth study. He also argued for increased security in the region, better communication with workers, and legitimate organizing.[10]

On February 10, 1955, in response to both Nguyên's December letter and another letter sent in January that focused more on the political leanings of workers, Diệm instructed Nguyên to investigate plantation conditions. Two weeks later, Nguyên reported that owners had not followed a November 1, 1954, law stipulating the timing and amounts of salaries, and that the most remote plantations were not providing adequate access to water. Most likely based on these findings, Trần Văn Lam, the South Vietnam government representative (Đại Biểu Chánh Phủ tại Nam Việt), sent a letter on February 23 to the province chiefs in Biên Hòa, Bà Rịa, Thủ Dầu Một, Tây Ninh, and Gia Định urging them to remind plantation directors to improve living conditions and protect the health of plantation workers. Lam listed a number of requirements based on a June 26, 1953, law, including a health unit with a nursing station, daycare for children, and a school for children between six and fourteen. The costs for these measures would be split between the plantation and the state, with the former paying for the building and materials and the latter paying the salaries of the teachers.[11]

In both his February 24 letter and another from March 1, Nguyên acknowledged many of the improvements to plantation conditions that had already been made by the French planters. This is surprising given RVN officials' critical attitude toward the rubber industry. When Nguyên visited plantations in Biên Hòa and Thủ Dầu Một provinces, he found that workers were well housed and fed, had access to health, religious, and educational facilities, and generally worked between eight and nine hours a day. He also argued that even though workers were demanding polished rice, the unpolished rice that owners provided contained more vitamins. Nguyên also wrote that the French plantation owners did not exhibit the same contempt for their Vietnamese workers that they had before 1954—a judgment that was contradicted by disparaging letters about Vietnamese tappers that had been written by planters. The workers themselves said that their lives were much better than before, and both Nguyên and Diệm agreed that those who wanted to break their contracts to leave the plantations and return to the north should be allowed to do so. Even DRV publications admitted that after 1954, plantation conditions improved as the rubber industry was integrated into a South Vietnamese economy.[12]

The main concern of Diệm's officials continued to be the workers' political leanings and their relationship to the communists. Communist activity varied by province, and according to Nguyên, early in 1955 communist agitators circulated actively in Thủ Dầu Một province, pretending to form legitimate organizations on plantations such as Lộc Ninh, then spreading out to other plantations. During Nguyên's earlier visit to the provinces, local officials claimed that in remote plantations the communists controlled 90 percent of the workers and in plantations close to population centers, they controlled 30 percent. Although the Geneva Accords had called for a regrouping of communists to the north, many cadre stayed in the south and continued to organize workers.[13]

In 1955, Trần Văn Lam articulated an RVN strategy for decreasing communist support when he argued for establishing labor inspectorates in the rubber-growing region, organizing health, youth, and sports clubs, explaining the contribution of citizens to government, and offering better security for citizens. To stabilize working conditions, the RVN government issued general rules (*tổng qui*) regarding labor in 1955 and a decree (*Nghị Định*) on November 17, 1955, concerning plantation workers' salaries. For nonspecialized male workers, the minimum wage was 30 đồng (the unit of currency in the RVN after the Indochinese piaster was taken out of circulation); for tappers, the minimum wage was 37 đồng; and for women and children between fourteen and eighteen years of age, the minimum wage was 24.50 đồng. While this structure continued the pay differential for women that was established during the colonial era, the decree did include a clause that allowed for equal pay if the quality and quantity of work done by a woman or a child was equal to that of a man. In addition, Lam argued that workers should be encouraged to form anticommunist organizations, a proposal supported by the government and plantation owners themselves, as it would reframe strikes as a question of personal rights rather than as a political movement. This position meant that organizing carried out by anticommunist labor unions received tacit governmental support.[14]

Trần Quốc Bửu was one of the most militant anticommunists within the labor movement leadership. Born in 1912 in Annam, Bửu received a French education but broke with his mandarin father over the question of French colonialism. In 1940, Bửu's anticolonial activism landed him in a jail on Poulo Condor, a famous breeding ground for Vietnamese communists. During his imprisonment, he befriended his cellmates, the nationalists Võ Văn Giáo and Trần Hữu Quyền, and an intense young communist named Lê Duẩn, who later became the secretary of the southern region. After the Japanese coup of March 1945, Bửu was released from jail and likely sent by the Việt Minh to work with the Cao Đài. Between 1945 and 1947, Bửu fought with the Việt Minh, but he eventually broke with the communists.[15] Native labor movements gained strength as the French colonial empire crumbled in the 1950s. When the Bảo Đại government issued new labor regulations in 1952, Bửu and Gilbert Jouan, a customs officer and representative of the French Confederation of Christian Workers, formed the Vietnamese Confederation of Christian Workers (Tổng Lien Đoàn Lao Công Việt Nam [CVTC]), whose membership rolls boasted as many as 300,000 workers during the mid-1950s. This union drew many of its members from the countryside, including the rubber estates.

In 1954, the CVTC began to receive funds from the American Federation of Labor and Congress of Industrial Organizations (AFL-CIO) (and perhaps indirectly the CIA) and decided to align itself with Diệm. This short-lived alliance ended in 1956, and many in the CVTC leadership sought a third way that avoided communist domination on the left and government control on the right. As the sociologist Tran Ngoc Angie has pointed out, many progressive Catholics also joined a third force that promoted democracy and people's

welfare within a unified Việt Nam.[16] During the early 1960s, the CVTC's ambivalent politics and the taint of foreign money rendered the organization
less effective as a political and social force in the RVN. After Diệm's assassination, the CVTC rebuilt its membership and changed its name to the
Vietnamese Confederation of Labor (Tổng Liên Đoàn Lao Động Việt Nam
[CVT]).[17] To NLF supporters, Bửu remained an outsider and a CIA agent.
To counter his efforts, revolutionaries formed a "Union of Vietnamese Plantations" that existed within the CVTC structure. Different interpretations
remain about the strength of the CVTC's leadership of the workers' movement during the 1950s.[18]

The distrust in government that workers had developed over the years
made relationships with the RVN difficult. These workers had a greater voice
as, thanks in part to the efforts of Bửu and others, the Plantation Workers'
Union grew in membership from 2,000 to 20,000 workers between the end of
1954 and June 1956. According to one study, no less than 28 strikes took place
between August 1954 and November 1959, with a median size of 600 workers,
lasting on average a little under three days. These strikes hit a range of plantations, including a one-day strike on September 11, 1955, involving 38,419 workers
and a strike for higher wages close to the lunar new year, or Tết, in 1956 involving 44,000 workers across all southern plantations. These strikes aimed
at improving living conditions and included demands such as the provision
of promised paludrine to fight malaria, an increased quality of rice and salary,
decreased working hours (including Sundays off), and the release of those
arrested for drunkenness. From the perspective of a U.S. embassy official,
these strikes arose from "anti-union maneuvers of employers, unfair firings,
and bad working conditions" and not communist infiltration.[19]

Workers' concerns centered on their bodies and ability to support their
families, and continued to drive unrest in the second half of the 1950s. Beginning in 1956, the workers and their unions began to push for a collective
agreement that would regularize conditions on plantations and allow workers
to negotiate through their representatives with owners. In May 1957, workers
of the Terres Rouges plantation sent a list of eleven demands to owners. Plantation management and workers reached a compromise on eight of the eleven
demands that involved salary, equal pay for equal work, health care, housing,
transportation, severance pay, and relocation funds while the RVN government
attempted to mediate the remaining disagreements. The first disagreement was
over the length of the lunch break. Management argued that the latex would
decrease in quality if it sat too long, so they gave workers only a thirty-
minute break rather than the two hours that was promised in their contracts.

Apparently workers would accept shorter lunches if they could return home earlier, but they objected to the management's arbitrary decision to change the times. They also pointed out that thirty minutes was not enough time to return to their houses and that if management wanted them to eat at the worksite it needed to build a shelter.

The second disagreement was over allowances paid for families, which management had reduced from eight to five đồng, with a chance for production bonuses. Workers pointed out that they were continuing to work the same number of hours and that there was no reason to reduce pay. In the end, the government negotiated a 1 đồng decrease, with a rise in production bonus from 0.40 to 0.50 đồng. A similar deal was made to end a strike on Terres Rouges during Tết 1958, where, with RVN pressure, workers and managers agreed on a base wage of 37 đồng, a bonus of 7 đồng per day, a Tết bonus, and an incentive bonus of 0.40 đồng per kilogram of dry rubber content. Finally, workers pushed for a reduction in number of trees tapped. Investigation by the labor inspector showed that the trees on certain lots had tougher bark and tapping heights, and recommended reducing the numbers for those tappers. Government officials also expressed concern about documented retaliation for workers who had spoken up about company abuses.[20]

Owing in part to its willingness to support workers' demands, tensions between the RVN government and French planters intensified. On August 31, 1957, an article appeared in the Vietnamese press about the attempts of the Labor Ministry to improve the standard of living of plantation workers. In a strongly worded letter, the vice president of the French Rubber Planters' Union argued that worker demands were threatening the future of plantations by pushing up the costs of production, which mostly came from salary and social provisions, and taking funds away from planting young rubber trees. According to the manager of Michelin's Phú Riềng plantation, a skilled tapper earned forty-one đồng per day and a skilled grafter received sixty đồng per day. The union stated that the 40,000 or so workers earned anywhere from 9,000 to 14,000 đồng, not counting the rice, water, electricity, medical care, and other advantages offered by the plantations. Thus, their standard of living was better than most of the nine million other South Vietnamese, who, according to the *Economic Bulletin* of the National Bank, earned an average of 5,043 đồng per year. Finally, the French argued that workers, encouraged by irresponsible labor union actions and tacit government support, adopted the unreasonable strategy of making as many demands as possible, knowing that only some of them would be met.[21]

In a letter to the RVN vice president and the economic minister, the labor minister Huỳnh Hữu Nghĩa scoffed at the position of the Rubber Planters' Union, arguing that it grossly misrepresented the situation. First, Nghĩa noted that the newspaper article was not an official ministry position, and if the article had been wrong, or even just too strident, the Ministry would have prevented its publication. Second, Nghĩa mocked the Planters' Union for claiming that the Ministry was behind worker agitation. If anything, the Labor Ministry had pressured tappers back to work. Third, Nghĩa showed that labor costs made up at most 25 percent of gross earnings. Moreover, the claim that plantation workers had a high standard of living was refuted by all evidence, not least of which was the difficulty planters had in recruiting tappers. Nghĩa asked the planters why, given their claim of not making a profit, they were staying in Việt Nam at all, and argued that French planters were not invested in Việt Nam or improving its economy. Nghĩa closed by saying that he was asking only that plantations follow the laws, which were essentially the same laws that were in place in 1927. In a more tactfully worded letter sent to the Planters' Union on the same day, Nghĩa covered similar objections and stated that the planters were overreacting to the article.[22]

Not surprisingly, tensions also existed between the CVTC and the government. On December 29, 1959, a strike began on Xa Cát and Lộc Ninh, plantations belonging to Caoutchoucs d'Extrême-Orient, as workers claimed that owners were not following agreements made earlier that year. Apparently the workers went hungry and the strike, which had been costing Caoutchoucs d'Extrême-Orient US$30,000 a day, ended when the CVTC failed to provide promised food. The CVTC official, Trần Hữu Quyền, argued that local officials had called on the police and military to suppress the strike. From the government's point of view, the claims made by the workers at Xa Cát were not justified, and many workers on both plantations wanted to resume tapping. The Bình Long province chief claimed that the CVTC leadership was pulling the strike forward.[23]

By the late 1950s, however, the RVN government began to seek ways to align its, the planters', and the workers' interests. Earlier RVN fears of communist infiltration of rubber workers' unions began to ease, as RVN officials were also potentially interested in stabilizing the plantation situation to carry out their rubber extension programs. On March 12, 1960, representatives of Vietnamese and French plantation management, the CVTC, and the Confederation of Vietnamese Workers' Unions/Trade Unions Confederation of Vietnam met at the Ministry of Labor to sign the Rubber Community

Convention/Collective Convention of Rubber Culture for Vietnam. The 92 pages and 257 articles of this convention detailed work conditions on estates (plantations with more than 500 hectares) and, more significantly, declared liberty of opinion and association, thus confirming the constitutional right of workers to strike and bargain collectively through unions. Speakers that day included the minister of labor and his cabinet director, François Schmitz, representing the plantations, and a representative of the CVTC. The cabinet director stated that unlike all previous labor agreements, this one emerged out of direct negotiations between the owners and the workers—650 hours of direct negotiations by one estimate—with the government serving as the intermediary. This meant that the resulting agreement was fair, free, and respectful of human dignity. The director also called for continued cooperation between the owners and the workers to help expand the rubber sector.[24]

A dispatch by U.S. embassy official Joseph Rosa highlights the significance of this agreement for Vietnamese labor history:

> The importance of the agreement, assuming it will eventually be put into force, cannot be over emphasized since it represents, on the one hand, a literal emancipation of the plantation worker, who once worked under a virtual system of indenture, and, on the other hand, a major step forward in Vietnamese labor/management relationships. It is undoubtedly this second point which has had and will continue to have the greatest impact since the convention is the first industry-wide collective agreement signed in Viet-Nam and is being accepted as a model for future such agreements. Already preparations are underway for the negotiation of similar agreements between banks and their employees, for petroleum workers, the power industry, and for the fish sauce (nuoc mam) and fish canning industries.[25]

On May 19 (incidentally Hồ Chí Minh's birthday), the RVN passed decree No. 66-BLD-LD/ND, which ratified the convention. With this decree, the rubber industry went from being one of the most backward to one of the most progressive in all of Việt Nam, serving as a model for other industries.[26]

During the early 1960s, competition between the CVTC and the left-leaning CVT continued. In the summer of 1961, the Caoutchoucs d'Extrême-Orient fired Nguyễn Văn Vỹ, supposedly for bad performance, including hitting workers, though it appears that he may have also been organizing workers. Vỹ sent a letter to Lê Đình Cự, head of the CVT, who then threatened to strike. Apparently the CVTC was prepared to strike if Vỹ was not fired. In a government-mediated solution, Vỹ was fired and then hired at a different Caoutchoucs d'Extrême-Orient plantation. Strikes continued on several plantations, with local officials often seeing the influence of workers' unions and the newly founded NLF.[27]

Transcending Colonialism

The trajectory of the rubber industry after Vietnamese independence in 1954 shows how local institutions such as plantations could redirect nationalist and community development projects into older patterns of the colonial situation. When Ngô Đình Diệm became president of the RVN, he brought autocratic rule and his brother, the ideologue Ngô Đình Nhu, with him. While Diệm's American advisers were unable to comprehend the coherence of his and his brother's behavior, they too deployed development ideology, nationalist rhetoric, and modernization theory. Rubber played a telling role in their vision as they were driven by their nationalism to create opportunities for Vietnamese in the industry. But facing tremendous fiscal pressures, and a continuing communist insurgency, they opted to encourage large-scale production of rubber over Vietnamese smallholders, and French-owned plantations continued to dominate rubber production until reunification in 1975.[28]

As Diệm consolidated his political and military power after his victory in the Battle of Sài Gòn in May 1955, he also faced questions about how to convert the colonial economy into a national economy that could absorb the massive influx of refugees from the north. The economic historian John Drabble has defined a national economy as "an integrated system operating within a defined geographical area under the control of a central government." This definition assumes that the economy is "subsumed within a de jure national state which further defines itself in terms of language, culture and citizenship." Diệm had limited tools at his disposal to create a national economy, so the rubber industry, which was still dominated by a few French-owned firms and their plantations, provided him with a precious resource. The economy of the RVN resembled in many ways the economies of other formerly colonized Southeast Asia countries, including the newly formed Indonesia and Malaya. Like the RVN, both of those countries inherited an economy that was heavily dependent on agricultural production, with a large percentage of export earnings coming from rubber. The state-directed, export-led growth of Japan after World War II challenged the liberal assumption that the state should not have a prominent role in economic activity. Diệm saw Japan as a model, and as nationalists, Diệm and his officials partly staked their political reputations on their ability to make the Vietnamese economy benefit its citizens. The legacy of French colonialism, along with the perception of Chinese and other "non-Vietnamese" dominance over the economy, made this issue politically explosive.[29]

In a September 1955 memo on economic policy discussions, a U.S. embassy official in Sài Gòn articulated the American view that the RVN leadership was overly nationalistic. "Naturally," he wrote, "discussion of these problems within the [Vietnamese] Government has taken a strongly nationalistic view vis-à-vis the French business interests in Sài Gòn." He went on to note that the Vietnamese "Supreme Monetary Council," which included Ngô Đình Diệm and other top members of the RVN government, advocated "a strong nationalistic position in economic policy" and was critical of the United States Operations Mission influence. While the Americans were unable to judge the capabilities of the Vietnamese government, they accurately recognized continuing tensions between French plantation management and the RVN government. RVN officials were hesitant to consider nationalizing foreign businesses; instead, these officials limited their objections to specific plantation practices. An unofficial paper handed to senator Mike Mansfield in August 1955 during his meeting with Diệm and his ministers stated, "Everyone knows that the French rubber plantations, not having confidence in the future of free Vietnam, practice bleeding to excess." While the French vehemently denied excessive tapping for short-term profits, many Vietnamese leaders assumed that the French were not interested in the long-term health of plantations. Even worse, French managers were not above trying to sell rubber to communist China, and, as security around plantations deteriorated, these managers often paid "taxes" to Vietnamese communists to keep the rubber flowing.[30]

Even though the Geneva Accords signed in 1954 called for the regroupment of military combatants, many former Việt Minh cadre remained in the south and continued to operate on and around plantations that were far from population centers. In a July 1956 letter, Trần Văn Minh, a military official responsible for security in the plantation region, compared his own task to that of colonial security forces. Minh recounted that the roughly 3,000 French-led partisans in the plantation region did a decent job of securing plantations but couldn't stop communist activity, and that the plantations called in the professional military when things got out of hand. According to Minh, the Army of the Republic of Vietnam had far fewer troops to control the area and counter NLF organizing, propaganda, and infiltration. The French, he concluded, had left a legacy of poorly selected and trained partisans.[31]

While some American officials were skeptical about the role of ex–Việt Minh agents in launching worker strikes, RVN officials agreed with French managers that communists were active in the region. In response, Diệm initiated a program called Tố Cộng, or "denounce the communists," in mid-1955. French planters had long claimed that communist agents were responsible for strikes, and in June 1956, four communists were supposedly identified as part of the CVTC membership at Lộc Ninh. To eliminate such agents, the RVN military launched multiple raids in provinces across the rubber region in the Trương Tấn Bửu operation, which was named after a famous general of the Nguyễn dynasty. Starting in July 1956, this operation dealt a severe blow to left-leaning organizations, and the military often used worker strikes as a pretext for carrying out its attacks. For example, on August 15, workers at the Phước Hòa plantation went on strike to increase hours and wages. This strike was legal, as the CVT had gone through the proper channels and received approval. A few days after the start of the strike, however, four or five labor organizers disappeared. Although the CVT leadership, including a probable communist operative named Lê Đình Cự (see chapter 4), protested their disappearance, the Army of the Republic of Vietnam commander, Mai Hữu Xuân, argued that the communist operatives in the organization had been uncovered and as a result had simply run away.[32]

A memoir by the Catholic rubber worker and Communist Party member Lê Sắc Nghi casts further light on Trương Tấn Bửu, confirming that it was devastatingly effective at eliminating the leadership structure of the Communist Party in the rubber region. Born in 1924 in Quảng Trị, Nghi, along with his father and his brother, left his mother and in-laws to work for a plantation in An Lộc after a broken dike had caused local hardships. The Catholic Church and a local priest facilitated their trip. Later, Nghi joined Military Unit 310 and used his cover as a Terres Rouges tapper to travel throughout the rubber region in support of the revolution. While many other leaders were jailed, tortured, exiled, and killed, including Nghi's own wife, who died in the Poulo Condor prison, Nghi drew on luck and his intimate knowledge of the plantation region to avoid capture. Despite these setbacks, the communists continued to engage in a cycle of activism and repression on plantations even before the 1960 formation of the NLF.[33]

After the formation of the NLF, RVN military and political leaders continued to struggle with the question of whether plantations should remain open. From their perspective, plantations provided a significant source of income for the NLF through the "disasters" of rubber theft and kidnappings. During

an August 1960 meeting between Bình Dương provincial officials and plantation directors, four steps to control the illegal sales of rubber were proposed.
First, plantation management would keep production logbooks and packing
slips to keep track of the creation and movement of rubber. Second, only
those in the industry could keep rubber. Third, the buying and selling of rubber would be limited to plantation owners, industrialists, and merchants.
Fourth, managers would have to keep a record of their travels. In a letter to
Diệm about the August meeting, the director of planning expressed his skepticism about the efficacy of the proposed steps. The director noted that large
plantations already kept statistics on rubber production and movement but
that smallholders were harder to regulate. Furthermore, even for large plantations, packing slips were easy to fake. The only effective means of control would
be to limit purchases and sales to those with licenses and to crack down on
those receiving stolen rubber (*kẻ oa trữ*).[34]

Government actions to limit smuggling were ineffective, and illicit activities continued into the early 1960s. For example, a French planter filed a complaint with the police stating that forty tons of rubber had been stolen in Tây
Ninh over several months in 1962. In a single incident on May 2, 1962, the
drivers of a truck that was carrying six tons of rubber to Sài Gòn were arrested
for smuggling. The minister of the interior, Bùi Văn Lương, noted with concern that some of the thefts were taking place in open daylight and were probably being abetted by local officials. He also recognized that many thieves
had been captured but that the courts had let them go, thus giving the impression that smuggling rubber was tacitly accepted. Measures such as
stamping rubber with company names and training officials to stop smug-
glers did little to stem the flow of illegal rubber.[35]

Kidnappings also presented a challenge to the industry. In December 1961,
two French plantation employees were captured by the NLF and were not
released until ransoms had been paid. Such NLF actions posed threats to
French and Vietnamese personnel and created hardships for Vietnamese
labor. Nguyễn Lê Giang, the cabinet director of the Ministry of Labor, expressed concern about the economic and social consequences of plantation
closures. When the Michelin plantation suspended operation of its Dầu
Tiếng plantation in 1962, workers' representatives requested that management either reopen plantations or give severance pay to workers as stipulated
by the 1960 Collective Convention of Rubber Culture. Michelin management argued that the convention did not apply in this case, because a lack of
security, rather than a lockout, was the cause of the closure. Similar discussions were taking place on plantations throughout the rubber region.[36]

Throughout the 1960s, the NLF and RVN sought to transcend plantations while still extracting advantages from them. The NLF collected taxes from workers and sought to reshape plantations to their needs. It left mature trees standing, focusing instead on destroying bridges and preventing grass cutting and fence building. The RVN responded with plans to link different provinces in the region through a security net and to increase security along roads and in agricultural resettlement camps. Plantations, not surprisingly, remained relevant to environmental diseases such as malaria for both sides. While novel techniques such as the use of DDT and synthetic drugs developed in the 1930s were introduced in both the north and the south, most medical doctors attempted to control malaria by employing colonial-era strategies. Memories of this era structured experience and approaches throughout Việt Nam as malaria knowledge was refashioned to fit wartime needs and political motivations, rather than specific techniques, separated efforts in the revolutionary north and the anticommunist south. As Pierre Brocheux has pointed out, even though most of the infrastructure of Việt Nam was destroyed during the twentieth century, foreign powers, especially the French, helped show "the path to follow, the methods and the know-how for pursuing modernization." Dominant wartime realities limited the options available to the Vietnamese, French, and American researchers as aid from China, the United States, and the Soviet Union helped shape knowledge production.[37]

Malaria remained a research priority in the RVN. A 1960 medical thesis at the University of Sài Gòn, for example, noted the high malaria rates in the coastal province of Bình Định, where rural health services had begun only a few years earlier. International organizations also instituted antimalaria projects during the late 1950s and continued them throughout the 1960s. In 1958, the WHO began its malaria eradication program in the RVN, and between 1960 and 1963, it carried out two DDT sprayings per year. These efforts reduced the rates of malaria from a high of 7.2 percent in 1958 to 1–2 percent in 1966. Synthetic drugs were also used to combat malaria. These drugs were divided into two main types: acridines, which include mépacrine (quinacrine), which had been used since 1930, and quinoléines, which include chloroquine (amodiaquine), proguanil, chlorproguanil, pyriméthamine, primaquine, quinocide, and pamaquine. After 1965, wartime conditions made it impossible to continue antimalaria activities, especially in the countryside, as DDT resistance in *Anopheles* and drug resistance in plasmodia added to the difficulties created by human violence.[38]

RVN medical doctors wrote of the successes of colonial antimalarial measures, while noting the high costs of such programs. In 1970, Đặng Văn Đang and Nguyễn Đăng Quế reviewed malaria rates between 1930 and 1944 using data from the Pasteur Institute. These rates varied between 15 percent and 31 percent and were consistently higher than the rates from 1958 to 1966.[39] Still, the authors spoke of the success of a few rubber plantations in dealing with the disease: "There were only a few plantations that if they succeeded it was thanks to large-scale organization. They cleaned up the gloomy places full of vegetation, filled in the muddy lagoons and ponds and applied the medical methods to thoroughly prevent malaria. The plantations mentioned above became prosperous but we had to pay dearly with much sweat and money." These experiences could offer lessons for the present, they continued: "The brilliant results of the rubber plantations with respect to economy and society mentioned above has shown us that if the development of medicine is implemented in an effective manner in the forested, mountainous or swampy regions, we can exploit a large expanse of land."[40] These statements paralleled those made during the 1930s, when techniques developed for malaria control on plantations were haltingly extended to the countryside. The production of malaria knowledge by Việt Minh and DRV medical scientists illustrates more clearly a postcolonial rationality. DRV and RVN medical doctors may have been dependent on the tools of empire, but they developed their own reasons to justify their use of these tools. As in the RVN, memories of the colonial period shaped understandings of malaria during the Vietnam War, with northern medical doctors judging their own actions and progress against these memories.[41]

North Vietnamese medical doctors, too, faced a monumental task. Left with an aging colonial infrastructure and the legacy of wartime disruptions that had drastically changed the disease enviroments of the north, the DRV desperately needed updated malaria knowledge. The Soviet malaria expert A. Y. Lysenko pointed out: "Most of the studies in malaria epidemiology in Viet Nam were performed in the years of colonial dependence of that country 15 to 25 years ago. The techniques of these studies, the goals toward which they were directed, and many of their conclusions have either become obsolete or have become inapplicable to a country which has won its independence and is developing a public health service. Besides this, during the 8-year war of resistance in Viet Nam, intensive migration processes occurred which of necessity introduced considerable changes into the regional epidemiology of malaria."[42]

Lysenko's quote highlights the importance of political contexts for public health measures and the role that migration played in transforming malarial ecologies. With the help of experts from the Soviet Union, the medical doctor Đặng Văn Ngữ investigated conditions in northern provinces such as

Thái Nguyễn as part of an intensive malaria survey that took place between 1955 and 1957. Armed with this research, Ngữ and Lysenko published several articles and produced an atlas of malaria in Việt Nam.[43]

In July 1957, Ngữ became head of the newly established Malaria Institute (Viện Sốt Rét), which in 1960 was renamed the Institute of Malariology, Parasitology, and Entomology. As part of his investigations, Ngữ established pilot stations to test the techniques developed in the Soviet Union and by the WHO, which had embarked in 1955 on a program to eliminate malaria worldwide. Between 1956 and 1961, the DRV, with Soviet aid, carried out research on a monumental scale: 3,000 locations including 646,277 people checked; 435,370 samples of blood tested; 319,087 houses checked for mosquitoes; and 168,084 water spots examined, which was 3.4 times the scale of the Pasteur Institute's investigations between 1927 and 1938. These investigations led to the creation of several programs. In 1960–61, DRV health workers prepared for antimalaria campaigns that were carried out starting the following year. During the relative peace of 1961–64, these campaigns achieved many successes but faced increasing hardships after 1965 due to renewed violence in the north.[44]

Ngữ and his medical colleagues used research from the malaria surveys to refashion techniques that were used to combat malaria on rubber plantations during the colonial period. In practice, malarial control involved heavy use of DDT and synthetic drugs when they were available, and manipulating environmental conditions when they were absent. Between 1958 and 1975, 10,633 units of DDT were used, making up 11,482 tons of 30 percent solution that protected 495 factories and fields. The largest amount of DDT was sprayed between 1965 and 1972, during the most intense period of fighting. At the same time, health workers were still distributing the synthetic drugs that had been developed during the 1930s. From 1962 onward, the DRV used Delagyl (choloroquine). Beginning in 1969, the DRV relied on pyrimethamine (daraprim), sulfamid slow-acting pills 3 and 2, and Fansidar, a sulfadoxine and pyrimethamine mix, which treats *falciparum*, albeit with serious side effects.[45] Despite the best efforts of the DRV and RVN, malaria control remained problematic, especially in southern Việt Nam, where much of the fighting during the Vietnam War took place. As part of its war effort, the DRV sent malaria experts to the south. A Malaria Institute publication noted that many comrades had become heroic revolutionary martyrs while combating malaria in southern Việt Nam. In 1967, Đặng Văn Ngữ was killed by a B-52 bomb around Huế, and one year later the minister of health, Phạm Ngọc Thạch, died during a mission to reduce the ravages of the disease.[46]

From Planters to Planners

Leaders of newly decolonizing states across Southeast Asia looked to increase indigenous participation in agriculture and industry, in part by tapping into the remnants of imperial networks. Furthermore, individuals and families, rather than plantations, became seen as the basis of health and good environmental management. The Federation of Malaya served as a model—and competitor—for rubber production in the RVN. Malaya had a clear lead in economic indicators such as gross domestic product, and the RVN's borrowing of Malaysian models is indicative of increased intra–Southeast Asian contact. Malaya also boasted a strong rubber smallholder sector, with smallholders producing over 40 percent (300,000 British tons) of Malaya's rubber exports in 1961. Although initially discouraged during the colonial period, smallholders benefited from the combination of midsized European plantations dispersed throughout the countryside, the activity of Chinese entrepreneurs, and official postcolonial support.[47] The lack of these factors in French Indochina, along with specific environmental conditions, resulted in a limited number of Vietnamese smallholders. Furthermore, planners in Malaya advocated for agricultural diversification, including palm oil, as they were worried about overproduction, the cost of replanting, and bark levels in the rubber industry. Just as cooperation and competition had structured relations between colonial powers, leaders of RVN and Malaya were open to participate in the "intellectual hegemony of science and technical reason." Throughout Asia, leaders sought to increase their connections to postcolonial networks as a means to strengthen their position relative to other leaders and to their own citizens.[48]

A comparison of the definition of "smallholder" in Malaya and Việt Nam provides further evidence of how little attention was paid to Vietnamese families who participated in the rubber industry. In Malaya, smallholders were defined as those holding less than 24.70 hectares (ha), while in Việt Nam, smallholders were defined as owning less than 500 ha. After the French military loss at Điện Biên Phủ in 1954, most French-owned plantations of less than 500 ha were transferred to around 400 Vietnamese smallholders. In 1956, one report claimed that all but one large plantation belonged to the French and all but three small plantations belonged to the Vietnamese, although later reports show a larger presence of French individual plantation owners who owned less than 500 ha. In terms of size and production, Vietnamese-owned plantations accounted for only 10–15 percent of the land that harvested *hevea* trees, and these plantations produced roughly the same percentage of rubber nationally. Many of the smaller plantations had been abandoned during the

First Indochina War, and the quality of the Vietnamese-owned trees and land tended to be lower.[49]

The violence of the two Indochina Wars further prompted the consolidation of the large plantations, or estates. Between 1943 and 1970, the number of smallholders remained approximately constant, but the average size of individual smallholder plots, and the total area dedicated to smallholdings, decreased by more than 50 percent. During the same period, the average size of medium plantations held steady, but the number of holdings, and the overall area covered by these holdings, declined by two-thirds. The biggest winners during this time were the large estates. While the number of estates decreased by one-third, the area held by each estate increased by roughly the same amount, leaving the total area dedicated to estates almost constant. Thus, estates, which controlled 64 percent of rubber lands in 1943, controlled 82 percent by 1970.[50]

The period stretching from 1954 through 1963 was a relative golden age for the rubber industry and its workers; the French had left and the Americans had not yet arrived in full force. By the late 1950s, rubber had become the leading export of the RVN, with its production peaking in 1961. Increases in productivity played an important part in this increase, and between 1957 and 1966, Vietnamese plantations produced more than one metric ton of rubber per hectare, a figure that compared favorably with the rest of Southeast Asia. In these years, the Vietnamese rubber industry faced three important questions: (1) What would be the role of smallholder production? (2) How would the issue of estate labor be addressed? and (3) How would French, Vietnamese, and American rubber experts best respond to the needs of a rapidly changing industry? New conceptions of Vietnamese nature and the actions of the leader of the First Republic of Vietnam, Ngô Đình Diệm, shaped the answers to these questions. Diệm not only had to create a national economy but also address southern critics who called for greater Vietnamese participation in the industry; locals wanted to be growers and owners rather than tappers and "coolies."

Promotion of Vietnamese Smallholders

In April 1957, the American agricultural economist Sheldon Tsu wrote a report justifying smallholder rubber in Việt Nam. Tsu noted that rubber had replaced rice as the primary export earner for the RVN, and he commented favorably on the quality of Vietnamese rubber. While he approved of the social services provided by French-owned plantations and urged that they not be taken from French hands, he gestured toward the many problems facing large and small plantations and tentatively suggested creating more smallholders. His recommendations for the Small Scale Rubber Community Development program were based on the principle that "'tappers should tap their own trees' as 'farmers should till their own land.'" This principle would "encourage land ownership, prevent concentration of land, and arouse the interest in taking good care of trees and rubber development." Tsu addressed a branding issue and called for the term "rubber community" rather than "rubber plantation" as a name for his proposed settlements to highlight "the independent, democratic and self-owned spirit and operation," which had supposedly been successfully created in South America, Malaya, and Liberia.[51]

Tsu specifically proposed the Further Expansion Program for promoting rubber communities that would be composed of 120–400 families owning about four ha of rubber trees each, for a total of between 500 and 1,500 ha. These rubber communities would raise livestock and other food crops to be self-sufficient. They would have centralized processing centers to ensure quality control and a shared marketing system, with profits going toward communal facilities, services, personnel, and welfare. Tsu called for promotional films and booklets, an Agricultural Development Finance Corporation, programs for the transfer of land ownership, rehabilitation, and expansion. He urged the United States Operations Mission to cosponsor these programs and provide an agricultural economist as an adviser. These rubber communities were to fulfill the "modern business principles" of "a. making profit, b. collective employment, and c. raising the standard of living."[52]

As Diệm had done with other foreign advisers, he took Tsu's proposal and transformed it into his own program, calling it Rubber Cultivation Development. More importantly, Diệm received a program that was originally meant to promote smallholders and reframed it largely for the benefit of foreign-owned plantations. During Diệm's July 1958 visit to SPTR, Lai Khê, and Dầu Tiếng, he signaled his decision to funnel money into research and the large plantations, encouraging planters to extend their lands or replace old trees with high-yield seedlings to set up the rubber industry for the future. With a natural cycle of roughly thirty years, most of the rubber trees on the large plantations were nearing the end of their economic life; the violence and lack of labor in southern Việt Nam since 1945 meant that almost all *hevea* had been planted before World War II. By laying his bet on large plantations, Diệm reproduced colonial prejudices that largely disregarded smallholders.

In mid-August, Diệm signed into law Decree (Sắc Lệnh) 414-KT and Governmental Decision (Nghị Định) 287-KT, which established the Rubber Cultivation Development Fund and its regulations. Under this decree, extension or replanting had to take place on properties that were adjacent to lands that

cultivated rubber in 1957. With security still a concern, opening new lands
was not desirable, so the law encouraged the replanting of already-existing
lands. Minimum qualifications included owning at least ten ha, with no less
than 5 percent of the land planted at the end of the previous year, and the resources to develop or replant at least one ha. These conditions essentially denied new rubber growers from taking advantage of the loan.[53]

A third stipulation stated that those who controlled between 10 and 100 ha
could ask for 5 đồng for every kilogram of rubber sold in 1957, with payments
divided into three installments: 60 percent for 1958, 30 percent for 1959, and
10 percent for 1960. Those with more than 100 ha could ask for 3 đồng for
every kilogram of rubber sold in 1957, divided into two installments: 60 percent
for 1958, 40 percent for 1959. While this condition appears to have favored
smallholders with a slightly higher per kilogram loan amount, the result was
quite the opposite. Most large plantations produced more rubber per hectare;
thus, basing loan amounts on the amount of rubber produced gave the largest
plantations a better rate. In addition, the money for smaller plantations was
distributed over three years rather than two. Interest was fixed at 2 percent per
year, and the loan had to be repaid in five years, beginning in year nine. The
borrower also had to pay 0.25 percent of the loan amount yearly to the National
Agricultural Credit Bureau (Quốc Gia Nông Tín Cuộc) to cover administrative costs, and a committee of six members would make loan decisions, with
the president casting a tiebreaking vote if needed. A representative of the Rubber Planters Union sat on the committee, with the nationality of the applicant,
either Vietnamese or French, determining the nationality of the union. The
cost of opening the account was 0.5 percent of the loan, with a minimum of
2,000 đồng, which meant that those who were replanting or developing less
than about ten ha paid a higher fee per hectare, and a larger percentage of
their loan, than those who were redeveloping large tracts of land. These
conditions effectively guaranteed that only those with large holdings would
benefit from the development fund.[54]

Many middle-class Vietnamese who owned rubber were politically active,
and in 1957, 180 planters formed the Vietnamese Rubber Planters' Association
(VRPA). The RVN government presented drafts of 414-KT and 287-KT to
the association only a few days before they were officially announced. When
the association met to discuss its terms, they critiqued the draft, noting that
the proposals favored large, French-owned plantations. They called for a loan
of 60,000 đồng for every hectare of extension, even for those who had begun
planting in 1958. They also called for loans for planters who owned less than
ten ha. Draft 287-KT was released on the following day with few changes, and

the VRPA responded quickly, both with private letters and in journals such as the *Saigon Trade/Commerce Bureau Weekly*. The editors of the weekly noted that basing the loans on production amounts disadvantaged Vietnamese growers. They stated that the growers were not being xenophobic, pointing out that in the past the colonial government had been quite supportive of French planters, citing its loan of ninety million piasters to rubber growers. The growers argued that "preferential rights to rubber replanting loans must go to Vietnamese."[55]

After a series of letters between various government officials and the VRPA, the two sides came to a tentative agreement. The RVN vice president, Nguyễn Ngọc Thơ, wrote that while the government fully supported the efforts of Vietnamese smallholders, it had to consider the big picture, citing concerns that the funds would quickly be exhausted. Thơ adopted Diệm's Personalist ideology and urged the planters to rely on individual effort to overcome any shortfalls. While Thơ considered ending new land grants, the government made two concessions to those who were already growing rubber: it agreed to provide more funds for Vietnamese planters and to lift the restriction of giving loans only to those who had grown rubber in 1957. This agreement created opportunities for new planters and for those whose plantations were slow to recover after the First Indochina War.[56]

A report from the Biên Hòa provincial head lists several Vietnamese who qualified for a loan. One such planter was Võ Thị Sen, who owned Gia Trập, a plantation of 26 ha of *terres rouges* in Long Khánh that had grown to 104 ha by 1965. Echoing earlier concerns about clean-weeding and aesthetics, Léon Morange, the SPTR director who was responsible for providing technical aid to small and medium planters, found that too much vegetation was growing in between Sen's *hevea* trees. Sen had otherwise taken good care of the plantation, and Morange recommended that she receive the second part of the loan. However, smallholders like Sen were an exception; by mid-1959, the program's biggest beneficiaries were large plantations such as Michelin, which had quickly filed for the loan, submitted reinvestment plans, and requested tax relief. According to *The Nation*, only 820,000 đồng had been lent to Vietnamese planters. Even given the 30,000 đồng minimum needed to replant one ha of rubber, this amount paid for less than thirty ha of extension. Meanwhile, foreign-owned companies received 142,000,000 đồng for 2,365 ha of extension, which worked out to 60,000 đồng per hectare.[57]

The RVN banking sector viewed rubber investment programs such as the Rubber Cultivation Development Fund favorably. In March 1960, the general director of the Vietnam National Bank, Nguyễn Hữu Hạnh, wrote a note in

French to Diệm arguing that natural rubber's future was bright and RVN plantations were well positioned to take advantage of this market. Hạnh also pointed out that the industry could eventually employ up to 500,000 workers and their families. Later that year, Hạnh offered a more restrained prediction in response to government skepticism about his projections. Moreover, other prominent RVN figures used the battles over Vietnamese participation in the rubber industry to challenge Diệm's rule. One such person was the rich and eccentric Nguyễn Đình Quát, who passed himself off as a representative of a world citizens' organization, was tried for tax evasion, and at one time served as president of the VRPA. In Quát's challenge of Diệm's rule, he appealed to a distrust of the French common among RVN elite, arguing that French planters were tapping trees to death and not replanting to maximize short-term profit at the cost of long-term health of the industry. He pointed to sudden spikes in latex production, though he did not consider increases that were due to years of tree resting, improvements through grafting, and the application of tree stimulants. Quát was not alone in his distrust of the French, as another anonymous commenter speculated that they were focused on short-term profits due to fears of nationalization.[58]

Quát believed that Vietnamese planters needed government assistance, and in a May 1956 letter, he pointed out that the colonial government had worked closely with French planters to overcome the difficulties of growing. After 1954, the number of Vietnamese planters had increased, yet they were losing money while French planters continued to make a profit. Quát urged Diệm to help loyal planters deal with rural insecurity and the low price of rubber. While Quát's analysis ignored some of the tension between the colonial government and French planters, he did correctly point to the timely legal and financial assistance that the colonial government had provided. An official from the Ministry of Agriculture responded to Quát's letter by offering technical advice, including tree selection and fertilizer. Although French planters had demonstrated a troubling lack of loyalty to the RVN, their plantations provided far too much in tax revenue for Diệm's government to eliminate them entirely, and debates over rubber ownership continued into the late 1950s, when the VRPA again raised the question at its 1958 congress.[59]

The most insistent criticisms of the Rubber Fund loans came from middle-class South Vietnamese. Early in 1960, the Hội Văn Hóa Bình Dân (HVHBD; known in French as l'Association culturelle populaire) held a series of three conferences on rubber. The first conference took place on March 13 in Phú Cường, located in the rubber-rich province of Bình Dương. Huỳnh Văn Lang, president of the HVHBD and a key figure in Sài Gòn's political and cultural life, delivered the opening speech. Both Lang and the HVHBD were interested in the question of popular Vietnamese participation in science, technology, and industry, and smallholder production was one means to increase involvement among the Vietnamese. The purpose of the conference, Lang stated, was to consider the rubber industry from the angle of production and commerce and to analyze its effects on society. The conference also intended to introduce Vietnamese planters to each other and to economists who could help frame the planters' efforts in an open forum.[60]

Hoa Văn Mùi, then general secretary of the VRPA and owner or part owner of several plantations, also spoke that day. Mùi and the association continued to fume over 287-KT, and his talk strongly criticized the program. He showed previous analyses proving that smaller planters, whom he defined as middle-class landholders investing in rubber, rather than individuals or families who owned a few trees, were disadvantaged compared with large plantations. Mùi argued that loan amounts were based on land possessions of 1957, a condition that the government had agreed to lift. In addition, he criticized Vietnamese planters' "traditional" production methods, which had not changed in over thirty years. The situation was exacerbated by the fact that the IRCV at Lai Khê had focused its efforts on issues relevant to estate rubber. Conference attendees asked tough questions about the contributions made by organizations such as the VRPA, and about the low number of Vietnamese members of the IRCV. Mùi responded with an elaboration of his pessimistic outlook for the prospects of smallholders; other speakers were more optimistic about the potential of Vietnamese smallholder production, but their rationales remained hazy.[61]

The second HVHBD conference took place on April 10, 1960, in the heart of rubber-growing lands in Xuân Lộc, the capital of the newly created province of Long Khánh. Both Vietnamese and French speakers employed the colonial-era rhetoric of the modernizing potential of rubber plantations. Yet, along with the potential benefits of rubber, speakers also raised issues related to plantation labor. Tăng Văn Chỉ, the RVN's delegate for the eastern provinces, said that the human aspects of industry, such as employer-employee relations, were important. The general labor inspector, Nguyễn Lê Giang, also reviewed the history of plantation labor, tactfully addressing the issue of colonial-era abuses. He cited key texts dealing with labor, including the Rubber Community Convention, which had been signed the previous month. While the Ministry of Labor failed to apply many of the provisions of this convention until the 1970s, Giang repeated the arguments that plantation labor conditions in the RVN were comparable to those of neighboring nations

and were better than the conditions endured by peasants living in the surrounding countryside.[62]

François Schmitz, the inspector general of SIPH, one of the nation's largest
rubber companies, offered numbers to support Giang's assertions. Schmitz's
data came from a study by the Economics and Humanism Center that assigned
scores to measures of human well-being at various population centers. Plantations, the study argued, measured up well when compared with places such as
the resettlement camp at Cái Sắn (though not as well as in Sài Gòn), and large
plantations did better than small ones. These numbers were skewed toward
the material organization of large plantations: they assigned high value to visible markers of modernity such as hospitals and row housing, regardless of the
practical contribution of such architecture in promoting well-being. Questions that day were once again pointed, and the audience asked about workplace accidents, death and birth rates, and the affordability of health care
for workers on smallholdings. The often-vague answers left the audience un-
satisfied, and the dialogue did not alleviate its apprehensions.[63]

Although the Rubber Fund program ended in the spring of 1961, discussion about how to support Vietnamese rubber growers continued. One
proposed solution was to reestablish the Rubber Bureau that had closed in
Việt Nam in 1948, though one skeptic pointed out that the main role of the
bureau had been to enforce the 1934 London agreement, which was no longer in effect. Over the next few years, Mùi's pessimism concerning smallholders was borne out. Some smallholders could obtain high-yielding breeds such
as PR107, PB86, GT1, and (the IRCV's own) OY1, but Vietnamese had very
few holdings of less than ten ha and estates still contributed the lion's share of
production. Furthermore, Vietnamese rubber lands held a relatively low percentage of trees that were in their productive years.[64]

Overstretched

The other program that Diệm did in fact use to create smallholders, and shift
responsibility for labor and health conditions to the individual and the family,
was the Land Development Program (LDP). Examining the successes and
failures of the LDP, also launched in 1957, is key to understanding the fate of
rubber in South Vietnam. Officially called Dinh Điền, a Sino-Vietnamese
term meaning "to nourish rice fields," the LDP focused on establishing populations under the control of the RVN. Ngô Đình Diệm envisioned the LDP as
a network of settlements in the southern lowlands called "dense and prosperous areas" (*khu trù mật*), which were meant to win the support of the local

population through political and economic means. These settlements also came to be known by locals as agricultural development centers and "agrovilles" that were organized under a General Committee (Tổng Ủy Dinh Điền [TUDD]). While historians have paid most attention to the agrovilles in the Mekong Delta, the LDP aimed to resettle both Vietnamese and ethnic minorities across four zones, three in the Mekong Delta and one in the central highlands.[65]

RVN officials saw family-grown rubber in the central highlands as a way both to resettle lowlanders and to fix uplanders in one spot. Political scientist Stan Tan has analyzed the Dinh Điền Thượng, or uplands portion of the LDP. He argues that RVN officials, to a greater degree than their colonial counterparts, sought to reform highlanders' agricultural practices and integrate them into the state. The central highlands' rolling hills and mixture of dense and open forests offered possibilities and challenges distinct from the Mekong Delta, and RVN officials needed a crop other than rice to settle both Montagnard and migrant Vietnamese on permanent plots. Initially, LDP planners viewed kenaf (a fiber) and flaxseed, rather than rubber, as both morally appropriate and capable of providing a living for program participants. Many RVN officials soon saw the environmental, economic, and social value of rubber. Rubber trees produced latex for upward of thirty years, which would fix residents to the land and provide a year-round source of income, and the technical division of the Land Development Commission published a pamphlet in Vietnamese in 1960 called the "Method of growing rubber." In its introduction, Bùi Văn Lương, the interior minister and member of the TUDD, promoted land selection for rubber, arguing that rubber trees had deep roots, didn't demand many nutrients, could help control erosion, and could be grown on relatively marginal land. Yet rubber was not without its drawbacks. The boom and bust cycle of the world market and the perceived threat of synthetic rubber presented difficulties for farmers who were dependent solely on this substance for a living. Moreover, *hevea* trees required an initial investment of several years before they would yield latex. LDP planners sought to overcome these drawbacks by promoting intercropping of annual cash crops to see farmers through the initial lean years.[66]

Resettlement in the central highlands involved negotiations among uplanders, lowlanders, and the RVN, all of which were complicated by the communist-led insurgency in the south. "By 1961," Stan Tan writes, "the focus on agrarian change to achieve the political and strategic objectives during the relatively peaceful years of 1955–1960 had given way to the priority of strategic deployment for insurgency warfare." As military and social control became the focus of the LDP, planners envisioned rubber production as a political and economic basis to resettle Vietnamese displaced by warfare. To create settlements that were loyal to Diệm's government, planners focused on communities and collectivities as opposed to individuals. For historian Ed Wehrle, the concept of *đoàn thể*, or the "search for better collective organization," helped unify much Vietnamese activity of the 1950s and 1960s. This concept shared many similarities with the idea of community development, and many Vietnamese and Americans initially highlighted these similarities. But as projects were implemented, the profound differences in these concepts became obvious. As Ed Miller has shown, Diệm had his own ideas of community development. Rather than a democratic ideal based on individuals acting freely and viewing themselves as independent agents, Diệm's vision emphasized hierarchical relations and obedience to authority for the good of the community—a system of development that required compulsory labor.[67]

In addition to the Americans, RVN officials considered other sources of international aid for the LDP. In June 1960, Huỳnh Hữu Nghĩa, formerly the labor minister, wrote to the general director of the Bureau international du travail (International Labor Organization) asking for advice on how best to integrate "aboriginal populations, semi-nomadic" into the RVN state and to resettle lowland populations to the highlands. Nghĩa cited the labor organization's experience in protecting aboriginal populations and providing technical assistance, particularly in Latin America. The RVN state also sought financial assistance from the labor organization, arguing that upward of 375,000 people could eventually benefit from a program based on giving two ha of rubber trees per family. RVN officials also cited the favorable environmental conditions of the region, including a dry season that suppressed diseases and relatively flat land that reduced the danger of erosion.[68]

French agronomists continued to advise Diệm's government about rubber. They highlighted the ideological importance of framing rubber through the lens of smallholdings and community development, and emphasized the differences between smallholdings and the plantation system in speeches associated with the opening of the Rubber Advisory Committee. On March 30, 1960, two weeks before the official start of the committee, a French tropical agronomist Philippe J. P. Richard presented his vision for rubber in a speech titled "Future of 'Small Holdings' in Rubber Agriculture in Vietnam." This speech, given in French, discussed a family holding that was worked only by a man and his wife, which included 1.5 ha of rubber, 0.2 ha of garden crops, and 2 ha of diverse crops, with a total of 3.7 ha to be farmed by the fourth year. The first part of Richard's speech focused on the policy and politics of smallholders, offering a comparative analysis of smallholder experience in Indonesia, Malaya, and Ceylon. Richard attempted to address the concerns of his audience about family holders, arguing that the experiences of neighboring countries showed that family plots could be successful. Việt Nam, with its "political, and above all moral, revolution," needed these plots.[69]

Richard did have concerns about family plots, including the poor treatment of soil. He admitted that social scientists did not pay enough attention to the limits of soils: "The economists give to the earth the quality of indestructibility that the tropical agronomists question or at least view as limited." Richard argued that the economists' assumption was based on soils in temperate climates, which did not behave like soils in the tropics, and pushed for the use of perennial trees to develop the central highlands. "The agrarian economic development of the High Plateau," he argued, "is strictly dependent on the development of perennial tree agriculture." These trees would protect the fragile red earth soils from the threat of erosion or of losing their structure and turning into a brick-like mass. He also argued for a diversity of crops that would shift over the years. With the appropriate technical, financial, and organizational assistance, Richard concluded, family holders in the central highlands could take care of soils, support themselves, and pay taxes to the nation.[70]

In the second part of his speech, Richard detailed his theoretical understanding of family holders and gave a lengthy defense of smallholdings, arguing that they were both healthy for the economy and beneficial for the individuals working them. Large capitalist collectives, including plantations, Richard argued, denied the humanity of workers. More broadly, he viewed the stock market as a reflection of human confidence and money as trust. He cited the writing of Pope Pius XII several times, including his statement that technological and scientific advances should not work to the detriment of families and smallholders. Richard also struck out against "the disciples of Marx," who, he argued, saw family holdings as a relic of an artisanal era, bound to disappear under collectivization. "The real problem that faces the rural world in light of modern evolution," he concluded, "is to create work communities where the human feels at home and where he does not risk dehumanization in losing his prerogatives to be free and responsible. Discipline is necessary but this implies neither the renunciation nor the alienation of the self."[71]

Richard's center-left Christian viewpoint on labor resonated with Diệm's Personalism philosophy (*Nhân Vị*) and aligned with his strong opposition to both communism and unfettered capitalism. These similarities did not prevent Richard from being disappointed with the execution of the LDP, and his evaluation of the program points to some of the factors that caused it to fail.

In January 1962, Richard wrote a letter to the general commissioner of agriculture and agricultural development that strongly criticized the management of the LDP. The program, he argued, was too top heavy and did not have enough people on the ground to carry out instructions. He said that more money was needed to recruit qualified professionals and that "the image we now have of Agricultural Development is only virtual and under these conditions cannot be used to predict the future." Throughout 1962, Richard continued to plead for more personnel. In a May 1962 letter to Tôn Thất Trình, a tropical agriculture expert and the general secretary to the minister of rural affairs, Richard argued that "the presidential ordinance n° 287 of August 18, 1958 constitutes the only real participation of the Government in *heveaculture*." Richard urged the RVN government to bring to fruition the project to create a National Office for a Rubber Cooperative that would "grant low-interest loans to cooperatives and to rubber-farmers associations leading to ownership . . . [and] satisfy the demands of modern rubber-growing techniques." Eventually, RVN officials grew tired of Richard's complaints and marginalized him.[72]

The use of *hevea* trees also led to resistance from French plantation managers. The agroville and community development (*phát triển cộng đồng*) programs drew potential labor away from the rubber industry, and in October 1960 the president of the Association of Rubber Planters in Vietnam, Jacques Polton, wrote to the RVN delegate for the Eastern Province at Bình Dương, Tăng Văn Chỉ, to complain about the use of plantation labor for government works. Polton discussed the construction of agrovilles, mentioning the use of factory workers from the Phước Hòa plantation in Phước Thành province to build the Khánh Vân agroville. In response, the RVN vice president urged provincial chiefs to resolve conflicts in favor of the plantation's owners. The RVN government also wanted plantation owners to help their former workers become growers by providing land, trees, supplies, money, and expertise. In July 1962, the minister of rural affairs wrote to provincial heads requesting information regarding the settlement of former workers on their own lands. In an exchange of letters, the planters agreed to let workers settle noncultivated lands in valleys and around streams while Diệm agreed to supplement this offer with public lands and resettlement money. When surveyed, around 9,000 workers expressed interest in this program, with over half coming from Long Khánh province. While SIPH offered seeds, stock, and expert advice, the rubber company claimed that it did not have the financial strength to help new planters.[73]

These claims seem to have been warranted. For example, the 650.5 ha Cầu Khởi plantation in Tây Ninh near the border with Cambodia, which

Stock sheet, Cầu Khởi. Source: author's personal collection.

employed 674 workers, recorded an overall decline in profits: in 1957, profits totaled US$117,904; in 1960, US$303,811; in 1961, US$107,047; in 1962, US$38,976; and in 1963, just US$21,045. Likewise, in the late 1960s, a U.S. military report noted that the average cost of production was 48 đồng per hectare, while the average selling price was 40 đồng. Plantation owners were reluctant to loan plows to would-be planters as they feared that the communists would destroy them. Their most pressing concern, however, was that Vietnamese smallholders would encroach on the land of their own plantations. Owners in other provinces offered a range of rationales, but they invariably said that they were not able to help. It was not completely the large plantation owners' fault that smallholders did not come flooding into the rubber market. A Vietnamese Institute for Rubber Research (IRCV) advising program that was initially paid for by the French Technical and Economic Aid to Vietnam met with interest; yet, as violence increased and smallholders saw few benefits to growing rubber, participation waned. The cost of defense favored large plantations, and the NLF ironically wound up boosting the very system that it claimed to be fighting against.[74]

The LDP also aggravated tensions in questions of land ownership. RVN land laws placed restrictions on foreign ownership and sought to collect taxes from squatters on rubber lands. A particularly sensitive question regarded whether government officials could be granted land to plant rubber in the province where they worked. In 1957, some RVN officials had been involved in scandals over land concessions, and the issue continued through the 1960s. Diệm made it clear that his priorities were to open land for rubber planting and to prevent corruption, especially land speculation that did not result in productive usage. Because of these concerns, all requests for land had to be submitted to the president, who would consider each case carefully to determine whether an official was acting on their own or as a front for others, and whether these officials had sufficient means to develop land or whether they were using resources from their office to develop land. This system did not prevent corruption, as it depended on local officials to verify whether land had been planted in rubber. Nor did it prevent Diệm's own brother, archbishop Ngô Đình Thục, from maintaining his lease on the Lê Văn Viễn plantation, which he had acquired from the government in 1957.[75]

Eventually, the government loosened rules on state officials applying for land grants. In 1966, a report argued in favor of the practice, stating that historically, when laws forbidding colonial officials from getting government land grants were enacted, officials were of a high position relative to the rest of society. By the mid-1960s, however, this was no longer true; officials had become poor because their salaries were fixed while inflation drove up the cost of living. The report made the case that officials should be treated like any other citizen who wanted to own land, and that the state needed to keep these officials on their side to fight communism. The only exception to this rule was that current officeholders should not be allowed to own land within their province.[76]

The LDP struggled in 1962, as corruption became a growing problem and the state did not provide settlers with the means to succeed. In December 1962, the National Assembly asked the minister of rural affairs a series of questions about the progress of the LDP. The first question, which assumed that the resettlement of northern and central refugees was going well and that rubber would soon be turning a profit, prodded the minister to discuss how the project would build rubber-processing factories for these farmers. The minister began by agreeing that the LDP had advanced the ideology of community development by settling more than 250,000 people, and gave a rather optimistic projection of 147,000 tons of rubber to be produced on 122,000 ha in 1971. By 1965, the minister hoped to have a factory with 2,400-ton capacity costing 20,000,000 đồng, with four to five factories built per year after that. The minister also reported on the forest products industry, the cost of pig meat, and other ways to make strategic hamlets economically viable.[77]

The Vietnamese-language press commented on the LDP. Articles in progovernment newspapers with titles such as the "Promise of Vietnamese rubber" offered encouragement and advice for those taking part in resettlement programs. Outside the RVN, however, coverage of the LDP was more criti-cal. One article titled "From the Rubber Plantation to the Highlands LDP" appeared in the February 20, 1962, edition of *Trung Lập* (*Impartial*), a paper published in Phnom Penh for the Vietnamese community in Cambodia. In the piece, Dư Cẩm Long reports on the lives of Vietnamese who lived on plantations in Cambodia and were thinking about resettling in the high-lands. Long begins by citing the saying "It is easy to go to rubber plantations and hard to return; males who go get sick from malaria, females who go re-turn with kids." He noted that many Vietnamese were returning to the RVN in search of "freedom" but warned against believing the promises of the LDP. He noted that not much help was given by the organization and that while plantations were not "heaven," in Cambodia people could live their lives in peace.[78]

The situation did not improve in 1963. An anonymous report written in French cited numerous problems with the system of land concessions, with almost all participants planting the rubber poorly or simply selling the trees for wood. The report pointed to corruption among officials, detailing how bribes were often needed for land purchases. Because plantations continued to produce most of the rubber in Việt Nam, the RVN government offered little support for smallholders and family holders and labor conditions remained an important issue. While the LDP sought to create smallholders, provide em-ployment for recent immigrants, and settle uplands with Vietnamese, in the end it accomplished none of these goals. Diệm's goal of enrolling the rubber industry in the RVN went unfulfilled; he was unable to implement his re-forms, and after his assassination, the LDP was terminated.[79]

The Natural and the Synthetic

Both material and symbolic factors influence the decisions of producers and consumers about what is "natural" and what is "synthetic." Proponents of the "natural" rubber (NR) industry emphasized the higher quality of NR while competing with rubbers labeled "synthetic." In 1970, rubber expert René Fabre noted that despite the competition from synthetic rubber (SR) "by its intrin-sic qualities, the natural [NR] had always kept in the eyes of many users . . . a 'margin of preference' " that could even compensate for higher prices. Fabre also explicitly tied world prices for rubber to "Third World" stability by noting that NR workers, who produced around 2 tons of rubber per worker per year, were more affected by low prices than SR workers, who pro-duced around 200 tons.[80]

SR was first developed in Germany, and its production expanded tremendously during World War II as the United States worked desperately to replace its access to NR, which had been blocked by Japanese occupation of Southeast Asia. By the late 1950s, S-, Butyl-, and N-type SRs constituted two-thirds of the rubber market in the United States and nearly half of the four million tons of rubber consumed worldwide. Moreover, new forms of SR, including polyisoprene and polybutadiene, presented direct challenges to the characteristics of NR; they duplicated the basic hydrocarbon molecule of NR, though the role of nonrubber constituents was still not fully understood. The International Cooperation Administration released guidelines meant to assist United States Operations Mission officials in their decisions about whether to assist rubber-production programs in countries such as Việt Nam. These guidelines projected global demand for rubber at approximately 6 million tons in 1980, with up to 4.5 million tons of NR. Some of the demand for NR would remain inelastic if it retained the unique characteristics that were desired by industry. However, the price of NR could vary, depending on demand, so it became critical that plantations lower costs per kilogram to maintain reasonable returns on investment, living standards for workers, and tax revenues for governments.[81]

To stay competitive with SR, NR researchers became dedicated to making gains in productivity. At the heart of this research in Việt Nam was the IRCV, the successor of the IRCI. The IRCV is absent from most histories of the RVN, but the postindependence fate of the institute reveals the ambiguities and tensions involved in the transition from colonial to nationalist rule. Many prominent guests, including Diệm and Sheldon Tsu, the agricultural economics adviser of the American fund, visited the IRCV in the late 1950s because of its function in the creation of scientific knowledge about NR. During the colonial era, sciences such as medicine and agronomy had become linked to both the civilizing mission and political and economic domination. The importance of scientific knowledge for the colonial project meant that the control of any type of science would be highly contested during the process of decolonization.[82]

From the 1890s to the 1930s, the availability of land at incredibly low prices throughout Southeast Asia encouraged plantations to expand into forested lands, and research was centered on extension techniques. Furthermore, during the colonial era, the division of research labor between Paris and Lai Khê meant that researchers in France had been responsible for "fundamental questions," while those in Indochina had been responsible for "applied questions." Following World War II, scientists in Việt Nam and other newly independent Southeast Asian nations had to carry out their own fundamental research and create the links between theory and experiment in increasing rubber production. By then, decades of rubber expansion and the creation of nationalist governments had begun to close the latex frontier, and without colonial control, free land and cheap labor were harder to come by. Researchers were forced to shift their focus to the question of getting as much latex as possible from each hectare of *hevea*. The lack of personnel, due in part to the threat of assassination, challenged their efforts even further.[83]

National independence reshaped the institute in other ways. In August 1956, its name was changed from IRCI to IRCV, with the last letter of the acronym changing from "Indochina" to "Vietnam." The Institute for Rubber Research in Cambodia (IRCC) was formed the following year. A new section in the IRCV was created in 1957 to improve smallholder yields and increase Vietnamese membership. Finally, the IRCV started to hire Vietnamese personnel in technical and research positions, and in 1958, a Vietnamese chemical engineer, Vũ Đình Độ, joined Nguyễn Hữu Chất, head of the microbiology division at the IRCV (after 1975 Chất became vice director of the RRIV). The institute also hosted its first Vietnamese intern, Nguyễn Phó Lu, a graduate of the Parisian Technical Institute of Agricultural Practice.[84]

In most respects, however, the IRCV continued to perform the same functions of the IRCI, but while the IRCV remained in the hands of the French well into the 1960s, its role as knowledge producer did not remain static. Starting in 1958, the Annual Report was divided into two sections, "fundamental research" and "research applied to growing and exploitation." The first section included findings related to anatomy and physiology, such as the relationship between bark and latex production under different tapping techniques. The overall goal of this research was to find possible limiting factors for latex production, whether directly through the formation of latex, or indirectly through damage to the bark or too much or too little tree growth. This combination of fundamental and practical research in Việt Nam and Cambodia attempted to increase the intensity of output per hectare of *hevea* through high levels of inputs (and costs), which was the dominant model of scientific agriculture throughout the world at the time. Much of the scientific research of the IRCV remained targeted at the needs of the large plantations, including the "physiological diagnostic" developed in the 1950s. These diagnostics studied the chemical composition of the *hevea* leaves and latex to help optimize the use of fertilizers and to understand the effect of soils on production. The studies began in 1954, and by 1959, close to 37,000 hectares of red and gray earth had been examined. In 1960, seventeen staff members carried out

2,628 diagnostics, of which 902 were for planters without their own laboratories, 62 were for Lai Khê, and 1,664 were for research purposes. Overall, diagnostics were carried out on about 40,000 hectares, largely consisting of plantation land, and they contributed to an increase in latex production from 1,500 to 3,000 kilograms per hectare per year on some plantations. They became such an important part of the IRCV's work that they necessitated a reorganization of the workflow in the laboratory.[85]

Determining how to increase the quality and quantity of latex production (measured as dry rubber content) through a "total treatment" was a complex task because it was almost impossible to control for all the variables. Researchers had to factor in the time of year that the fertilizer was applied, the weather, the effects of stimulation through the injection of micronutrients such as iron, copper, and boron, density of planting, tapping methods, and the selection of individuals. Researchers also experimented with the plant growth hormones 2,4-D and 2,4,5-T, the same compounds that formed the herbicide commonly known as Agent Orange. A surprising degree of individual responsiveness to treatment remained, even among cloned trees. PR107 clones, considered one of the highest producers, with an average of two tons of latex per hectare per year, could release anywhere from one to three tons of latex, depending on their stock and their individual placement. In addition, the IRCV considered the availability of macronutrients, micronutrients, carbohydrates, and other limiting factors in the metabolism of latex. Finally, researchers studied the relationship between microbial infection of *hevea* and the amount of latex produced. Initial research suggested that higher levels of microorganisms were correlated with greater levels of latex, though this relationship was not necessarily causal. Scientists suggested that "the factors tending to increase the productivity have for a secondary effect to produce a latex more susceptible than the others to be infected." It appeared that the trees bred to be latex machines were also more likely to be sick. D. H. Taysum at the Rubber Research Institute of Malaya even proposed using *hevea* as a model organism to understand how plant immunity worked.[86]

French researchers sought to make themselves relevant in a newly independent Việt Nam by holding conferences on the rubber industry and presenting information on growing and process techniques. In 1960, the head of the IRCV's Phytopathology and Nursery Division gave a talk on replanting. He spoke about new varieties of *hevea*, both grafted and seed grown, and presented research on the production and consumption of rubber. IRCV researchers continued to study plant diseases and served as technical advisers. These French experts had to both prove their worth to the Vietnamese government and

B — ORGANISATION DU LABORATOIRE D. P.

1° *Schéma.*

Le plan des laboratoires actuellement en service à Lai-Khê pour l'exécution des D.P. a été reproduit dans les figures 1 et 2. L'établissement de ce plan a été guidé par les impératifs suivants :

— Eviter au maximum les déplacements inutiles de personnel.

FIGURE 1
Laboratoire D.P. — Circuit feuilles

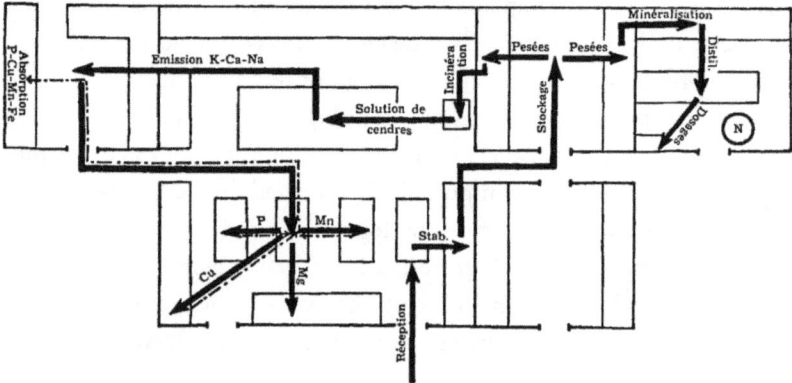

FIGURE 2
Laboratoire D.P. — Circuit latex

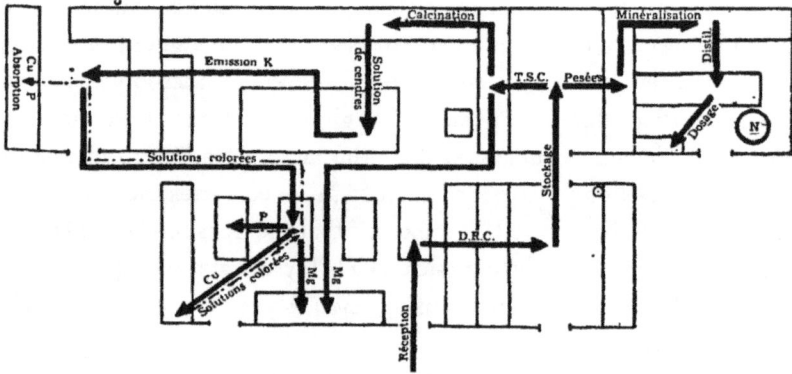

IRCV work flow, 1958. This diagram shows the growing importance of exact diagnostic techniques aimed at getting the most out of each rubber tree. Source: IRCV, *Rapport annuel de 1958.*

claim relevance to scientists in France. Eventually, some of the intellectual production concerned with the rubber industry was taken over by Vietnamese in both the north and the south. In 1961, Nguyễn Phó Lu wrote a trainee report on the technical and social organization of several plantations, and in 1963 Lâm Văn Thương completed a student thesis on the PR107 high-yield clone. During the 1970s, students at the National Institute of Administration undertook studies of the organization of the industry at both the provincial and national levels.[87]

Many experts also called for the industrial manufacture of rubber products, and in both the RVN and Cambodia, attempts were made to establish businesses that processed rubber into sandals, bicycle tires, and automobile tires. Speaking at the Association of Vietnamese Engineers and Technicians, the French head of the newly formed Rubber Consumption Division of the IRCV, argued in favor of this type of production. Việt Nam, he pointed out, exported latex rather than rubber-manufactured goods, while importing 5,400 tons out of the 6,600 tons of rubber goods it consumed. Việt Nam was importing 1,500 tons of bicycle tires at a cost of 70 to 80 million đồng, but he believed that, with proper investment, companies in Việt Nam could be manufacturing those tires and creating rubber boots for Asian rice farmers. He noted that the Vietnamese domestic market had ample room to grow; it consumed only 600 grams of rubber per person per year, whereas the world average was 1 kilogram, and up to 10 kilograms in the United States by 1959.[88]

Arguing for increased industrial production in Việt Nam challenged the long-standing model of colonial economy that placed the creation of raw materials in Việt Nam and industrial manufacturing in France. The rubber industry also divided intellectual labor into two locations; knowledge creation about the production and treatment of latex took place in Southeast Asia, while the research on industrial uses of rubber was performed in Europe and America. The head of the IRCV reiterated this point in his talk to the Vietnamese engineers and technicians. The planters in Indochina and the IRCI had taken the lead in researching tree yields and treatment methods, while French researchers and the Paris-based IFC had worked on different uses for rubber in the manufacturing process since 1936. After 1954, this division was no longer sustainable and the IRCV took on the mission of researching both improved latex production techniques and manufacturing uses for rubber. The laboratory and the field were brought together as the IRCV was generating "knowledge that could permit a better understanding of practical problems or that serve to create hypotheses that Agricultural Experimentation could then verify."[89]

Plantation workers, owners, government officials, and activists were not the only ones who had to negotiate the tricky waters of development in an independent South Vietnam; the IRCV faced attacks from the NLF as well. In 1963, the vice president of the IRCV reviewed thefts and sabotage that had occurred over the previous three years. These acts cut into the profits that the IRCV earned from selling rubber and hampered its ability to carry out tapping and growing experiments. In a desperate ploy, he sent a letter to the general population that appealed to the interests of science, arguing that shutting Lai Khê down would harm everyone.[90]

THE POST-1954 fate of the rubber industry raises again the question of why Việt Nam was unique in terms of the degree of de facto exclusion of smallholders from rubber production. Part of the answer lies in technological momentum, or the way in which human decisions become layered in assemblages, both material and symbolic, which become difficult to alter. In other words, once "solutions" have been found to "problems," as defined by understandings of *hevea's* biology and symbolism surrounding rubber, they become settled. Another part of the answer comes by placing rubber in the context of community development projects. Rubber plantations occupied an ambiguous place ecologically in the foothills and rhetorically in community development. Plantation space, high production levels, and control by transnational capital marked them as emblems of high modernism. However, plantation owners had been emphasizing some of the themes of community development since the 1930s. These owners highlighted the plantations' building of hospitals, schools, churches, and pagodas. While management and labor relations on plantations prior to 1954 ran contrary to the ideas of community developers, these relations were not substantially different from Diệm's approach to the countryside. The extreme violence brought to bear on the Vietnamese society and environment by the American war machine ensured that little changed even after the end of Diệm's rule.

CHAPTER SEVEN

Militarizing Rubber

It became necessary to destroy the town to save it.
—U.S. major to Peter Arnett, 1968

The distinction between people and their environment is mainly academic. They are both integrant parts of the same nature. They are closely related and react continually on one another.
—Phan Quang Đán, 1975

By the end of the Vietnam War, concepts of governance based on management of the relationship between people and their environment had entered the Vietnamese vocabulary. In January 1975, Phan Quang Đán, a South Vietnamese politician who had trained at Harvard's School of Public Health, self-published a book titled *The Republic of Vietnam's Environment and People*. In the epigraph to the book, cited above, Đán calls into question the artificial division between humans and their environments. In a more political vein, he wrote: "The ruthless but efficient Nature's equilibrium which has preserved life on our planet should be gradually replaced by a rational and well planned harmony between man and his environment." While this quote seems to have socialist undertones, Đán was not a communist. His book reflects, instead, the awareness of the embeddedness of humans in nature and the dominance of developmentalist ideology. This chapter examines the interactions of the Vietnam War and plantations and returns to some of the long-term environmental and social consequences of rubber for Việt Nam.[1]

Warfare had an important, if sometimes difficult to evaluate, impact on the physical and biological environments of Việt Nam. In terms of rubber, political and social violence kept plantations from expanding and encouraged the continued production of rubber on large plantations. Plantations, in turn, influenced the fighting of the Vietnam War, or the American War as many Vietnamese called it. Plantations were liminal spaces that brought together nature and culture, north and south, peace and violence, and rubber production remained an important ideological and material battleground from 1963 to 1975.

As the U.S. military increased its presence, it sought complete control of the environment via herbicides and weather modification programs. Because plantations were economically valuable to the French and south Vietnamese, the U.S. military was limited in its ability to spray herbicides, drop bombs, and command plantation space. Plantations offered additional advantages to insurgents because they existed outside the control of the RVN, offered cover from airplanes and access to Sài Gòn, and were positioned on international borders.[2] Plantation space was, from the U.S. and the ARVN perspective, at the heart of the Corps Tactical Zone Three (III Corps), one of the four military regions. For the communists, the former War Zones C and D and the Inter-zone for the southeastern region (*Phân liên khu miền Đông*) became War Zone VII. The ramp-up in violence caused by the U.S. military's full entry into the war in 1965, which included massive herbicide sprayings, bombing runs, and procurement of plantation space, eventually made rubber production in southern Việt Nam impossible.

Burning Rubber

Throughout the twentieth century, Southeast Asia's forests and unruly jungles were home to antistate and nonstate peoples as well as states in the making. Plantations offered a simulacrum of these unruly places, and the edges where plantations met with jungle served as a setting for political and social violence, or "militant tropicality." Planters, RVN officials, NLF guerillas, rubber experts, and workers struggled to control plantation space, and these overlapping interests largely protected plantations from complete destruction. Certain forms of mobility were built into the tropical geography of plantations, as rubber tappers could travel around plantations but not off them. Workers poured latex into and out of buckets, and rubber moved out of southern Việt Nam by road, by river, and by air. Rubber trees allowed for both limited horizontal and vertical visibility, providing protection from the air and a terrain that allowed greater mobility by foot. The necessity for circulation and mobility of plantations made them vulnerable to sabotage; roads could be cut and boats on rivers could be ambushed. A new militant tropicality emerged as plantations were destroyed, communist militancy expanded, and the United States lay waste to the Vietnamese environment.[3]

In the early 1960s, the NLF sought to improve the plantations' usefulness for guerilla warfare. Its activities focused on cutting down young trees to prevent plantation expansion while leaving strategically useful older trees standing. As the DRV negotiated an uncertain situation in the south in the early 1960s, its publications about plantations painted a more nuanced picture than they had during the First Indochina War. Nguyễn Phong's *French Capitalists and the Problem of Rubber in South Vietnam*, published in the DRV in late 1963, adopted a greater focus on the economic aspects of the industry. The end of colonialism meant that the book lacked a clear-cut enemy of the Vietnamese people; the analysis of the rubber industry was more detailed and complex, and the ideological message was less explicit. The author pointed out that French plantations continued to dominate production, with the French generating almost 300,000 metric tons of rubber between 1955 and 1959, while Vietnamese-owned plantations produced a little over 40,000 metric tons in the same period. French plantations were also the main recipients of aid that was meant to encourage plantation growth and smallholder ownership.[4]

The book had been written in the spring of 1963, and, with the French politically and militarily neutralized, the "American-Diệm regime" posed a greater threat to the DRV's reunification project. The French had become a potential ally, as they could serve as a counterweight to the United States in international affairs. The DRV tried to drive a wedge between the United States and the French, and Phong tried to ease fears of nationalization under communism by suggesting that the French might play a role in the Vietnamese economy. The United States had not yet sent its military into the south, and Phong, given the possibility of intervention, tried not to strike too strident a tone. Finally, Phong chose not to address the subject of the Plantation Workers' Union, founded in 1955, that competed for the loyalty of rubber workers.[5]

Diệm's assassination in November 1963 and the subsequent founding of the Second Republic of Vietnam led southern society into tense and uncertain times. After Diệm's assassination, a twelve-man Military Revolutionary Council took control of the government, eventually forming the Second Republic of Vietnam. The Council initially proved inept at facing the threat posed by the NLF, and communist organizing expanded. The following spring, General Nguyễn Khánh and others who were disaffected with the Council launched a coup and successfully took control of the government. As part of the effort to increase security in III Corps, the counterpart to the DRV's and NLF's War Zone D, the Council considered the question of whether to close certain plantations or to nationalize the industry. Workers mostly fought to keep plantations open, and the RVN leaders decided to keep plantations open and to leave them in the hands of the French.

The lack of government control made plantations a security concern for the state, and incidents of sabotage on remote plantations increased during the months before and after Diệm's assassination. One option for military and government officials was to simply close plantations; yet, when the owner of the Trần Văn Phòng plantation petitioned to close a section of his plantation after the disappearance of several workers, local officials refused. After further negotiations, local military and labor officials and the owner agreed that the plantation could close temporarily but would open earlier than usual the following year, and that workers would receive some severance pay and daily rations of rice. Because of plantations' importance for security, RVN officials generally sought to help them through mechanisms such as tax breaks.[6]

RVN officials did express concern about some aspects of the rubber industry, including the "tax" paid by French management to the NLF, either as protection money or as ransoms for kidnapped Frenchmen. In his survey of the history of rubber, René Fabre counted 100 French planters killed out of a total of 250 present in Việt Nam during the First Indochina War. Rather than outright assassination, the NLF often took French plantation managers hostage. A former planter named Michel Michon recounted how he and other managers were often captured and exchanged for cash, while recalcitrant Vietnamese plantation workers were often simply killed. After changing its strategy from sabotage to extraction, the NLF could provide a kind of protection for the plantations. Michon recalls that in the late 1950s, an ex–Việt Minh cell assumed responsibility for his plantation by searching for and punishing noncommunist thieves who had supposedly stolen rubber coagulum. The NLF also redirected the money and supplies that were sent to the plantations in exchange for rubber. Medicine was particularly valuable, and South Vietnamese authorities sought to stem its flow from plantations to the NLF.[7]

In the spring of 1964, Jean Chateau, the director of the Quản Lợi plantation, was kidnapped. Chateau was held for three months before the SPTR gave the NLF three million đồng for his release. Yet, the RVN police had a difficult time documenting the extent of French "support" for the NLF. French payments often took place outside the republic's borders, and the police suspected that the French were funneling payments to the NLF through their representative in France. Even payments made within RVN space were difficult to track. In the case of Chateau, Trần Gia Huấn, a former Việt Minh nurse turned RVN informant, gave information that suggested an inside job and cast suspicion on Chateau's chauffeur. Huấn, who had been living in Quản Lợi for four years, also offered to give information about French anti-RVN activities. RVN police, however, were suspicious of Huấn's motives. A northerner who had moved south, Huấn was a former Cần Lao party member with close ties to the NLF, and a national police director concluded in a May 1964 letter to the prime minister, General Nguyễn Khánh, that Huấn was trying to misdirect government intelligence. While RVN officials remained suspicious of the French, the police director argued that Huấn was aiming to get his enemies in trouble.[8]

In early 1964, rumors circulated that the RVN government was considering nationalizing the rubber industry. Nationalizing rubber may have provided an opportunity for improving workers' lives, as it had with the oil
industry in Mexico. RVN policy also encouraged the sale of existing plantations to private Vietnamese owners or to the state. Yet, given the instability of
the RVN government at the time, full-scale nationalization was unlikely. The
interior minister urged General Nguyễn Khánh, who had seized control of
the Military Revolutionary Council at the end of January, to ease French fears
by publicly announcing that plantations would not be nationalized. The interior minister was less concerned with French interests than he was with the
idea of plantation management inciting its workers to protest to blunt the call
for nationalization. The minister of labor, Đàm Sỹ Hiến, also weighed in about
the possible connection between rumors of nationalization and worker unrest. In a July letter, Hiến summarized a report that had asked owners, labor
organizations, and local labor officials about possible links and found none.
Instead, the report argued that worker unrest was due to lack of safety, the
intransigence of owners during negotiations, and NLF influence. To address
worker demands, Hiến suggested increasing security in the region, providing
rice for workers, and reducing taxes on plantations. He also suggested putting
the kibosh on rumors of nationalization.[9]

Rubber workers' activity increased, and in the fall of 1964 plantation workers launched a wave of strikes. At 10 A.M. on September 30, drivers on
Michelin's Dầu Tiếng plantation refused to transport latex until workers' demands were met. The plantation director, R. Desplanques, and the local Labor
Ministry representative attempted to flee by airplane, but the workers allowed
only Desplanques to leave. Earlier that day, 300 workers on the Bình Ba and
Xuân Sơn plantations of the Gallia company in Phước Tuy province had presented a list of demands, including the end to military conscription and ARVN
firing drills in the villages, increases in salary and rice provisions, and a reduction in the number of trees to be tapped. The strikers briefly held the assistant to the director of Gallia hostage but released him later that morning. On
October 1, an airplane took the labor official who was being held on Dầu
Tiếng, along with a representative for the workers, to the provincial capital
to negotiate. Meanwhile, workers on the Lê Văn Viễn plantation submitted
demands for 50 percent higher salary and the 933 grams of rice per day that
was promised in their contracts. Finally, workers on the three remaining
Gallia plantations also struck and called for the involvement of the CVTC.[10]

French planters and many RVN officials argued that the communists were
instigating these actions. Desplanques, for example, reportedly claimed to have

evidence that an NLF company stationed in villages 2 and 4 of Dầu Tiếng had urged workers to strike. Đàm Sỹ Hiến, the minister of labor, following reports from local labor officials, agreed with Desplanques. Hiến also stated that the "VC" (Việt Cộng, a pejorative name for the Vietnamese communists) had forced workers at Quản Lợi, Cầu Khởi, and Hội Cao su Tây Ninh to strike, or at least prevented them from going to work, and that the RVN police knew about NLF involvement in strikes that had taken place on September 22 and 23 in Long Khánh and Biên Hòa. Hiến explained that the NLF knew that strikes were inevitable because the demands of workers would never be met by owners. Hiến argued that losing the plantations would be bad for RVN society, economy, and politics and recommended trying to work with labor organizations to avoid worker actions. Workers, however, stated that they had grievances that were unrelated to the NLF. As historian Ed Wehrle notes, the recently renamed CVT called for a general strike on September 21 to protest then head of state Nguyễn Khánh's August ban on strikes and meetings. With Trần Quốc Bửu in jail, the leaders of the Plantation Workers' Federation, including its general secretary Phạm Văn Vỹ, sent a letter on November 11 to General Khánh and other RVN state leaders explaining that workers had their own grievances and were not motivated solely by communists.[11]

In the spring of 1965, Khánh himself was ousted by a U.S.-backed coup, and air marshall Nguyễn Cao Kỳ came to power, eventually taking the position of prime minister. Building on the Vietnamese nationalist sentiment that had been expressed by many Saigonese elites since independence, the Kỳ government threatened to break diplomatic ties with France. Vietnamese leaders did not want to favor the "ex-colonialists," an attitude that was particularly worrisome for French-owned plantations. Many in the RVN government distrusted French planters, both because of continuing colonial attitudes and because of their perhaps involuntary cooperation with the NLF. A U.S. report at the time noted, and dismissed, "the popular belief that the planters are pro-Gaullist and therefore in some respects anti-American." The report surmised that the French wanted to keep plantations going for profit and because of "traditional and professional reasons." In the end, French interest lay with the RVN since "the French realize that they have some chance of retaining their interests under a non-Communist government in Vietnam but that they would be quickly supplanted without compensation by a Communist regime as occurred in North Vietnam after 1954," which finally occurred in 1975.[12]

After coming to power, Prime Minister Kỳ ruled with a heavy hand, and in May 1965, as part of his move to suspend civil liberties, he required all union meetings to seek official approval at least two days in advance. Bửu and the CVT were successful at courting Kỳ, however, and he and the RVN leadership were sympathetic to the rubber workers. On behalf of the workers, the CVT requested raises from 10.5 to 12 percent to meet the increased cost of living, which would push salaries to over 100 đồng per day. Plantation management responded that they were open to such increases, but because of the economic situation, they could agree to only a 2 percent raise. The owners then used the labor demands to bargain with the RVN for lower taxes on rubber. RVN officials tried to work out tax schemes that would benefit Vietnamese smallholders, but their attempts were fruitless. Workers, through the CVT, also requested the rice rations that were agreed to in earlier negotiations, which had been withheld on the pretense that rice surpluses were being funneled to the NLF. The RVN's Interministerial Committee, which oversaw plantation affairs, proved more resistant on this point, and offered salary increases instead, asking the workers to sacrifice for the nation. As negotiations dragged on, the Federation of Plantation Workers, which was affiliated with the CVT, issued a statement pointing to the suffering of plantation workers due to cost-of-living increases. With worker salaries stuck between thirty-eight and fifty-seven đồng, the Federation of Plantation Workers announced its intentions to call a one-day strike on August 1 on all rubber plantations to call attention to the plight of its workers.[13]

Rebounding NLF

The landing of U.S. Marines at Đà Nẵng in 1965 marked a return to condemnations of colonialism and neo-imperialism as the DRV attempted to equate American presence in South Vietnam with French colonialism. One of the best-known publications from that year was Trần Tử Bình's memoir *The Red Earth*. Although scholars have incorporated Binh's description of his life on Michelin's Phú Riềng plantation during the late 1920s into their historical narratives, they have paid less attention to how the book reflected the moment in which it was published. By 1965, Bình had risen to the rank of general in the DRV military. Like other memoirs, such as Nguyễn Thị Định's *No Other Road to Take*, *The Red Earth* was meant to bolster the spirits of the southern NLF and remind North Vietnamese about why they were being called on to sacrifice their sons and daughters to the war effort.[14]

Diệp Liên Anh's 1965 book *White Blood, Red Blood: The Damned Lives of Rubber Tappers* similarly portrays plantation life as hellish for its workers. Although it is unclear whether Anh had firsthand experience of plantation life, he demonstrated detailed knowledge of the industry and the red earth region. Anh conducted a nuanced analysis of the difficulties of Vietnamese capitalists in their attempts to run small plantations, and cited several *ca dao*, or folk songs, of rubber workers. He styled his book as a piece of reportage, a genre made famous in the 1930s when journalists such as Vũ Trọng Phụng exposed the abuses of the colonial system. *White Blood, Red Blood* includes a series of vignettes that illustrate different abusive aspects of the plantation system that were in place before 1945. Anh's book is of interest because it was published in Sài Gòn, even though Anh's rhetoric resembles DRV-produced works on plantations. He argues, for instance, that "conditions under the French colonial regime resemble those under the puppet governments of the Imperialists, [and] plantations are hellish worlds." Anh praises the revolution, but he focuses more on anti-French than anti-Diệm rhetoric, concluding with a promise of an exposé of plantation conditions under Diệm (this work seems to have never been written).[15]

Both Bình's and Anh's works speak to Vietnamese perceptions of the link between nature and nation. Bình recounts how workers perceived plantation environments and Northern migrants such as Bình often considered the southern forests as strange and terrifying environments. Accustomed to the flat plains of the Red River Delta, where most wild animals had long been expelled, Bình found the dark forests and the harsh cries of the primates disconcerting. Yet Bình expresses ambivalence toward plantations as he attempts to incorporate those spaces into a national landscape, arguing that despite the loneliness of the "forests and mountains" of southeastern Việt Nam, they "were still my country."[16]

Anh's book describes plantations as beautiful but deceptive places. At one point he exclaims: "The rubber tree is very strange!" Anh noted the way that the existing forests shielded the young *hevea* from the wind, even as the old trees were cut down to make room for plantations. Anh argues that from the outside, the plantations, with their French-owned villas and quaint-looking villages, looked very beautiful. Years of labor during the colonial period had turned forests, with their poisoned springs, malaria, and wild animals, into well-tended gardens. But this beauty, Anh argues, was only one of form, and was meant to please only the owners and their henchmen. Anh explains that the phrase "Rubber grows well in this place, Every tree is fertilized with the body of a worker" indicated that the high death rates of workers paralleled the expansion of rubber plantations. Anh covers the standard stories of manager abuse, including rape, lack of medical treatment, and poor living conditions.[17]

The plantations retained their strategic role in fighting during the 1960s. As Anh wrote of the First Indochina War, the plantations were in "a key region for military posts that included . . . many strategic transport routes and turned into strategic locations for the military and the economy." The plantations offered many strategic advantages for the Việt Minh's successors, first the NLF and later the Provisional Revolutionary Government (PRG). Moreover, the area around the Michelin-owned Mimot plantation near the border between Cambodia and Việt Nam at the end of the Hồ Chí Minh trail became one of the locations of the Central Office of South Vietnam (COSVN). This location enabled the COSVN to coordinate actions of the People's Army of Vietnam (PAVN) in the north and the People's Liberation Armed Forces (PLAF) in the south. During the devastating American B-52 bombing raids of the 1960s, plantations provided convenient cover for the NLF, as bombers were initially discouraged from targeting the estates. The proximity with Cambodia provided additional advantages, as the Cambodian government under Sihanouk turned a blind eye to the presence of DRV troops.[18]

U.S. military leaders were aware of the advantages that rubber plantations offered in terms of landscapes, infrastructure, supplies, and finances, and that rubber trees were useful for the NLF in ways that tea, coffee, and cinnamon were not. As one U.S. report on rubber, tea, coffee, and cinnamon plantations put it: "The nature of the terrain and foliage of rubber plantations is well suited to VC military operations. The tree canopies deny Allied aerial observation of VC activities. Under the canopies, rolling terrain and open fields of observation and fire provide ideal conditions for ambushes along lines of communications, or for prepared defensive positions. Secondary routes interlacing the area simplify control when withdrawal from contact becomes desirable or necessary."[19]

PLANTATION INFRASTRUCTURE AND MATERIAL proved useful as well. Medical supplies, heavy machinery, communication equipment, trucks, and other material associated with making rubber were all available to the NLF. It even made use of plantation-owned airplanes, and on "at least one occasion . . . may have conducted an aerial reconnaissance of an area bombed by B-52's." Food from plantations was perhaps one of the most valuable resources for the NLF. Workers and cadre grew crops such as manioc, sweet potatoes, vegetables, and fruit in their gardens, outside strategic hamlets, along the perimeters of plantations, and on the sides of roads. An NLF report for activities on the Bến Củi plantation lists the following numbers: "Manioc, 19,350 plants, Sweet potatoes, 40,212 meters. The quantity of other crops such as Tu [a kind of

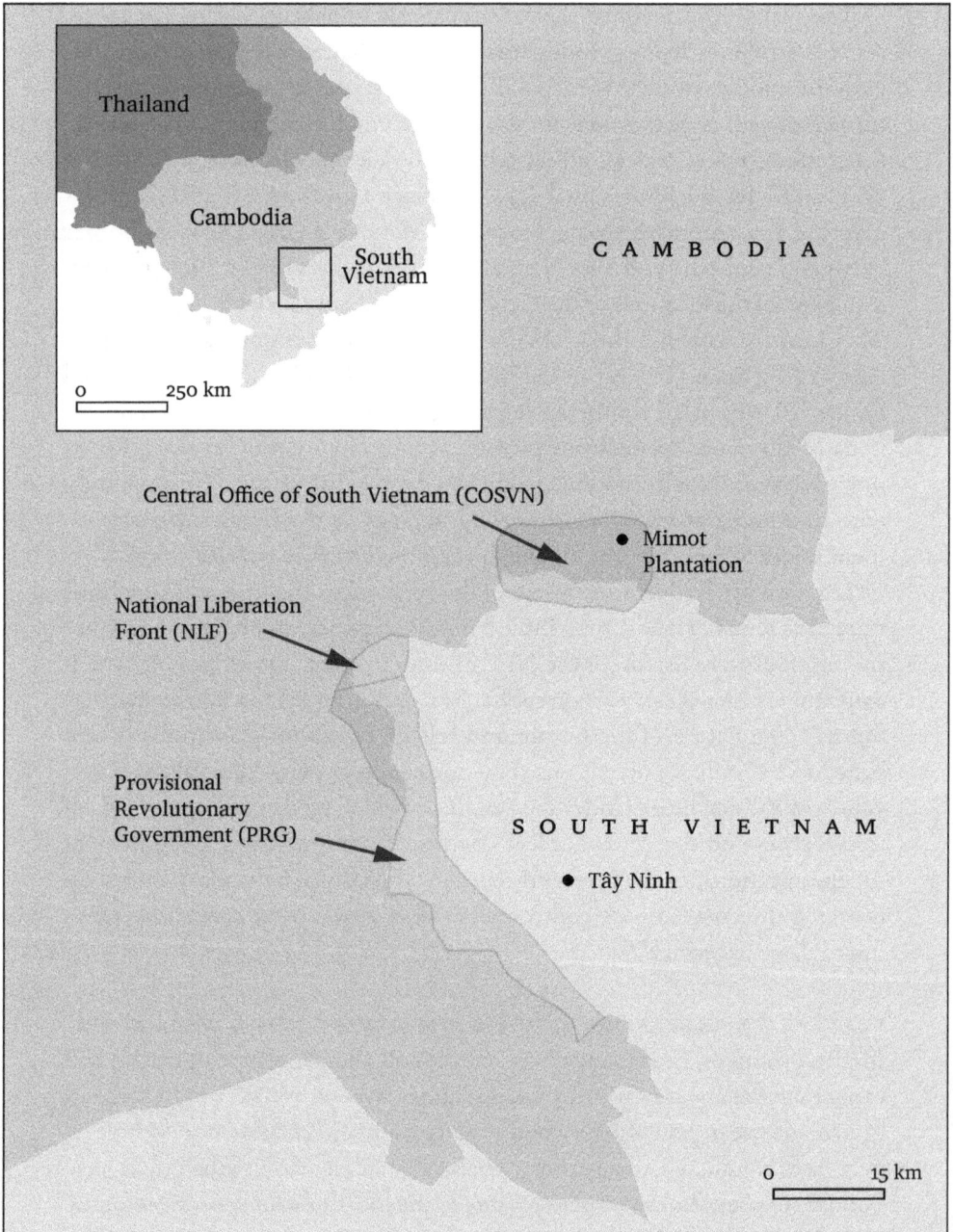

Thailand

Cambodia

South
Vietnam

0   250 km

C A M B O D I A

Central Office of South Vietnam (COSVN)

● Mimot
Plantation

National Liberation
Front (NLF)

Provisional
Revolutionary
Government (PRG)

S O U T H   V I E T N A M

● Tây Ninh

0   15 km

COSVN. This map shows the approximate region where the NLF and the PRG operated.
It also shows the region in which the COSVN moved around. Note the proximity to
rubber plantations. Source: Trương, Chanoff, and Doan, *A Vietcong Memoir*, 162. A similar
version of this map appears in Aso, "De plantations coloniales à la production socialiste,"
and I thank Frédéric Fortunel and Christophe Gironde for permission to reprint that map.

potato], Mon [a kind of potato], Cu Chuoi Tay [banana root] etc . . . is unknown. 79 percent of the families produced vegetables. Generally speaking, every house grows a fruit tree." In a Vietnamese version of a victory garden, local cadre even named a furrow of sweet potatoes: "Teenagers' anti-American sweet potatoes furrow," referring to the NLF units composed of the teenage children of plantation workers, though apparently "this was not popularized among the workers." From 1965 to 1966, the NLF of Bến Củi also produced 150 bushels of rice, and during the first six months of 1966 the unit was self-sufficient in food. Food of the workers also found its way into NLF hands, and the requisitioning of rice, either voluntary or involuntary, from plantation workers became such a problem that RVN authorities allowed the distribution of only 400–600 grams of rice to workers, with the rest paid in cash. On the Bình Sơn plantation in 1969, local inhabitants needed permission from the hamlet chief to buy food, with the allowed amount based on the number of family members and estimates of one week's worth of consumption.[20]

According to the U.S. military, plantations provided tremendous financial benefits for the NLF. One military report estimated that "taxes on plantations probably provide the NLF with their second most important source of revenue in SVN [South Vietnam], the first being the agricultural tax." The report estimated that "in 1964 in Bình Dương Province the NLF collected 20,531,612 Vietnamese piasters [đồng]; taxes on rubber plantations accounted for 14,699,139 Vietnamese piasters." Individual plantations could provide a windfall, as "in 1963, Hang Gon Plantation . . . in Long Khanh Province, paid 2,500,000 Vietnamese piasters in taxes to the NLF, and in 1964 and 1965 the plantation paid a total of 7,200,000 Vietnamese piasters." At the high end, the military report pessimistically estimated that "the total estimated tax receipts from rubber plantations in 1965 . . . [were] 119,584,000 Vietnamese piasters," including 85,984,000 from land/rubber tax and 33,600,000 from taxes on individual workers. If plantations were unable to pay their taxes, NLF members would tap the trees themselves and sell the latex back to the owners, which happened on the Vên Vên plantation in 1964. Some documents also suggest that different rates were applied to owners of different ethnic origin (i.e., French, Vietnamese, and Chinese). While most of the tax burden fell on the owners, individual workers also contributed. On the Bến Củi plantation, "each month in 1965, the workers contributed an amount from 1000 to 2000 [Vietnamese] piasters. This year [1966], as of June, the total amount of this money was 13000 [Vietnamese] piasters." On the Lộc Ninh plantation, the NLF asked for 50–100 đồng, the same as union dues. For much of 1970, however, the NLF was unable to collect taxes and so it asked the people to buy food, sugar, milk, and medicine instead. Workers on the An Lộc plantation complained that the NLF even "forced the people to sell them their dogs," making nightly visits to the hamlet easier.[21]

The noncommunist labor unions presented the greatest competitor for organizing workers. Regarding unions, the NLF adopted two strategies: infiltration of existing unions and the formation of new unions. The U.S. military report noted that

> whether attempting to infiltrate a legitimate union or attempting to establish a union, some of the same tactics are used. Workers are warned, for example, not to pay dues to the Vietnamese Confederation of Workers (CVT).... They are told it is controlled by Americans and does nothing to improve the position of workers. In addition, existing union leaders may be eliminated. In July 1965, on Cau Khoi Plantation ... in Tay Ninh Province, a local union leader was assassinated. If by these various techniques the VC can gain control of a union, the benefits are obvious: union members can be propagandized, recruits obtained, dues collected, and pressures placed on management.[22]

The early 1960s presented multiple challenges for the NLF, including legal unions, the efforts of Diệm, and a general lack of experience. An NLF report for Bến Củi reported that "at first, the situation was difficult because our cadre's background was still weak. Our cadre could not keep close touch with the people to satisfy their requirements and aspirations, thus, the result is not worth mentioning." This report continued that "in 1965, during the days when our area was not liberated, the situation of our unit was very difficult. About 20 cadre of our unit had to live in privation due to the small contribution of the people who were then living under the enemy control." And on the Lai Khê plantation, "the local (RVN) Administrative Personnel and Public Security Agents have, with surrenderers [sic], identified our agents. Discouraged people and workers have struggled weakly against the election of the enemy National Assembly." Furthermore, "a labor union was organized with 176 members of which only 17 members remain, now, due to the enemy terrorism and resettlement." Finally, an NLF leader noted that "our soldiers fought fervently in the beginning. Gradually they became negligent because the leadership was directed now by the district, now by the province. For a while the Command Committee of the American Base Unit lacked determination. During 5 months there were no Command Committee conferences, Party conferences or Group conferences."[23]

Americanization

Some critics of the U.S. intervention into Việt Nam argue that access to rubber provided one motivation for the American War. In 1960, François Schmitz mentioned that U.S. leaders were afraid of being cut off from natural rubber supplies, and that this fear drove their push for synthetic rubber production. But this understanding does not match the U.S. government discussions of the time. Việt Nam played a part in the regional economy, but its rubber was never considered vital to U.S. interests. Instead, plantations were key sites of political struggle and military strategy. While the first half of the 1960s saw increasing threats to production estates, thanks to their greater resources and value for the Free World forces (consisting of soldiers from the United States, the RVN, and their allies), the NLF, and the French management, the estates could survive the vagaries of war more easily than smallholders.[24]

With massive U.S. intervention in 1965, security concerns in III Corps came to the fore, and in 1966 the American embassy and the Military Assistance Command, Vietnam commander suggested the official selling and closing of the Lộc Ninh plantation near the border with Cambodia. In July 1966, the General Staff held a meeting with representatives from the Ministries of National Defense, Economy, Labor, and Communist Refugees, along with III Corps staff and Bình Long province officials, to study the American suggestion. The meeting members did find some beneficial aspects to closing Lộc Ninh, including cutting off funds flowing to the NLF, avoiding worker contact with NLF cadre, and making military operations easier. The members pointed out many drawbacks as well. From a military and political perspective, closing the plantation would mean losing a military base and source of intelligence, while the NLF would continue to use the area as a corridor for providing military supplies from Cambodia. From an economic and financial perspective, closing this plantation would cause distress among the population. Replacing the rubber produced on Lộc Ninh would take twelve years and an investment of 735 million đồng simply to cover the value of the trees and severance pay for laid-off workers. From a social perspective, closing this plantation would mean having to resettle 18,000 people, which would cost at least 36 million đồng. Finally, from an administrative perspective, government activity in Lộc Ninh district and Bình Long province revolved around the activities of Caoutchoucs d'Extrême-Orient's three plantations. The closure of these plantations would have effects at the district and provincial levels. They concluded that it was better to keep the plantations open and to more carefully control the flow of resources to the NLF by increasing the military presence. When U.S. leaders decided to intervene in Việt Nam, this insecurity became their problem.[25]

When the U.S. military increased its involvement, it sought out advice from "counter-insurgency" experts who could speak to "the connections between the physical and political environment of war." One such expert was Robert Thompson, who was largely responsible for the successful British effort against communist guerillas during the Malayan Emergency (1948–60). Drawing on the Việt Minh's experience during the First Indochina War, communist gueril-las operating on Malaya's rubber plantations had sought to take power from the colonial government. The British successfully isolated the guerillas, who were predominantly ethnically Chinese, from the population at large using strategic hamlets. During the Vietnam War, according to Bowd and Clayton, Thomp-son "criticised the U.S. military establishment for not coming to terms with the truly interwoven nature of the political, military, psychological and physi-cal (jungle) dimensions of guerrilla warfare." Seeing parallels with the Emer-gency, Thompson argued that the leaders of the U.S. military were not "able to see the woods for the defoliated trees."[26]

While Thompson's observations may have held true when U.S. Marines first landed at Đà Nẵng in March 1965, After Action Reports from major Free World forces military operations in III Corps, such as Crimp (January 1966), Mastiff (February 1966), and Junction City (February–May 1967), show an awareness of the nature of the plantation, paddy, and jungle terrain and to a de-gree the "psychological physical and dimensions" of the fighting, if not yet an understanding of how to effectively combat PLAF and PAVN forces. Accord-ing to one After Action Report, Operation Mastiff was designed "to locate and destroy VC forces and base camps in the vicinity of the MICHELIN Rubber Plantation." Because of NLF demands, plantation managers moved their head-quarters to Sài Gòn and suspended operations at Bình Dương, the Dầu Tiếng plantation, on October 10, 1965. The NLF, which had controlled the region between the plantation and Cambodia since 1954, wanted to produce rubber but apparently lacked the technical personnel to run processing equipment. Instead, they poured latex into holes in the ground, resulting in "VC latex" that was unmarketable, likely due to its poor quality. The report also stated that the eighty-eight square kilometers of rubber trees on the Dầu Tiếng plantation "offer good concealment from air observation." Furthermore, "the ground under the canopy is generally clear of under-growth providing good mobility along existing roads and trails." The report also noted the mixed en-vironment of jungle and rice paddy, plus river access to Sài Gòn.[27]

Between 1965 and 1968 the NLF was quite active on many plantations. During these years, according to rubber workers in Bình Long province, the NLF often met with workers in hamlets and organized musical shows, weapons exhibitions, and political discussions. Anti-RVN and anti-U.S. banners held in the U.S. National Archives and Records Administration (NARA), with the bark of rubber trees still on them, attest to political activity. One reads "With the Americans, [Nguyễn Văn] Thiệu, [Nguyễn Cao] Kỳ, and [Trần Thiện] Khiêm still around, the people are bleeding, suffering hardship, and poor."[28] Militarily, the 272nd NLF regiment operated in the rubber region, fighting and winning battles at Đồng Xoài (June 1965) and Ông Thành (October 1967). PLAF and PAVN forces also took advantage of the Long Nguyen secret zone near Michelin's Dầu Tiếng plantation, where they rested and prepared for attacks. To complement aboveground movement, NLF forces created an underground system of tunnels, rooms, and even hospitals in the rubber zone, including the famous Củ Chi tunnels in Tây Ninh province. The region's red clay held its shape for the "NY Subway system" and thus shaped fighting.[29]

The first order of business for the U.S. military was to extend its presence into the rubber region, which was largely a no-go zone for the ARVN. Starting in late February 1967, ARVN marines and U.S. military units, including the First Infantry Division, or the "Big Red One," the First Cavalry Division, and others, undertook a joint action called Operation Junction City. One main goal was to find and destroy the COSVN headquarters, which Military Assistance Command, Vietnam suspected was in Tây Ninh province. At the time, the region was a mixture of plantations, rice paddies, and "two and three-canopy jungle" interspersed with clearings that offered "a safe haven for Viet Cong forces." Lasting until May, the operation unfolded in two major phases: phase I put forces near the Cambodian border to block NLF retreat, and phase II landed forces near "enemy lines of communication in eastern War Zone C" for search-and-destroy missions. Called "the largest operation of the Vietnamese War to date" in the After Action Report, "it demonstrated the U.S. capability to enter areas which were previously Viet Cong sanctuaries and to conduct successful search and destroy missions." The military also viewed Junction City as crucial to future activity, as "actions taken during the operation to facilitate reentry into War Zone C will further decrease the VC tendency to regard the area as impenetrable." However, the operation failed to eliminate the COSVN.[30]

A second major aim of Junction City was to reopen National Highway 13 (Quốc Lộ 13, the old Route coloniale 13), which U.S. soldiers called "Thunder Road," for South Vietnamese traffic. As during the colonial era, this road served as one of the key links between Sài Gòn and the plantations. In addition to direct military action, the U.S. and RVN militaries employed psychological warfare, or psyops, to make National Highway 13 available for the movement of South Vietnamese people and goods. Junction City's After Action Report describes leaflets dropped during the operation, including several that targeted specific NLF units, with information meant to lower morale and increase defections. One leaflet was "aimed at the people of SVN, showing two pictures of Rt. 13." This leaflet appealed to "South Vietnamese compatriots [*đồng bào*]" to help keep National Highway 13 open by emphasizing the importance of the road for transporting goods and food and for keeping prices down. This leaflet also mentioned plantations, arguing that "unless Route 13 is open, the rubber plantations cannot stay open." The front of the leaflet asked, "brothers and sisters, which National Highway 13 do you want?" with two images below: one of a well-built road, with a street light, a trilambretta, a Vietnamese family walking, shops, and an RVN flag; and the other of a road overgrown with plants and full of holes, with a straw hut and an armed NLF cadre approaching menacingly from the right. Such propaganda attempted to enroll South Vietnamese support for the RVN by tying modernity to capitalism and the Free World.[31]

Rubber workers became an important target of pacification programs in III Corps, and in September and October 1967, ambassador Ellsworth Bunker and commanding general of Military Assistance Command, Vietnam, William Westmoreland exchanged letters about how to preserve plantations while also denying their resources to the NLF. Preliminary studies had given them four alternatives to consider: (1) forced abandonment and destruction of plantations, (2) selective destruction through defoliation, (3) nationalization of plantations to prevent French collaboration with the NLF, and (4) joint RVN/U.S. inspection and supervision of plantations to prevent collaboration. Options 1, 2, and 3 were "rejected as being too costly and impractical in terms of refugees created, indemnification to be paid and lack of trained GVN manpower to run the plantations." Thus, the preliminary study recommended active U.S. assistance along with RVN/U.S. inspection and supervision of plantations. In response, Ambassador Bunker advised against inspection as seeming to involve the United States in "pre-nationalization procedures" and instead recommended placing base camps near plantations and creating a committee to implement pacification procedures. Westmoreland viewed placement of troops and civil pacification near plantations as a waste of resources, but he agreed to the formation of the Mission Interagency Committee on Plantations (TARA), which held its first meetings in November and December 1967.[32]

Tết 1968 shattered the illusion that Free World forces had neutralized the NLF threat, as PLAF and PAVN troops infiltrated Sài Gòn, making use of paths, including "Serge's Jungle Highway," that traversed plantations. TARA's activities had also been suspended, but a combination of plantation directors' complaints "about extensive tree cuttings and damage to plantation infrastructure during U.S. military operations" and Westmoreland's changing tactical objectives in War Zones C and D granted plantations "a new lease on life." TARA met in June and emerged with a secret report recommending a U.S. policy regarding plantations. The result of nine months of research, this report saw a hopeful future for natural rubber in postwar Việt Nam and the world despite the current threats. It argued that "the basic objective of U.S. policy should be the preservation of the plantations and their infrastructure." In practical terms, this meant that the United States/Free World forces must try to avoid damaging the plantation, either through felling or slashing of trees, damaging or destroying processing plants, or calling down artillery, mortar, or airstrikes without the consent of authorities. Tree felling for defensive positions should be limited, and road clearing, another area of activity, was discussed at length. The committee concluded that no more than one hundred meters should be cleared from each side of the road and that clearing should be limited to main lines of communication.

The report also recommended a set of procedures for any clearance. The military should allow plantation managers time to remove equipment and clear trees when possible. Officers had to submit any request for clearing to the Civil Operations and Revolutionary Development Support (CORDS) province senior adviser ten days in advance and wait for approval from the American ambassador or his designee, who had the sole authority to permit destruction or expropriation. All clearing, whether approved or not, had to be justified in After-Action Reports that would be forwarded to appropriate authorities, including the economic and political counselors at the U.S. embassy in Sài Gòn. U.S. military and civil advisers were told to urge their Vietnamese counterparts to do the same. The complexity of these procedures hints at the seriousness with which civil and military leadership took this issue.[33]

As the report circulated for comment, some officers raised objections, including the distance for road clearance and the appropriate lead time for notification. But the recommendations largely stood as proposed and were adopted as U.S. policy. The pacification side was more difficult to carry out, and U.S. military and civilian leaders talked at rubber workers instead of listening to them. In this respect, Robert Thompson was right. After the shock of Tết, the U.S. military built command centers and fire support bases on plantations to control the wider region. In addition to the command center at Lai Khê research center, Michelin's Quản Lợi plantation was retooled into a U.S. military base and airfield. During the "savage retreat" of the early 1970s, U.S. and RVN violence ended rubber production and threatened to destroy plantations as B-52 bombing runs collapsed tunnel systems and Rome plows, and herbicides such as Agent Orange flattened rubber trees. Companies such as Michelin even attempted to sue the U.S. military for damages. As a TARA report mordantly noted, "The U.S. has been . . . both the savior of the plantation economy and the ultimate cause of its deterioration," a plantation equivalent of destroying the town to save it.[34]

The decreasing numbers of workers reflected the drop in production: Vietnamese plantations employed roughly 30,000 workers and their dependents in 1961; that number decreased to 24,000 in 1964 and to 11,000 in 1969. Northeast Cambodia offered a relative haven for growing rubber, and many companies shifted their personnel and resources to this region in the second half of the 1960s. Michon recalled being transferred to Cambodia in 1964 after he was married, as Việt Nam was considered unsafe for families. Michon does not provide details of his life for the years 1964 to 1970, but he generally considered them to be peaceful and productive. Plantations often transferred Vietnamese personnel in addition to their European managers. Those from the rubber industry that were active during this period can still recall Khmer phrases and write some Khmer words. This solution was temporary, however, and when Lon Nol took power through a coup d'état, Cambodia too was drawn into the destructive violence of war. By the early 1970s, rubber production in Việt Nam was slowly restarting and planners began to look ahead to a postwar future.[35]

Vietnamization

While Tết Mậu Thân in the spring of 1968 was a symbolic victory for the NLF, it was a tactical disaster. After Tết, Phước Hòa plantation workers reported no NLF infiltration into the village and that Regional Forces and Popular Forces and People's Self-Defense Force units patrolled day and night. A Free World forces' inventory showed thirty families with relatives in the NLF in 1968 but only seven or eight in 1970. Since 1969, workers on Lộc Ninh had little contact with the NLF and said its activities were limited to discussion of the eight points for peace, threats for those cooperating too enthusiastically with the RVN, and prevention of workers from joining the People's Self-Defense Force. Steady U.S. pressure during Operation Phoenix also lessened NLF influence.

On the Bình Sơn plantation, workers reported that the 274th NLF Regiment had received heavy casualties during Tết but that it was rebuilding with local NLF forces and preparing to elect a hamlet chief and chief of state for the PRG. The CORDS survey named the leading NLF in the area, including Huynh Van Quan, a man named Lien, and Nguyen Van Chinh (the chief of the NLF Quartermaster Section), his wife Nguyen Thi Long, and daughter Nguyen Thi Hong. The most important NLF cadre was Hoang Van Xuong, who had received an RVN draft deferment so that he could move freely. With these cadre named, they could be dealt with; in the margins of the report, an unknown U.S. analyst wrote "phoenix now has and is working on."[36] In 1970, ARVN and U.S. forces destroyed several NLF sanctuaries in Cambodia, which, according to the workers, led to decreased activity on the Lộc Ninh plantation.

For the NLF that remained, and for the incoming PAVN forces, plantations retained their importance, even after Tết. A TARA report on the plantations noted in 1968 that "there are strong indications . . . that the Communists desire the survival of the plantations and have therefore reduced their tax demands and harassment of plantation managers." As with the U.S. military and the RVN, during interviews between 1969 and 1971, many rubber workers noted that the NLF told them not to cut the rubber trees down. In short, most combatants in the Vietnam War saw usefulness in the plantations.[37]

Although some plantations were still making healthy profits, studies of the rubber industry from the 1960s noted the unfavorable exchange rates between the Vietnamese piaster and the U.S. dollar, which had been set up by the RVN state to increase its purchasing power. The rubber industry was forced to pay for labor and material in the relatively costly đồng, yet received fewer đồng abroad due to the Vietnamese piaster's expensiveness. The price of rubber on the world market fell by half between 1960 and 1967, decreasing from US$0.71 to US$0.37 per kilogram. Productivity per hectare also fell in absolute terms, and this problem was compounded by the rising productivity of planters in surrounding countries. Part of the decline in productivity came from a shift in emphasis to productivity per worker, as the labor force was scarce in relation to land. Part of the decline was also due to aging trees on the plantations, as little replanting had been performed during the 1960s. These factors combined to reduce rubber export earnings from US$48 million in 1960 to US$9 million in 1969. Even with this reduction, rubber still constituted nearly 80 percent of total RVN exports by the end of the decade.[38]

Plantations remained valuable for reasons other than profit. Following NLF attacks on the Lộc Ninh plantation in 1967 and 1968, vice president

Nguyễn Cao Kỳ and the deputy minister of interior Lê Công Chất visited the plantation. Later, the minister of welfare came to inaugurate three buildings built by Father David, who tended to the important congregation of Catholic workers. Such activity supports the conclusions of a 1968 report from TARA "that the GVN is closely interested in the plantations not only for revenue losses or population dislocations which the GVN budget could scarcely sustain; but as a valuable national economic asset." One U.S. report estimated that plantations employed between 15,000 and 20,000 workers, which meant that 100,000 people depended on plantations for food, lodging, schools, and medical care. Plantations were particularly important for some provinces, such as Bình Long, where 50 percent of the people depended on them for food. For these reasons, the report noted that the prime minister at the time, Trần Văn Hương, "has issued explicit directives to stop all rubber tree cutting." As a signal of support for plantations, the RVN reduced its 40 percent rubber export tax to a nominal 1 percent and did not authorize the requested 40 percent wage increase requested by the rubber workers until late 1970.[39]

Immediately after Tết 1968, the RVN police arrested many CVT leaders, who were later released under American pressure. Meanwhile, declines in profitability and the closure of plantations because of insecurity continued to hurt workers and their families. The chief of Pacification Studies Branch, Office of Program Coordination, CORDS, requested a series of surveys between 1969 and 1971 that gathered the views of hundreds of people living on or near the major plantations about the RVN government, the NLF, their living situation, and what they thought about the future. These interviewees came from all classes and age groups, from Vietnamese plantation owners to ex-workers to the poorest day laborers who also farmed to survive. They included animists, Buddhists, Protestants, Catholics, and members of the Cao Đài sect, as well as Vietnamese, Cambodians, and Montagnards.

Rubber workers felt caught between the RVN and U.S. militaries during the day and the NLF presence at night. Many rubber workers said that before 1954, they had been supportive of the Việt Minh. As a survey of the rubber workers on the An Lộc plantation in Long Khánh put it, "according to the interviewees, before 1954, the plantation workers were mistreated by the French people, and consequently, a great number of them joined the Việt Minh to fight against the French during the period from 1946 to 1954." Attitudes toward the communists began to shift during the First and Second Republic as the workers "realized that the VC are different from the Viet Minh." Although some rubber workers with children in the NLF continued to be sympathetic toward its cause, most workers resented NLF demands for taxes and its disruption of their lives. Rubber workers at the Vên Vên plantation noted that NLF cadre required poor workers, including tappers and day laborers, to give one day's wages and well-off workers, including managers and office personnel, to give one basket of rice per month. As workers at the Phước Long plantation put it, the NLF "not only terrorized but also stole vegetables, destroyed roads, bridges, and caused trouble for the workers." Respondents in Bình Long province said that "the VC policy was not suitable to their way of living," with others noting the contrast between clever lectures of the NLF and its actions. The presence of U.S. and RVN interviewers may have encouraged workers to give negative evaluations of the NLF, but workers did not spare the RVN and Free World forces from criticism, which suggests that responses were honest.[40]

In a report prepared by Biên Hòa Province Advisors in 1969, workers at the Bình Sơn rubber plantations complained of the abuses of Thai and local Popular Forces troops. A thirty-four-year-old female shop owner said, "I like the soldiers because they work hard and are sometimes short of money. This is the case of my son. That is why I understand them. I buy one kg of fish for 350 đồng but I sell it to them for just 300 đồng. However, they misunderstand me. Recently they bought a carton of Salems and just paid me 50 Vietnamese piasters. If they do not have enough money they should let me know or ask me for I am willing to lend to them. But if they buy, they must pay me. When I cannot accept such a low price for cigarets [sic], they threaten me. This make [sic] me very frightened."[41] She also witnessed a family that needed treatment for wounds received from stray bullets fired by soldiers and noted that they had molested a fifteen-year-old girl. Many Thai forces spent their time in coffee shops and getting drunk, and a forty-year-old male plantation worker complained: "It seems to me that the Thai force only receives orders to operate defensively; they never conduct any sweeps or patrols." Thai forces were a double-edged sword; they also contributed to a thriving black market that provided much-needed rice, canned food, and other goods for the local economy. U.S. soldiers, especially officers, were generally viewed positively, but the Americans were blamed for causing disastrous inflation in local prices. For example, retail prices for food on Bình Sơn were 500 đồng/kg for pork; 400 đồng/kg for pork, thigh; 200–300 đồng/kg for pork, chest; and 38–400 đồng/kg for fish. Eighteen respondents at Michelin's Dầu Tiếng plantation blamed the Americans for creating rich Vietnamese who spent money without regret. "They don't bargain or think when they buy something, therefore, the cost of living rises."[42]

Almost all workers complained about the cost of living. In 1970, rubber tappers on the Vên Vên plantation took care of 300 trees per day, which meant work from early morning until noon. They received 108 đồng a day, plus between 725 and 900 grams of rice for each family member. In addition, they could earn bonuses by reaching certain collection goals and for longevity. When the RVN government finally enforced a provision for a 55 percent pay raise along with new incentive plans on October 1, 1970, salaries on many, though not all, plantations increased. A male worker on the Terre Rouge plantation in Bình Long optimistically calculated that adding up his new 155 đồng daily wage, 4 đồng bonus for each kilogram of dry rubber, and the 6,000 đồng rice provision for his wife and five children, he could earn 11,500 đồng, a salary that was higher than that of ordinary soldiers and equal to the average government official. He added that his entire family could earn 20,000 đồng a month. Yet, due to runaway inflation, most workers complained of poverty. As a forty-two-year-old male on the An Lộc rubber plantation said: "With 155 [Vietnamese] piasters per day, what do you think we can buy? Therefore, we have to tap more latex than we are supposed to. If we tap 50 more liters of latex, we can earn some extra money. If not, we will have to grow sweet potatoes or wheat along the road in order to make some extra money." Another worker on Vên Vên noted that before this mass inflation, 5,500 đồng could take care of a family of five, but now this amount covered the costs for only two-thirds of a month. Vietnamese directors, supervisors, office employees, and managers could individually make from about 5,000 to almost 80,000 đồng a month, including allowances. Even so, with rampant inflation, a bicycle cost over 9,000 đồng, a major outlay for all but the highest-paid employees.[43]

Wages were similar on plantations run or owned by Vietnamese. On the Phước Hòa plantation in Bình Dương, which was rented by a group of Vietnamese, including Hoa Văn Mùi, tappers received a daily payment of 157 đồng plus 933 grams of rice, enough for two modest meals. In 1969, the Vietnamese Rubber Planters' Union, and its president Mùi, recognized the need to increase salaries but argued for postponement as plantations were suffering. The Vietnamese-owned Bukarr Rubber Company, which did not belong to the Planters' Union, paid its grass cutters 250–400 đồng per day and its tappers 500 đồng per day at the end of 1970. Colonial categories continued to play a role in wage discrepancy, as ethnic minorities and women were paid less than their Vietnamese male counterparts, with Montagnard and Cambodian women earning 250–300 and men earning 300–400 đồng per day. While these numbers seem higher than those paid on French-run plantations, they were considered market price for labor and most likely did not include provision of rice. Workers also complained that "there are no laws to protect the workers as in the French plantation system." Making an astute observation about the lives of rubber workers, one forty-eight-year-old man on the Quản Lợi rubber plantation said, "The plantation gives the workers just enough to live on, but not enough to allow them to ever improve their lot, to give their children a complete education before putting them to work, or to escape from the system of paternalism in which they find themselves caught."[44]

In response to the complaints of the workers and in recognition of their importance to southern society, the RVN government took a few steps. First, RVN officials worked to increase loyalty by stopping NLF infiltration, improving security on the plantations, and holding elections for village and hamlet officials. Second, the Labor Ministry agreed to enforce the provisions of the Rubber Community Convention of March 12, 1960, between representatives of the plantations and the various workers' unions. Third, the RVN started many development programs. Workers at the Terres Rouge plantation were apparently pleased with RVN assistance, including animal-raising People's Common Activity Groups. The animal-raising programs on the SIPH plantations of An Lộc in Long Khánh were successful, with seventeen cow-raising cooperatives, two goat-raising cooperatives, and two pig-raising cooperatives spread across several hamlets. A fifty-year-old man at the An Khánh plantation said, "Our 1969 cow raising cooperative has been very successful. One calf was born, and we hope to use it to carry wood and bananas and possibly to plough."[45] Respondents on other plantations, however, often viewed government aid as benefiting the rich rather than the poor. On the Lộc Ninh plantation operated by Caoutchoucs d'Extrême-Orient and located next to National Highway 13, the self-development program had a bad reputation. A fifty-one-year-old worker explained: "The government gave three pigs to Mr. Ta Van Xuan and three pigs to Mr. Ngo Van Nho two months ago. According to the government, the six pigs cost 80,000 Vietnamese piasters and when the pigs were big, the recipients could sell them and keep the profits. But it was obvious that the pigs were not worth 80,000 Vietnamese piasters to begin with."

On the Phước Hòa rubber plantation, most of the chickens and cows had died. Participation in such activities was, moreover, compulsory. A male latex worker said that "there were many required lectures or meetings conducted by specialized cadre from Information or Pacification and Development departments. If the workers did not participate in these activities they would be listed as pro-Communist." He also noted that the meetings interfered with daily life and that "the workers wished that the government would organize the meetings for them on Sunday or in the evening from 1700 to 1800 hours."[46]

Workers complained about other hardships, including not having Sundays off and the lack of medical care and schooling for their children. On certain plantations, noncommunist worker unions played a role in fighting for improved labor conditions. On the Vên Vên plantation, the pacification report noted that in 1967 some cadre from Tây Ninh Labor Services came to talk about organizing a Vên Vên plantation workers' union, but nothing came of it. On the Terre Rouge rubber plantation, 50 percent of the workers said that the Confederation of Trade Unions helped workers. One forty-five-year-old man said, "The Confederation of Trade Unions has instructed the workers on labor laws and helped them fight for their labor rights. Thanks to the union, the workers now know how to work together for their common good and they are now respected by the directors who no longer mistreat the workers." These rights included getting two Sundays off a month, a condition that mattered to Catholic workers. Other workers wanted improved medical care. A forty-five-year-old worker on Terre Rouge said, "The laborers can become sick easily, because they are poor and they work hard. It is shameful that a plantation that employs thousands of workers does not have a permanent doctor, but only has nurses who dispense medicine. The workers wish that the plantation would observe its agreement to provide one doctor for every two thousand workers. They have made this request through the union to the board of directors of the plantation, but so far no decision has been reached. At present, the plantation has only one doctor who takes care of the workers one day a week."

SUCH MEDICAL SITUATIONS were common on plantations, and having a Vietnamese medical doctor did not necessarily improve matters. For the union's work, laborers on the Terre Rouge voluntarily paid fifty đồng, about a third of a day's wages, as dues. Yet, the unions could be problematic as well. On Terre Rouge, 30 percent of respondents said they knew there was a union but didn't know what it did, while 20 percent didn't know there was a union. On the Lộc Ninh plantation, sixty-two respondents felt that prior to 1970, the union collaborated with owners, although they did believe that it was struggling for better salaries at the time of the survey. A further twenty-nine respondents were fearful of backlash for working with a union. Despite, or because of, the war, worker action continued. In January 1969, workers on Michelin's Trí Tâm plantation illegally struck but returned to work voluntarily after the strike leaders had been arrested and the local government requested the workers to return to the plantation. In February 1971, the workers of Michelin struck again when 120 of 200 workers were fired. RVN officials urged the courts to treat strikers leniently, as southern society needed all the cohesion that it could muster.[47]

Many rubber workers described their prospects as bleak, with 60 percent of Terre Rouge workers reporting that a worker's life was not prosperous and that they stayed on only because they lacked an education and professional skills. A thirty-two-year-old male returned from Cambodia and said he tried to work as a bus driver and a mason before going to work on the Terre Rouge plantation. Most people were desperate for any kind of work, and in July 1969, workers in Bình Sơn sent a letter to the district chief at Long Thành asking him to try to reopen the plantation. In May, a workers' representative contacted the Labor Ministry to intervene, but the owners' representative did not return to the plantation. Some workers collected latex on their own and sold it for 15 đồng per kilogram, earning up to 240 đồng per day. Others turned to collecting bundles of firewood, which could earn them up to 500 đồng per day, while trading with Thai forces could bring in up to 600–700 đồng per day. People supplemented their income with gardens and catching fish and frogs.[48]

Development has served as a key justification for the American War; clearly, the fighting of the war served to deter development, at least regarding the IRCV rubber research station at Lai Khê, which suffered directly from the U.S. military presence. In October 1965 the U.S. Army 3rd Brigade, First Infantry Division, began seizing parts of Lai Khê to build a base. According to the IRCV director Jean-Paul Polinière, between 1965 and 1968 the 3rd Brigade didn't pay rent, paid below-market rates for electricity, and vandalized buildings. Moreover, the IRCV had to rent facilities on a plantation in Long Khánh province. In 1969, Tôn Thọ Xương, the director of Lai Khê rubber Experimental Center, estimated that the damages to that point came to over 69 million đồng in felled trees and an overall loss of 150 million đồng, with over 300 workers laid off. Of that total, only 6.5 million đồng had been paid, and Xương requested the 20 million đồng that had been promised in 1969 to pay off delayed wages and other debts. The U.S. military, however, refused to pay rent to the IRCV, citing the terms of the Pentalateral Agreement of 1950, under which the RVN was to freely supply any land and buildings needed to U.S. military. Xương appealed to the U.S. ambassador, and eventually the United States agreed to a USAID loan of 13.5 million đồng, insisting that this was aid and not an indemnity. By the late 1960s, the French had also become unwilling to support the IRCV, and when Polinière requested 15 million đồng from the planters' association, it gave him only 5 million đồng, blamed him for trusting the U.S. military, and stated that long-term research was of little use.[49]

Rubber production and exports in the RVN, 1960–1972

			Rubber exports		
Year	Planted area (ha)	Production (metric tons)	Weight (metric tons)	Value (USD)	% of total exports
1960	109,470	77,560	70,118	48,000,857	57
1961	122,720	78,140	83,403	43,831,829	63
1962	135,630	77,870	74,497	37,917,229	68
1963	142,770	76,180	68,926	33,479,914	44
1964	134,700	74,200	71,630	33,299,171	69
1965	129,660	64,770	58,161	26,010,886	73
1966	126,340	49,455	44,899	22,035,000	80
1967	115,735	42,510	37,704	13,264,050	81
1968	105,730	34,000	29,247	9,705,913	83
1969	104,950	27,650	20,831	9,394,725	79
1970	105,800	33,000	23,601	8,851,187	77
1971	103,200	37,500	30,358	9,500,025	76
1972	83,300	20,000	19,638	5,671,549	36

Note: Weights of rubber exports for 1971 and 1972 vary significantly in sources.

Source: Phan, *The Republic of Vietnam's Environment and People*, 345, 383, 385; Đặng, *Kinh Tế Miền Nam*.

Because of declining revenues of rubber exports and the military appropriation of land, the IRCV fell heavily into debt; in 1969, the institute had an annual revenue of thirteen million đồng and a debt of forty million đồng. Despite these difficulties, the IRCV retained its potential as a significant center for agricultural research in the RVN, as noted by UN and USAID studies. A 1971 agreement for a second USAID loan of ninety million đồng justified this support by noting that despite destruction over the past few years, "the RRIV remains a repository of technical expertise, experience and equipment, unduplicated elsewhere in Vietnam. . . . Loss of the RRIV [i.e., IRCV]," the agreement continued, "through bankruptcy or diminution of the services which it can perform through loss of operating funds would be a significant loss to Vietnam's post-war economic development potential at a time when the country's economic development is receiving increased emphasis." When the loan was signed in January, the institute's personnel had shrunken to forty-four Vietnamese, including a few scientists, and three Europeans. The agreement argued that this reduced staff and "limitation of monies have limited the importance of its contribution to Vietnam's urgent present needs."

The second USAID loan was meant to tide the IRCV over until the RVN state could fund the institute, but in return Lai Khê had to sign over any rights to damages caused by the U.S. military occupation of the grounds that had lasted from October 1, 1965, to March 19, 1970.[50]

To survive, the IRCV attempted to reinvent itself as an organization that was capable of a range of scientific and technical research and could funnel international development aid into the RVN. Jean-Paul Polinière noted that there was "an argument for expanding the scope of the RRIV from a rubber research organization to an industrial research institute concerned not only with rubber but also with other industrial areas." The IRCV carried out research for the United Nations and in 1970 rebranded itself as the Technical Service Institute for Agriculture and Agro-Industries (Viện Dịch-vụ Kỹ-Thuật Nông-Nghiệp và Kỹ-Nghệ Nông-Nghiệp). This new institute would not only research rubber but also investigate corn, peanut, sorghum, and soybean and would provide contract-based studies, leaving cutting-edge rubber research to the better-endowed Rubber Research Institute of Malaysia. The Technical Service Institute would also provide a conduit for personnel to interact with the RVN government, and serve as a documentation center that would bring both foreigners and nationals up to speed on current issues. Through these functions, the Technical Service Institute could contribute to the "integrated development of rural areas." However, the institute was unable to realize this new vision, and rubber remained the focus of most research. The IRCV's chemist also suggested that the institute switch its research focus from aiding plantation growth to improving the industrial production of rubber goods.[51]

Studies carried out on the rubber industry in the early 1970s, particularly after the 1973 Paris Peace Accords, were focused on postwar recovery. Although the RVN was never able to achieve complete security, and abductions of French and Vietnamese plantation personnel continued into the 1970s, most were sanguine about the future of natural rubber and predicted relatively high levels of production once the fighting ended.[52] The U.S. embassy economic section estimated full postwar production of Việt Nam at 200,000 metric tons, or 8 percent of the world total. Such numbers formed a happy medium whereby the rubber industry could make up a significant part of Vietnam's economy without affecting global prices. Near the end of 1972, Walter C. Tappan, an agricultural specialist for the U.S. Agency for International Development, met with representatives of banks, plantations, smallholders, and the government and argued that rubber, along with other crops, would be a worthwhile investment. In the fall of 1973, based in part on

Tappan's work, the Asian Development Bank president recommended US$84,000 to help the industry as part of an eight-year postwar redevelopment plan.[53]

Pulling Out

With the growing global environmental consciousness of the 1960s and 1970s, experts began to think about plantations' effects on nature as the environment became a category of analysis. Rubber's association with automobiles, the deforestation caused by plantation extension, and more recent images of tires burning or as breeding grounds for mosquitoes all suggest a strong role in environmental damage. Yet, quantifying the impact of rubber plantations on air, water, soil, animals, and other aspects of nature into numbers is difficult due to a lack of research. The environmental impact of plantations, as measured by rates of deforestation, seems to have decreased during the Vietnam War. Continuing the work of the colonial forestry department, the RVN government kept statistics on forest cover. Because of the threat of violence and the lack of labor, plantation owners rarely engaged in clearing work. When labor was available, planters focused on replanting young trees. Smallholders also performed little clearing work because the threat of communist attack kept would-be deforesters at bay. Not until peace in 1975 were plantation forces released, but by then the country was so ravaged that people were simply trying to survive, and large-scale extensions did not resume until the late 1980s.[54]

One research topic of the late 1950s that helps shed light on the environmental impact of rubber was the relationship between the circulation of water and the production of latex, or the "economy of water." This question was particularly important for lands where a distinct dry season might limit production. Patrice Compagnon, the research director of the IRCV at the time, argued that since the red and gray soils of the region retained water, the distinct dry season was beneficial for rubber because it limited tree diseases, a belief that had been promoted since the colonial era. Not only did drier weather directly inhibit the growth of microorganisms, but such a climate was different from the Amazonian home of *hevea*, thus releasing the tree from its parasites. The highest production came at the end of the rainy season, from mid-November to mid-January. Such evidence suggested that, "all other conditions being equal, production can be found to be influenced by the ration of the water absorbed and conveyed toward the leaves to the water lost by evapotranspiration."[55]

Recent studies of rubber trees have shown their high demand for water and their potential to dry out soils. When grown in areas with rainfall throughout the year, this use of water is not a problem. On drier lands, however, rubber trees can cause severe water deficits. The processing of latex into rubber also consumes large amounts of water. More recent material flow calculations of the amounts of chemicals used as fertilizers, pesticides, and agents in the processing of latex into rubber have raised questions about water pollution and the impact of rubber in tires. A 1972 USAID report on water quality in Southeast Asia included little discussion of rubber plantations, although it did mention acetic acid made from rice and used in coagulation along with agricultural fertilizers as potential sources of water pollution.[56]

Plantation owners also responded to the growing environmental awareness of the 1960s by marketing their rubber as "natural" as opposed to "synthetic." They made a distinction between rubber that was produced using hydrocarbons derived from petroleum products (dead plants) and rubber made from latex produced by living trees. In this new terminology, "natural" rubber was a renewable resource, unlike "nonrenewable" petroleum. With growing concerns in the 2000s over climate change, those in the natural rubber industry argue that rubber trees are potential carbon sinks. While natural rubber begins with latex from trees, planters' distinction between natural and synthetic is tenuous. Every other part of the process of manufacture of rubber, whether synthetic or natural, used nonorganic chemicals. The process of making natural rubber required acetic acid to coagulate rubber and petroleum to ship rubber over long distances. Furthermore, rubber lands took up large swaths of forest and, until the 1950s, leaned more on extensive than intensive agricultural methods. From planting and collection to processing and final use, humans intervened in every step in the production of natural rubber.

As plantations became naturalized, the destruction of plantations also came to be viewed in terms of environmental impact. In addition to studying the destruction of rubber trees by Rome plows and explosives, scientists studied the effects of herbicide sprayings on plantations. In 1969, the State Department sent Fred Tschirley, assistant chief of the Crops Protection Research Branch, U.S. Department of Agriculture, to South Vietnam to study the effects of herbicides. From mid-March to mid-May, Tschirley traveled around South Vietnam and Cambodia, with "no constraints" and "all the help and cooperation that was possible" from military and civilian officials. Tschirley saw little chance of permanent harm to the climate and soil, concluding that a single application of herbicides resulted in a nearly complete destruction of mangrove forests, which might take twenty years to return to their original state. He noted that multiple applications might result in serious disruption in semideciduous forests, that herbicides did not harm animal life—and may even be correlated with an increase in fish—and that there was little chance of direct toxicity to humans. The greatest danger of ecological change came from invasive bamboo, which was hard to eradicate once established. Still, he suggested that scientific study of the effects of herbicides continue during and after the war, and that a checkerboard pattern of spray be followed.[57]

As part of his study mission, Tschirley traveled with Americans and Cambodian agronomists to study effects on fruit and rubber trees in Cambodia. The team's report found that defoliation resulted from direct spraying rather than drift, and that the extent of damage depended on the kind of rubber and fruit tree. The various rubber tree hybrids and clones responded differently to herbicides, with PR107, one of the highest yielders of latex, being particularly affected. While most rubber trees seemed to grow their leaves back, the extent of effects could be determined only by latex yield.[58]

In the late 1960s and early 1970s, the violent competition between communist and anticommunist forces resulted in daily tensions on plantations. To a much larger degree than during the First Indochina War, the NLF and PRG turned the plantations into spaces of revolutionary governments. For example, Trương Như Tăng, a Paris-educated lawyer, eventual minister of justice with the PRG, and son of a Vietnamese plantation owner, wrote of the many NLF activities taking place on plantations, including votes and "People's tribunals," which sentenced and executed Vietnamese plantation personnel. Michel Michon, a French manager with the SIPH, recalled with horror the swift, though not necessarily arbitrary, justice carried out by these tribunals. To Michon's eyes, NLF members were experts at using terror to control local populations and were much more disciplined than RVN troops. The NLF could control remote plantations even with limited numbers. As a worker on Michelin's Dầu Tiếng plantation put it: "Only one or two [VC] could control a great number of plantation workers when they are in a deserted area." And according to a worker in Phước Long province, "The workers hate the VC very much but they do not dare show their hatred because they are afraid they will get hurt or killed by the VC."[59]

In the early 1970s, some observers thought the RVN could withstand northern and NLF attacks, and in 1971 one Quản Lợi worker said "the VC are exhausted and discouraged, and that the GVN is winning the war." Yet many respondents to a survey conducted by CORDS saw only a bleak future. On An Lộc, Catholic workers were afraid of a coalition government with communists, while many Buddhists just wanted the war to end. Some respondents deflected the questions by saying they were not "soothsayers," though many had little hope for an eventual RVN victory. One respondent articulated the ways in which ordinary Vietnamese were buffeted by global forces when he described the American war as "an ideological war of great nations so nobody can forecast the future. The United States is a great nation which should be able to defeat the VC in one or two days, yet it has failed."[60]

NLF bombing attacks on CVT offices and the assassination of around sixty officials increased the anticommunist sentiment among the CVT's leaders. Phạm Văn Vỹ and other leaders of the Federation of Plantation Workers maintained pressure on RVN officials and plantation management to fulfill earlier agreements on rice provisions and to repay the patriotic sacrifices of the workers. Plantation closures presented a grim situation for workers and their families, who depended on their salaries for a living, and labor unions struggled for unpaid back wages and severance pay that the transnational plantations did not turn over. Planters, too, faced many problems, including lack of access to markets to sell their product.[61]

After the 1973 Paris Peace Accords brought about a temporary peace, the government became more receptive to planters' demands and both Vietnamese and French planters requested government subventions to support the rubber industry. Various associations and groups also tried to take advantage of the lull in fighting by investing in rubber. A women's association, for example, tried to buy abandoned plantations in the central highlands in 1974, stating that it would use profits to run its hospital for the poor, and the Cao Đài and a Buddhist association wrote to the RVN state inquiring about purchasing public property.[62]

Overall, however, prospects remained uncertain. Because of security concerns, RVN officials denied Michelin a 1973 loan request, and in February 1974, Hoa Văn Mùi, still president of the Vietnamese Rubber Planters' Union, noted that even the few Vietnamese rubber growers who had restarted activity were having trouble because of the closure of National Highway 1 to transport rubber. Later that spring, Trần Quốc Bửu wrote to the prime minister about the plight of labor, though his letter mentioned only 1,500 workers and their families who were still dependent on rubber plantations for their salaries. As PAVN forces advanced in the south in 1974 and 1975, one plantation after another closed due to lack of security, and the few remaining workers left the region. In May 1974, the labor minister reported that the Courtenay plantation in Long Khánh had closed as ARVN forces had withdrawn, leaving an estimated 30 percent of the plantation's trees destroyed. At the end of

Liberation đồng. This bill shows rubber harvesting and heralded the celebration of rubber production under socialism. Source: author's personal collection.

March 1975, the labor minister reported that Dầu Tiếng had 350 workers who received regular monthly wages when Michelin had lost contact with the plantations around March 11. Finally, on April 29, one day before the fall/liberation of Sài Gòn, Americans evacuated many CVT leaders, leaving rank-and-file members to deal with the new regime.[63]

In 1974, the DRV seemed to be in a strong position, and as the Hà Nội government readied for material ownership of the southern plantations, it also worked to reappropriate their collective image. In one propaganda poster from 1974, two women, most likely representing north and south Việt Nam, pour latex from buckets that flows in a stream leading to industrial production. The slogan at the bottom of the poster encourages workers to "produce a lot of rubber for industry." A year later, a twenty-cent (*xu*) bill of the "liberation" *đồng* (dated 1966 though printed in 1975) replaced the currency of the previous regime. This bill depicts a working landscape of rubber tappers loading latex onto a truck, a sign of the mechanized and industrialized Socialist Republic of Vietnam (SRV). In June 1975, the Khmer Rouge attempted to restart latex production on the Cambodian plantations, even as they were destroying other signs of capitalist modernity, including money, banks, and electricity. The protean nature of plantations meant that communists could co-opt rubber for their own vision of modernity. This reappropriation stood in stark contrast to the ravages of the Vietnam War, which had brought much destruction to the plantations. Lê Sắc Nghi recounts the role of B-52 bombings and Agent Orange in this decimation, and one of the former plantation workers I

"Produce a lot of rubber for industry." This poster shows the complete shift in meaning of plantation rubber production from exploitative colonialism to progressive socialism. Source: author's personal collection.

talked to was deeply saddened by the loneliness he felt. He told me: "And [during the war] many left. When the Americans attacked here, they showered the area with bombs and no more houses were left at all. There was an old house near here, with a large tree in front; the tree is gone and all of the houses are new."[64]

TECHNOPOLITICS DURING THE Cold War is often viewed in the blinding light of the atomic bomb and other "big" sciences. This chapter has advanced discussions of Cold War science by examining an industry and governmental actions that were not particularly costly, which relied on techniques and ideas elaborated in the first half of the twentieth century, and whose practitioners faced questions of race and science in former colonies. The rubber industry was left in French hands well into the 1960s, with science caught between empire and nation. It is not surprising that the postcolonial rulers of the RVN, the DRV, and the SRV attempted to gain legitimacy through a cosmopolitan concept of modernity and access to scientific knowledge production. Gyan Prakash has shown how in India the nationalist leader Jawaharlal Nehru viewed science and technology, as practiced by the West, as the path to an independent, modern nation-state. Similarly, rubber production, representing the coupling of scientific knowledge to agriculture, became part of a Vietnamese national identity. Thus, rather than too much science, Đặng Văn Vinh, the former aide to Yersin on the Suối Dầu plantation in Nha Trang, concluded that the greatest limitation of French scientific work was the failure to train the Vietnamese.[65]

The role of the rubber industry in the Vietnam War also shows the influence of French transnational corporations, and institutions dedicated to agricultural research, in this hot war. Well after 1954, French scientists and businessmen continued to dominate industrial research. Industry leaders had a difficult time adjusting to the postcolonial situation even as they professed to deeply care for Việt Nam and its inhabitants. The complexity of the situation was compounded by the fact that these industry scientists worked for transnational companies whose interests were not always aligned with those of the government.

The plantations were also affected by the decision to conserve some environments and destroy others, due to both geopolitical and local forces. While revolutionary forces were unable to tame all environments, they managed to transform plantation landscapes into national spaces. The rubber industry, with its discourse of improvement, was an important agent in various political projects, ranging from *mise en valeur* and the civilizing mission to the planning projects of anticolonial nationalism and socialist internationalism. Even as rubber production dropped to near zero by the 1970s, leaders of both South and North Vietnam looked to the postwar future of the rubber industry as a source of economic power. Not until the 1990s did southern rubber landscapes recover from the trauma of thirty years of war. This recovery was celebrated with a spate of histories of the rubber worker movements along with company museums. Rarely do these histories consider the role of violence in transforming colonial landscapes into national ones. Today, with the expansion of Vietnamese rubber companies into Cambodia and Laos and nationalistic outcries over Chinese-controlled bauxite mining operations, hybrid landscapes should be considered potent tools for Vietnamese nationalism, expansionism, and environmental change.

Conclusion

Đồng Nai is one of the big provinces of the country. . . . [It] has land well-suited
to growing short-term and long-term industrial crops on a large scale, including
some trees that have the strategic value of tropical agriculture.
—Lê Duẩn, 1977

In 2001, General Trần Tử Bình (1907–67), probably best known for his role in
the worker strikes on Michelin's Phú Riềng plantation in 1930, was posthumously awarded the Hồ Chí Minh medal for service to the nation. Six years
later, in 2007, the Labor Publishing House reprinted Bình's famous memoir
Phú Riềng Đỏ, which has been translated into English as *The Red Earth*. Bình's
recent glorification is representative of how the colonial rubber plantation experience lives on in official Vietnamese memory. It is a memory that equates
the hardships suffered by Vietnamese workers under French colonial rule
with a celebration of the heroic struggles of the Communist Party in organizing a worker revolution that led directly to the emergence of Việt Nam as an
independent, socialist nation in the fall of 1945.[1]

The Việt Nam of 2001 was a very different place from that of 1945, or even
that of 1991, the year of the conference on Alexandre Yersin that opened this
book. By the 2000s, more than a decade of *đổi mới*, or renovation policy, had
drawn Vietnamese society into a neoliberal world. This process was reshap-
ing the landscape of Việt Nam yet again, as rubber plantations gave way to
industrial parks for transnational companies such as Fujitsu and Nike. Yet
memories of plantations and the colonial past remained. Immediately after
the fall/liberation of Saigon in 1975, leaders of the SRV such as Lê Duẩn had
emphasized the national importance of industrial crops, including rubber.
With the disruptions of the communist takeover, the difficulties created by
the U.S.-led embargo, and the decrease in international socialist support,
Vietnamese faced a severe shortage of food and goods in the years after 1975.
The expertise of many workers who had served under the French and the
Americans was wasted, as they either were sent to reeducation camps or were
simply unable to find employment. Many plantation skills, as former rubber
workers pointed out, could be learned only through demonstration and
physical activity, and without qualified workers, the industry had to be rebuilt
almost from scratch.[2]

In the 1980s, a series of histories outlining the heroic contributions of rubber workers to the communist movement began to appear. On first read, these celebrations of labor appear to fit into well-worn channels of communist thought. Yet, these publications also reflect a problematic past in which the interests of the workers were not always aligned with those of the Communist Party. With the DRV victory over the RVN in 1975, Hanoi's perspective became hegemonic in Việt Nam, and individuals and communities learned that claims to state resources had to be made via the discourse of a revolutionary past. Reflecting the new socialist sensibility, the Vietnamese term for "plantation" changed from "*đồn điền*" to "*nông trường*," terms meaning "agriculture" and "field" that together mean "state-run farm."[3] At times, tappers have felt slighted in the SRV, and even on the Đồng Nai plantations, which had been particularly successful in retaining skilled managers through the transition to socialist rule, some workers expressed nostalgia for the sound organization of the plantation system that existed under the French.[4]

Not surprisingly, those who had supported the DRV, either by moving north or through activism in the south, retained high positions in the industry. After the end of the American war, Lê Sắc Nghi became the general director of the Đồng Nai rubber company, and Đặng Văn Vinh and Nguyễn Hữu Chất became director and vice director, respectively, of the RRIV. Vinh, like many other French-trained agricultural agents and technicians, decided to fight on the side of the Việt Minh, and during the 1960s, he played an important role in the attempt to develop rubber plantations in the DRV. In the years after 1975, Vinh played an important role in international relations, and under his leadership the RRIV helped reestablish cooperation with other rubber-growing Southeast Asian nations. A 1980 letter from Chất convinced Prime Minister Tố Hữu (1920–2002) to authorize the RRIV to contribute US$10,000 to the International Rubber Research and Development Board at a time when foreign exchange was scarce. While the rubber industry was once a contested space of colonialism, it was by the turn of the twenty-first century recovering from its post-1975 slump to again become a productive sector of southern Vietnam's economy. Narratives about the recent success of plantations emphasized the good lives led by workers and the role of the party in the skillful management of the industry.[5]

By the 2000s, a more nuanced evaluation of the French presence became possible. The economic historian Đặng Phong, for example, argues that the sciences, both natural and social, were areas in which the French "objective spirit" (*tinh thần khách quan*) and "humanistic character" (*tính nhân bản*) were sometimes displayed. Phong also discusses the limitations of projects to increase agricultural effectiveness, arguing that these projects failed for three reasons: (1) the people's lack of capital, (2) the traditional ways and conservatism of peasants, and (3) the French scientists themselves, who, despite their goodwill (*thiện chí*) and heart (*tấm lòng*), could not deliver benefits because they remained outsiders and hierarchical, could not be "with the people" (*không hội nhập vào đời sống cộng đồng của nông thôn Việt Nam*). While production capacity greatly increased under the French, living conditions did not improve much for the peasants.[6]

In France during the colonial period, the plantations generally had a positive cultural valence, and this view is still promoted among certain sectors. For example, Michelin, which once owned large plantations, continues to sponsor works such as *Des hévéas et des hommes: L'aventure des plantations Michelin* that treat plantations as places of adventure. Even in more complex novels such as Erwan Bergot's *Sud lointain* (translated as either *Distant Skies* or *The Distant South*) and Erik Orsenna's *L'Exposition coloniale* (Colonial exposition), plantations often appear as places of danger and opportunity where young Frenchmen could make their fortune.[7]

Yet some popular French cultural productions do portray plantations as places of hardship and suffering. Probably the most internationally visible recent French cultural production that touches on rubber is Régis Wargnier's *Indochine* (1992). In this film, which is inspired by the story of Madame de la Souchère, Wargnier re-creates the world of a female plantation owner. Wargnier tries to capture a "romance for Indochina" with his lush scenes and the "blues" of his landscapes. Yet *Indochine* is more than just "a symptom of the current French fad for things exotic," as the literary scholar Panivong Norindr has argued.[8] The opening scene begins with the voice of the plantation owner Eliane (Catherine Deneuve), which informs viewers that they are witnessing events from her perspective. The audience is asked to question *her* fantasies about plantations. Although in several scenes Eliane imagines herself in the position of caretaker and mother, recalling colonial rhetoric of "protectorates" and civilizing missions, she is also denied many signs of motherhood, such as biological children, and is presented as existing far from the ideal of French femininity. In an early scene Eliane is shown whipping a Vietnamese man who had deserted her plantation. The filmic representation of the beating is relatively mild when compared with the historical record, but upon hearing the worker say to Eliane, "You are my mother and my father," viewers are called to ponder this strange relationship between parent and child. While invoking the colonial-era rhetoric of protectorate and tutelage, Wargnier considers the violence that could easily slide into moral and physical abuse.[9]

Retreaded Tires

Projections for a 100,000-hectare expansion of rubber production in Cambodia and Laos over the next few years show that industrial agriculture still exerts a powerful impact on the Indochinese peninsula. Researchers have begun to publish their findings on various aspects of this expansion, including the role of international nongovernmental organizations in mitigating the possible and actual social and environmental consequences that are tied to this capitalist expansion. Yet for the most part, the involvement of Vietnamese rubber companies remains poorly understood. Vietnamese-controlled science and technology have played a growing role on the Indochinese peninsula. These companies have invested the capital (machinery and saplings), provided the knowledge, found the markets, and sometimes provided the labor. It is no accident that rubber production has provided an important conduit for the investment of Vietnamese capital, for, in the words of a former Michelin plantation employee, "Vietnamese know how to grow rubber!"[10]

The expansion of Vietnamese rubber into Cambodia and Laos also reflects global trends, as the production of natural rubber has grown substantially over the past few decades. In the early 1970s about three million metric tons were produced, rising to seven million in 2001 to between twelve and fourteen million tons today. Most rubber is still consumed in the United States, East Asia, and Europe, and most rubber is produced in Southeast Asia, though other production centers have operated in Africa and South America. In Malaysia, long the industry's leading producer, rubber has been replaced by palm oil. Currently, production rates range from one to two tons per hectare, which means rubber trees themselves occupy anywhere from six million to fourteen million hectares.[11]

The environmental legacy of rubber plantations is still controversial. Sensing growing environmental awareness, especially climate concerns, managers have attempted to redefine plantations as "forests" rather than "agriculture" and receive credits under carbon-counting programs.[12] Not surprisingly, most historians' evaluations of rubber's impact on the environment are negative and tend to focus on the deforestation that occurred during the process of extending plantations. Rubber's roots in imperial expansion and industrialization and more recent images of waste tires have imputed rubber with damaging effects. In Việt Nam, some experts have raised concerns about an overemphasis on rubber, with one observer arguing that when comparing profits from cashew nuts with those from latex, "one could speak of an error in *hevea.*" Others have continued to raise concerns about the transnational movement

Advertising sign. This sign from the late 2000s reads "All for the cause of industrialization and modernization. Rich people, strong country, just, democratic, and civilized society." Source: author's personal collection.

of plant diseases related to *hevea*, and South American Leaf Blight has often been discussed in relation to bioterrorism.[13]

PLANT AND ANIMAL MOVEMENT, coupled with human migration in Việt Nam, resulted in mixed environments that were composed of people, plants, and animals from across the globe. Plantations served as experimental sites for the European powers and helped transform agricultural and medical practices and theories. These colonial enclaves served as foci for agronomists and medical researchers, colonial officials, private enterprises, and social activists, generating practices and discourses that anticipated and helped bring about postcolonial development and nation-building projects. Finally, as rubber knowledge became an object of sharing, competition, and modeling across colonial empires in Southeast Asia, it helped create the material and mental networks that gave substance to that region. As such, rethinking the history of plantations contributes to historians' on-the-ground understandings of how plants, animals, and people were incorporated into, and helped shape, imperial and national political rule.

Biên Hòa had been the capital of a rubber-producing province; yet, when I taught English there in the late 1990s and early 2000s, I saw few traces of the rubber that once covered the land. Instead, I saw an industrial landscape and groups of mostly female workers walking home from factories after their shifts that had begun early in the morning. I did not realize at the time that Justin Godart, who had called plantations "forests without birds" during his visit to Indochina in 1937 as part of the Guernut Commission, had probably witnessed similar scenes as groups of mostly male workers walked along dusty plantation roads. Then, as now, this region of southern Vietnam was an industrial landscape, and although I thought that the work that these factories brought to the city was important, I felt uneasy. I was concerned about the sweatshop conditions and the chemically polluted lands and waters of Biên Hòa.

Like Godart, I spent a limited time in Biên Hòa. I also consider myself both a critic of what is occurring in Việt Nam and yet utterly dependent on the labor and the environment of that society. For Godart, rubber, among other products, helped his world move. Apparel and computer chips help run my own personal world. As during the colonial era, twenty-first-century global systems bring jobs and ill health, material development and pollution, to places such as southern Vietnam, while leaving those in France and the United States insulated from these effects. Neither of us knew, exactly, what we could do or whether we as outsiders should do anything other than to report our observations. Despite the upheavals of the twentieth century, the contingencies of history, the lost opportunities for change, and the activism, from a certain angle, the nature-culture worlds of the rubber plantations are the world we, Godart and I, and you live in.

Notes

Introduction

1. There are several hagiographies of Yersin, including those by Bernard, *Yersin*, and Mollaret and Brossollet, *Alexandre Yersin*. For critical work, see Au, *Mixed Medicines*, and Jennings, *Imperial Heights*, 14–34 passim.

2. Quote from Bernard, *Yersin*, 140. See also Đặng, "Etudes sur la contribution," 6. For more on the plantation, see chapter 6 in Bernard, *Yersin*, 133–64.

3. Nguyễn, "De l'organisation de l'hopital psychiatrique"; Bernard, *Yersin*, 141; Đặng, "Etudes sur la contribution," 7, 9.

4. There are now many surveys of Vietnamese history. See, for example, Goscha, *Vietnam*, Kiernan, *Viet Nam*, and Taylor, *A History of the Vietnamese*.

5. See proceedings from the conference on Yersin published as Coll., *Alexandre Yersin*. These proceedings include a slightly edited version of Đặng's paper on Yersin.

6. Đặng, "Etudes sur la contribution," 7; Bernard, *Yersin*, 139. For Pasteur, see Latour, *The Pasteurization*. I have written Vietnamese names with diacritics, unless they are not clear from the source material.

7. Đặng, "Etudes sur la contribution," 9, 11.

8. Ibid., 8, 10, 11.

9. Au, *Mixed Medicines*, 42–45. For Vernet's work on rubber in central Việt Nam, see Vernet, "L'hévéa brasiliensis" and Vernet, "Observations sur l'hévéa."

10. Dean, *Brazil*, 7–23, and Jackson, *The Thief*. For a focus on supply, demand, and marketing, see Barlow, Jayasuriya, and Tan, *The World Rubber Industry*, 15. For an extended analysis of rubber and the Columbian exchange, see Mann, *1493*. For the roots of dependency in the nineteenth century, see Weinstein, *The Amazon Rubber Boom*. See also the work of Barham and Coomes, e.g. "Wild Rubber"; Aso, "Le caoutchouc au dix-neuvième siècle." For data on *hevea*, see U.S. Department of Agriculture, Natural Resources Conservation Service, "PLANTS Database," accessed January 4, 2011, http://plants.usda.gov, and Voon, *Western Rubber Planting Enterprise*.

11. For the quote about rubber, see Davis and Brother, *Illustrated and Descriptive Catalogue*, 3. Catalogue from the Gutta Percha & Rubber Manufacturing Company of Toronto, Canada, 1896–97. For more on gutta-percha, see Hunt, "Insulation for an Empire." For a reflection on early French fascination with rubber, see Chevalier, "Le deuxième centenaire"; Chevalier, "Les premiers découveurs français."

12. For a history of the bicycle, see Herlihy, *Bicycle*. For the growth of a bicycle culture in the Midwest, see Gant and Hoffman, *Wheel Fever*.

13. See Weinstein, *The Amazon Rubber Boom*, and Hecht and Cockburn, *The Fate of the Forest*. For production numbers, see Tully, *The Devil's Milk*, 67. For biographies of

the individual barons, see Serier, *Les barons du caoutchouc*, 49–145. For price data and the cost of the Opera House, see Tully, *The Devil's Milk*, 67, 69.

14. For the Devil's Railway, see Tully, *The Devil's Milk*, 70–71. For this evaluation of India rubber supplies, see Davis and Brother, *Illustrated and Descriptive Catalogue*, 4. For a historian's evaluation of environmental effects, see Dean, *Brazil*.

15. For UK legislation, see Tully, *The Devil's Milk*, 57.

16. Morel, *Red Rubber*, 45. For an extended treatment of the Belgian Congo, see Hochschild, *King Leopold's Ghost*.

17. Ghosh, *The Glass Palace*, 172; Amrith, *Crossing*, 117; Kaur, "Rubber Plantation Workers," 19.

18. Scott, *Weapons*. A vast literature on plantations in the U.S. South and the Caribbean deals with topics such as the creation of race that are relevant to rubber plantations in Việt Nam. A starting point for this literature is Thompson, *The Plantation*, which originally appeared in 1932 as a dissertation at the University of Chicago.

19. See Mintz, *Sweetness and Power*; Richards, *The Unending Frontier*; and Curtin, *The Rise and Fall*. This section draws on Aso, "Les plantations."

20. Uekötter, "Introduction," 11; McCook, "Ephemeral Plantations," 87; and Tsing, *Friction*, 6. See also Evans, "Between the Global and the Local."

21. Quote from Chambers and Gillespie, "Locality in the History of Science," 230. See also Anderson, "Introduction: Postcolonial Technoscience," and Latour, *Science in Action*.

22. See Bunker, "Modes of Extraction."

23. For recent work by historians, see Brocheux, "Le prolétariat des plantations"; Murray, *The Development of Capitalism*; Murray, "'White Gold'"; Bonneuil, "Mettre en ordre"; Kalikiti, "Rubber Plantations"; Boucheret, "Les plantations d'heveas"; and Slocomb, *Colons and Coolies*.

24. Manderson, *Sickness and the State*, and Monnais, *Médecine et colonisation*. In the U.S. context, see Stewart, *"What Nature Suffers to Groe"*; Nash, *Inescapable Ecologies*; Mitman, *Breathing Space*; and Andrews, *Killing for Coal*. The only writings to consider the environmental history of this region in even a tangential way are the colonial-era *Monographies* and their Vietnamese equivalents.

25. Johnson, "On Agency," and Shaw, "A Way with Animals," 8.

26. Mitchell, *Rule of Experts*, 19–53.

27. Cooper, *Colonialism in Question*, esp. chapter 5. For questions of modernity and its variations in East and Southeast Asia, see Bayly, "Racial Readings of Empire," and Barlow, *Formations of Colonial Modernity*. For hygienic variations, see Rogaski, *Hygienic Modernity*.

28. Latour, *We Have Never Been Modern*, 48. See also page 104 "There are only natures-cultures, and these offer the only possible basis for comparison," as quoted in Thomas, *Reconfiguring Modernity*, 219. By using a notion of an absolute "Modern" and locating it in the West, Latour's work does not allow for multiple modernities located in many physical and abstract landscapes. Furthermore, while I agree with Latour that nineteenth- and twentieth-century moderns do not represent absolute breaks with the

past, changes to the environment were new in scope. See McNeill, *Something New under the Sun.*

29. Thomas, "Modernity's Failings," 727–28.

30. Murray, " 'White Gold,' " 42, 45. See also Zinoman, *Colonial Bastille*, 2001.

31. Takahata, *Pom Poko.*

32. Thomas, *Reconfiguring Modernity*, 225.

Chapter One

1. For Southeast Asia, see Adas, *The Burma Delta*; de Jesus and McCoy, *Philippine Social History*; Boomgaard, Colombijn, and Henley, *Paper Landscapes*; Boomgaard, Henley, and Osseweijer, *Muddied Waters*; Boomgaard, *World of Water.* For fisheries, see Butcher, *The Closing of the Frontier.* For social and environmental changes related to U.S. plantations, see Stewart, *"What Nature Suffers to Groe."* For midwestern food markets, see, for example, Cronon, *Nature's Metropolis.*

2. The contemporary geographical extent of Việt Nam roughly came into being after the 1802 unification of Đàng Trong and Đàng Ngoài, a long-standing division of two Vietnamese-speaking societies in the north and south. For a discussion of the history of *đồn điền*, see Tạ, *Đồn Điền.* For histories of southern Vietnam under the Nguyễn, see Li, *Nguyen Cochinchina*, and Choi, *Southern Vietnam.* For smallpox vaccines, see Thompson, "Mission to Macau," and Thompson, *Vietnamese Traditional Medicine.* For the story of *ficus elastica*, see NAVN1 AFC 456, Notice 14 janvier 1901, which mentioned one "giant that measured 16 meters in height" that had been growing in a Vietnamese court official's garden for about twenty years.

3. Nick Cullather argues that technology is a form of discourse in "Miracles of Modernization." See also Adas, *Machines*; Headrick, *The Tools of Empire*; and Headrick, *The Tentacles of Progress.* Although at the time, most states saw the separation of humans from nature as leading to dominance of both, Latour argues in *An Inquiry* that ever-greater entanglement has led to increased exploitation and potential disruption.

4. Uekötter, "Introduction," 22.

5. This chapter uses the term "Montagnards" as an analytical category to describe the peoples distinguished from the larger groups of the Kinh (Vietnamese) and Khmer, peoples identified as lowland rice growers, regardless of whether individual Kinh lived in the lowland and grew rice. This chapter uses the pejorative "moïs" and "phnong" derived from the Vietnamese and Khmer words for "barbarian," and "autocthone," or native, when citing the terms' usage during the colonial period. It identifies smaller groups when possible and relevant. The word "ethnic minority" is commonly used today but implies a bounded nation-state that was less relevant before 1954, and the term is thus anachronistic when applied to the colonial period. Early missions record the first impressions of these people, later called the Stieng, the Phnong, the Choma, and so forth.

6. Han and Li, *Nan-Fang Ts'ao-Mu Chuang*; Borri and Baron, *Views of Seventeenth-Century Vietnam.* For an early Western-language publication, see Loureiro, *Flora*

cochinchinensis. For a comparison with the Atlantic world, see Schiebinger, *Plants and Empire,* and Schiebinger and Swan, *Colonial Botany.*

7. Trịnh, *Histoire et description de la basse Cochinchine;* Sterling, Hurley, and Le, *Vietnam,* 26. See also Baudrit, "Monographie d'une rivière cochinchinoise"; Gerber, "Coutumier Stieng"; Mạc, Bạch, Cong, and Nguyễn, *Địa Chí Tỉnh Sông Bé.* As Philip Taylor has shown for the Khmer lands of the Mekong Delta, the settlement of the south is not only a Vietnamese story. Taylor, *Khmer Lands.*

8. See, for example, Brenier, *Essai d'atlas statistique.*

9. Pierre, *Flore forestière de la Cochinchine,* and Pierre, *Notice sur une espèce d'Isonandra.* For his obituary, see Anon., "Miscellaneous notes, XXIII." See also Anon., *Annales du jardin botanique,* and Chevalier, *Catalogue des plantes du jardin botanique.* In the archives, see NAVN2 VA.8/117, l'étude de Anglès, "Le jardin botanique et zoologique de Saigon de sa fondation à 1930." As Hazel Hahn has recently shown, the Saigon botanical garden's cultural and touristic functions came to overshadow its scientific work in "Botanical Gardens of French Indochina, 1860s–1930s."

10. John Phan, personal communication; Thomas, *Reconfiguring Modernity,* 32–34.

11. Hoàng Phê et al., *Từ Điển Tiếng Việt.* For an excellent consideration of agriculture in communist China, see Schmalzer, *Red Revolution, Green Revolution.*

12. McElwee, *Forests Are Gold.*

13. Việt Nam stretches between approximately 8 degrees and 23 degrees N latitude. See Gourou, *L'utilisation du sol,* where Gourou called Vietnamese civilzation "la civilisation du végétal." NAVN1 AFC 116, Administrateur de la province de Tay Ninh à Lieutenant-gouverneur de la Cochinchine, Rapports économiques des années 1898–1909, 137–38.

14. Map from NAVN2 M.7/73, "Etablissement du programme de colonisation," 1930. For Southeast Asian forests, see Kathirithamby-Wells, *Nature and Nation;* Vandergeest and Peluso, "Empires of Forestry." For the extensive literature on forests in South Asia, see among others Rajan, *Modernizing Nature.*

15. For the colonial era in Việt Nam, see Thomas, *Services forestiers français;* chapter 1 of McElwee, *Forests Are Gold;* and Cleary, "Managing the Forest."

16. For background on Bến Cát, see Nguyễn, *Từ Điển Địa Danh;* de Laprade, *Simples renseignements,* 6, Bên Cát listed as "Canton de Bình Hưng, avec chef-lieu Ben Cat (An-Phuoc)." In 1891, the name of the area was changed to Bình Hưng and it now belongs to Bến Cát district, Bình Dương province. See also NAVN2 VA.2 041, Situation politique de l'arrondissement de Thu Dau Mot, Rapport au sujet du poste de Thi Tinh, 1877. For forest reserves, see NAVN1 AFC 627, Délimitation et aménagement des réserves cadastrées en Cochinchine, 1895–1914.

17. As Richard Grove has shown in *Green Imperialism,* these theories derived in part from early colonial experience on islands.

18. Buchy, "Histoire forestière de l'Indochine," 223. Originally from Anon., *L'Indochine forestière* (Saigon, 1942), 16.

19. Grove, *Green Imperialism.* For French tradition of forestry, see Scott, *Seeing Like a State;* Davis, *Resurrecting the Granary of Rome,* 78–79.

20. For the colonial state's approach to resource management, including selling the right to cut forests, see NAVN1 643, Réserves forestières en Cochinchine - Travaux exécutés : prestations dans le cantonnement Tay Ninh, 1910; NAVN1 642, Vente par adjudication publique des coupes de bois situées dans les Forêts de Baria et Tay Ninh, 1907–1912. For issues of peasant resistance related to uses of nature, see inter alia Karl Jacoby, *Crimes against Nature*; Guha, *Unquiet Woods*; and Scott, *The Art of Not Being Governed*. NAVN1 AFC 656, Instance introduite par Depierre, garde forestier à Trang-bom (Bien Hoa) contre Hangouvart (Hangouwart) au service de M. Posth, marchand de bois, pour menace dans l'exercice de ses fonctions, 1905–1910; NAVN2 4804, Bien Hoa, Requête de M. Mettel gérant de la Société Générale des Plantations, de Donnai contre l'agent du Service Forest Pochont; NAVN1 AFC 663, Rapport, M. Chapotte, Inspecteur adjoint eaux et forêts, chef p.i. Service forêts Indochine à M. chef circon-scription forêt Cochinchine, Hanoi 7 septembre 1904.

21. Thomas, *Services forestiers français*, 112.

22. Baillaud quoted in Thomas, *Services forestiers français*, 46, originally from Bail-laud, "Le régime forestier," 203–7. For more on BIF, see Thomas, *Services forestiers fran-çais*, 56.

23. Because of the reorganization of the Forest Service that started under the GGI Paul Doumer and was carried out by *arrêtés* of December 22, 1899, February 5, 1902, and January 15, 1903, much management of the forests of Indochina took place at the protectorate or colony level. For the contest over wood in Cambodia, see NAVN1 AFC 116, Administrateur de la province de Tay Ninh à Lieutenant-gouverneur de la Cochi-nchine, Rapports économiques des années 1898–1909.

24. Thomas, *Services forestiers français*, 62–63, figure 11, and 54–55.

25. Ibid., 88, 89, and 114. See also Chevalier, "Premier inventaire des bois," 743.

26. For an overview of forests to the 1920s, see Couffinhal, *La situation actuelle*.

27. The classic Marxist text in Vietnamese about the situation in rural Việt Nam dur-ing the colonial period is Nguyễn, *Những Thủ Đoạn Bóc Lột*. For the land situation in the Mekong Delta, see also Brocheux, *The Mekong Delta*, and Tai, *Millenarianism*. NAVN2 3900, Mode d'aliénation des terrains domaniaux ruraux incultes. Concession de la gratuité. Concession accordées par les Administrateurs, 1881–1885. For more on Nguyễn dynasty laws, see Choi, *Southern Vietnam*, and Tana, *Nguyen Cochinchina*. See Boudillon, *Le régime*, 8–10; types of land on, 14–15; see also Pasquier, *L'Annam d'autrefois*; Feyssal, *La réforme foncière*; Gueyffier, *Essai sur le régime*. Quote from Ly, *Le régime des concessions domaniales*, 52.

28. Boudillon, *Le régime*, 60–61. Categories included (1) rural red earth, given, and (2) urban/semirural gray earth, sold. See also Couzinet, *Les concessions domaniales*.

29. The GGI published a few laws related to land—e.g., in NAVN2 IB.23/124, 5 sep-tembre 1888, 15 octobre 1890—concerning concession of land but these laws largely left in place previous arrangements. Goucoch quote from *Arrêté* Goucoch 27 janvier 1896, *Journal officiel de l'Indo-chine française*, 30 janvier 1899, 224, cited in Ly, *Le régime des concessions domaniales*, 111. For official support of plantations, see provincial economic reports from NAVN1 AFC 624, Essais de culture des plantes caoutchoutifères dans les

réserves forestières de Kompong Thom (Cambodge), 1905–1906 and Rapport by M. Bordenave, chef de la division forêts de Kompong Thom, 23 août 1906. NAVN1 GGI 4052, Rapport sur le fonctionnement de la Direction de l'agriculture, des forêts et du commerce de l'Indochine, 1909, 61. NAVN1 AFC 643, Rapports sur les travaux exécutés au moyen des journées de prestations dans les divisions forestières des cantonnements de Baria, Bien Hoa, Tay Ninh, et Thu Dau Mot, 1909–1911. NAVN1 643, Lettre, Chef service forêts Indochine à Administration province Tay Ninh, 16 novembre 1907, au sujet des forêts.

30. Doumer, *Situation de l'Indochine française*. Conseil colonial, Séance du 24 septembre 1909, extrait, rapport 4711 au sujet achat par la Société Suzannah des terrains dont elle est concessionnaire. NAVN2 M.71 4853, Vente de gré à gré au profit de la société agricole de Suzannah 2,714 ha, Bien Hoa, An Vieng Concession, 1910.

31. *APCI*, no. 10, 12 décembre 1910, Actes officiels: *Arrêté* réglementant les aliénations des terrains destinés à la culture du caoutchouc (terres rouges), 50. Vu la délibération du Conseil colonial dans sa séance du 3 octobre 1910; Le Conseil privé entendu, *Arrêté*: Article 1er. Les terrains domaniaux disponibles, allotis ou non allotis, dont la cession est demandée en vue de la culture du caoutchouc, peuvent être aliénés par voie de concession gratuite, de vente aux enchères publiques ou de gré à gré. Article 2. Les terrains domaniaux non encore aliénés, connus sous le nom de terres rouges et situés dans les provinces de Tay Ninh, Thu Dau Mot, Bien Hoa et Baria, sont réservés à la culture du caoutchouc . . . . Art. 19. L'aliénation à titre gratuit ou onéreux des terres destinés à la culture du caoutchouc est faite sous la condition essentielle de leur mise en culture. The article goes on to discuss the terms of granting the land—essentially, the land had to be given over to the growing of rubber. Signed by Gourbeil on 13 octobre 1910. See also: *Arrêté* du 20 octobre 1910 sur aliénation des terrains domaniaux; *Arrêté* du 15 septembre 1910 du GGI et circulaire de Lieutenant-gouverneur en date du 21 octobre 1910; Boucheret, "Les plantations d'hévéas," 919–21.

32. For land concession battles, see *APCI*, no. 1, 12 juillet 1910. For pressure on Vietnamese authorities, see Procès verbal de la 6e réunion du comité, 3 octobre 1910. Vœu au sujet de l'hostilité des villages vis-à-vis des nouveaux planteurs, 18.

33. *APCI*, no. 4, 30 août 1910, 28. For Crémazy's overview of the rubber industry, see Crémazy and Bazé, *L'hévéaculture en Indochine*.

34. *APCI*, no. 18, juillet 1912, 304.

35. *APCI, Annales*, 1913, 28.

36. For land grant conditions, see NAVN2 VA.2 041, Rapport de l'administrateur de Thu Dau Mot sur le statut des Moïs de la région de Hon Quan et sur les différents survenus entre la Société des plantations d'hevéas de Xatrach et l'administration, 1912.

37. Under the colonial system of 1903, there were four types of property: public doméin, state lands, private/colonial, and local. Tạ, "Concessions agricoles," 79–87. See also Deroche, *France coloniale*. NAVN2 M.71 4853, Vente de gré à gré au profit de la société agricole de Suzannah 2,714 ha, Bien Hoa, An Vieng Concession, 1910. Lettre 2113, à Lieutenant-gouverneur de la Cochinchine, 5 octobre 1904, au sujet Cazeau demande

d'autorisation de culture. Lettre, M. Carrière à Administration Bien Hoa, 1 août 1904, au sujet de la mise en culture de terrains anciens boisés et contigus à la forêt de Bien Hoa. Estimates for exchange rates are based on Boucheret, "Les plantations d'hévéas," 812; "Discours prononcé par M. A. Klobukowski," *Bulletin économique de l'Indochine*, 29 octobre 1910, 22–23; and "Historical Currency Converter," http://www.historicalstatistics .org/Currencyconverter.html.

38. For the police post at Xuân Lộc, see NAVN2 M.71 4326, Goucoch au sujet de l'incendie de sa plantation 1914, Lettre 1880, M. Cuniac, Administration de Bien Hoa à Goucoch, 20 mars 1914, au sujet de l'incendie sur la plantation de caoutchouc de l'Indochine, 9. Planters wanted their own security apparatus. For the quote about the "transformation" of woodcutters, see NAVN1 AFC 116, Rapports économiques des années 1898–1909 sur la situation économique dans la province Tay Ninh, Administrateur de province à Lieutenant-gouverneur, 6 mars 1908. Even though the Indochinese government structure was under French control, rebels and bandits (sometimes one and the same) continued to attack French forces when they had the opportunity. These outlaws and anticolonialists often made their bases in places such as the mangrove forests of the Mekong Delta and the forested hills of the Tonkin-China border. General insecurity was also a problem, and many attacks without an anticolonial motive occurred. This issue is discussed in Scott, *Seeing Like a State*, and Biggs, "Managing a Rebel Landscape."

39. NAVN2 Services locaux 362, Rapport du Goucoch sur la situation politique du pays, 1917, No. 312, 1er trimestre, 9 mai 1917. Thomas, *Services forestiers français*, 47. Locals were most often the target of these attacks. *APCI*, no. 18, 7 août 1911 Police rural, 373. M. Le-Phat-Tan, a property owner in Baria and member of the chamber of agriculture, also grew rubber trees. He wrote to the president of the association of the need for rural police.

40. For roads in French colonial Africa, see Freed, "Networks of (Colonial) Power."

41. Del Testa, "Automobiles," 64.

42. Thủ Dầu Một was the best connected of the eastern provinces, with 90 kilometers of major roads and 436 kilometers of smaller roads. See Cochinchine française, *Rapport au Conseil colonial*, 1880, 492–93, for table of roads; 485, for the quote about roads; 449, for information on automobile services and Vietnamese use of them. For more on a "pioneering ethic," see Biggs, "Managing a Rebel Landscape."

43. Cochinchine française, *Rapport au Conseil colonial*, 1910, VII; telegraph network in Cochinchine française, *Rapport au Conseil colonial*, 1880, 24–25; Anon., *Bien Hoa en 1930*, 222, 227; Cochinchine française, *Rapport au Conseil colonial*, 1909, xxxii. For the idea of mobility and power in colonial Cambodia, see Edwards, "The Tyranny of Proximity."

44. Fall, "Investissements publics et politique économique," 238–46. In fact, Albert Sarraut, the GGI at the time, had wanted 200 million francs for his *mise en valeur* program, but the Chamber of Deputies was reluctant to invest in such a large expenditure after seeing the poor results of Doumer's 1898 loan. See also Morlat, *Les affaires politiques*, and del Testa, "Paint the Trains Red." As late as 1907, the administrator of Thủ

Dầu Một had to argue for the importance of creating a faster way to connect the province with Sài Gòn and Chợ Lớn. See NAVN1 AFC 117, Rapports économiques de la province de Thu Dau Mot, Administrateur de Province à Goucoch, 1898–1909.

45. Other centers of rubber production, such as Lộc Ninh, were linked by weekly buses to Sài Gòn and to Hớn Quản. Cochinchine française, *Rapport au Conseil colonial*, 1914, 179. For the Goucoch's thinking, see his 1918 annual report where he wrote about the Colonial Council's vote for funds that would "permettent aux colons d'étendre leurs cultures spéciales (caoutchouc, canne à sucre, etc.) dans des terres nouvelles." NAVN2 Services locaux 363, Rapport du Goucoch sur la situation politique du pays, 1918.

46. NAVN2 M.71 4290, Lettre 918, Administrateur de Bien Hoa à Goucoch, 7 mars 1913. The land sold for 1.50 piasters per hectare. This company was later the subject of a 1933 complaint from its workers who were not being paid—see Boucheret, "Les plantations d'hévéas," 425, Annex, 1006-09. For railroads, see NAVN2 M.71 4307, Bien Hoa Vente de gré à gré de terrains domaniaux 191 ha et 243 ha sis à An Loc, concession de la Société des Plantations d'An Loc, 1908. In this file, the engineer of the Public Works (Travaux Public [TP]) argued that the Goucoch needed to be careful not to concede too much land surrounding train stations, as this would inhibit other traffic. This mistake had been made at the Dầu Giây station with the Suzannah concession.

47. NAVN2 IIA.46/223, Main d'œuvre, dossiers relatif à la règlementation de la main d'œuvre indigène et javanaise en Indochine; Affaires diverses concernant la main d'œuvre des plantations caoutchouc en Cochinchine, 1905, 1910-18.

48. For critique of planters, see NAVN2 VA.2 041, Route de Chala, délégation de Hon Quan dans la portion traversant la plantation de Xa Cam, 1914–15. For planters' objections to paying costs of governing, see NAVN2 VA.2 041, Création d'un village annamite dans la région Moïs, *arrêté* du 30/12/08, Thu Dau Mot, 1908. In administrateur de Thu Dau Mot à Goucoch, 11 août 1915, the administrator argued that planters refused at all costs to have villages on their land because of the added costs and obligations to the plantations.

49. Anon., *Bien Hoa en 1930*, 224; del Testa, "Automobiles," 64–65.

50. For rusting bridges, see Chollet, *Planteurs en Indochine*. A small-gauge railroad line was used instead, and one can still see the old track bed. Đỗ and Nguyễn, interview transcript, 44; Phạm, interview transcript, 11–12.

51. Bonneuil describes how plantation labor was quite consciously subjected to the dictates of Taylorism in "Mettre en ordre."

52. *Bulletin du Syndicat des planteurs de caoutchouc de l'Indochine* (*BSPCI*) no. 7, octobre 1918. This planters' journal ran from 1911 to 1941.

53. Begon, Harper, and Townsend, *Ecology*, x, originally from the introduction of Krebs, *Ecology*.

54. See Lowe, *Wild Profusion*.

55. Worster, *Nature's Economy*, 203, 204. The definition of synecology is taken from Shelford, *Laboratory and Field Ecology*, 2.

56. Sharples, *Diseases and Pests*, 2.

57. On the rise of a "biological perspective" in nineteenth-century Germany, see Nyhart, *Modern Nature*. For ecology in Germany's nineteenth-century colonial empire, see Cittadino, *Nature as the Laboratory*.

58. Tilley, *Africa as a Living Laboratory*. See also Aso, "How Nature Works."

59. Miller, *Nature of the Beasts*, 2–3.

60. For a relatively early study of rubber in Indochina, see Spire and Spire, *Caoutchouc en Indochine*.

61. Quoted in Barlow, *Natural Rubber Industry*, 119.

62. Thomas, *Reconfiguring Modernity*; McNeill, *Something New under the Sun*.

63. For the phrase "garden in the machine," see Russell, "The Garden in the Machine." Biological knowledge is a reminder that plants are more than the sum of their moving parts.

64. For an excellent study of related groups on the Cambodian side of the border, see Guérin, *Paysans de la forêt*. For the source of *plasmodia falciparum* in China, see Packard, *Making of a Tropical Disease*. James Scott has recently argued that these groups stuck to the highlands as a survival strategy, using the malaria they brought with them as a barrier against the lowland states of the rice-growing deltas. Scott, *Art of Not Being Governed*.

65. See Hickey, *Sons of the Mountains*.

66. For example, the border farther south between Hà Tiên and Kampot was settled in 1905. For the Brao, see Baird, "Making Spaces," and Baird, "The Construction of 'Indigenous Peoples' in Cambodia."

67. Ernest Outrey was highly involved in questions of colonization. In 1907, he considered questions of colonization by Tonkinese in the Mekong Delta. NAVN2 IIA.46/311, Main d'œuvre dossier concernant la colonisation des terres incultes de la Cochinchine par la main d'œuvre tonkinoise 1907, 1908 (see also NAVN2 IIA.46/224) E.g. Lettre, Outrey à Goucoch, 25 septembre 1907, colonisation tonkinoise.

68. See, e.g., Azémar, "Les Stiengs de Brolam"; Carrau, "Du commerce et de l'agriculture"; and Gautier, "Voyage au pays des mois." NAVN2 IA.7/221, Mission: reconnaissance dans l'Hinterland moï: Dossier mission Patté. (province de Thu Dau Mot) 1903–04.

69. Patté, *Hinterland Moï*, 2, 188.

70. NAVN2 IA.7/221, Mission: reconnaissance dans l'Hinterland moï: Dossier mission Patté. (province de Thu Dau Mot) 1903–04. Lettre, Administrateur de Thu Dau Mot, Outrey, à Goucoch, 24 janvier 1905.

71. NAVN2 IA.7/221, Mission: reconnaissance dans l'Hinterland moï: Dossier mission Patté. (province de Thu Dau Mot) 1903–04. Lettre, Administrateur Thu Dau Mot, Outrey, à Goucoch, 24 janvier 1905. For more on Núi Bà Rá, see NAVN2 IIB.54/013, (1) Domaines, ouverture à la colonisation de Phu Rieng, demandes de concessions, cartes de régions, rapports, 1927. Nui Bara à ouvrir après établissement du poste à la Yumbra.

72. Cochinchine française, *Rapport au Conseil colonial 1909*, viii–ix, for good relations with Stieng but not with "moïs insoumis." See also Loubet, *Monographie de la*

province; Baudrit, "Monographie d'une rivière cochinchinoise," 12. Baudrit was one of those who wrote of Patté's mission as a success.

73. Loeng was not captured and killed until 1935. Loubet, *Monographie de la province*, 19–28; Guénel and Guerin, " 'L'ennemi, c'est le moustique.' " For the work of Henri Maître, see Maître, *Les régions moï*, and Maître, *Les jungles moï*. See Jennings, *Imperial Heights*, for a recent discussion of Maître's work. NAVN2 VA.8/125, (2) a/s pénétration et de l'organisation des région moï, 1921–22, Lettre, Goucoch Cognacq à GGI, 29 novembre 22, a/s pays moï, 2.

74. NAVN2 VA.8/125, (2) a/s pénétration et de l'organisation des région moï, 1921–22, Lettre, Administrateur Bien Hoa, a/s pays moï.

75. Quote from Cleary, "Managing the Forest," 279. For fire in Montagnard societies, see works of Georges Condominas and Gerald Hickey. For discussions among foresters in the French and British empires, see Rajan, *Modernizing Nature*.

76. NAVN1 AFC 663, Répression des délits de "rãys" constatés en pays moïs 1904. M. Chapotte, Inspecteur adjoint eaux et forêts, chef p.i. du service forestier de l'Indochine à M. chef de la circonscription forestière de Cochinchine, Hanoi 7 septembre 1904.

77. Allouard, *Pratique de la lutte*; Consigny, *Considérations sur les feux de brousse*. Montagnard labor maintained more than symbolic value into the late 1920s, and in 1928 the Société agricole et industrielle de Cam-Tiên listed "moïs" and Cham among its 600 "free" laborers (i.e., those without a contract). NAVN2 Goucoch IIB.56/029, Procès verbal de visite de la plantation de Cam-Tien, 1928 janvier 14. For Sipière discussion, see "Ray.–Dangers d'incendie," *APCI* 22 (1911): 641. Chamber of Commerce report in NAVN4 3113, Vœux de chambre mixte de commerce et d'agriculture de l'Annam, Touraine, 1932.

78. Cleary, "Managing the Forest," 274.

79. For more on the two schools of thought regarding plantation clearing, see Bonneuil, "Mettre en ordre," 303. See also the 1917 unpublished manuscript of Georges Vernet, who directed. Yersin's plantation in Nha Trang, NAVN2 N5.59. See Clarence-Smith, "The Rivaud-Hallet Plantation Group."

80. For more on Henry, see Coll, *Hommes et destins: Dictionnaire biographique d'outre-mer*; Tome V: Yves-Marie Henry (1875–1966), by André Angladette, 245. For quote, see Henry, *Terres rouges*, 34. For Consigny quote, see *Considérations sur les feux de brousse*, 1, 8.

81. Rollet et al., *Les forets claires*.

82. Gourou, *L'utilisation du sol*, 494.

83. Đặng, *100 Năm Cao Su*, x. For work in English defending swidden, see, inter alia, the writing of Hickey, "Memorandum for Record: Montagnard Agriculture and Land Tenure" of April 2, 1965. Hickey writes: "Using the swidden method has the advantage of leaving tree roots in the soil which helps to retain the structure. If a plow were used in these circumstances there would be grave danger of having the top soil wash away." From Salemink, *The Ethnography of Vietnam's Central Highlanders*, 243.

84. See Goscha, *Vietnam or Indochina?*

85. For recent work on the rubber plantations of Cambodia, see Slocomb, *Colons and Coolies*, and Boucheret, "Les plantations d'hévéas." De Lachevrotière was the first to grow rubber in Cambodia, according to the *Annuaire* produced by the Indochinese Rubber Planters Syndicate (SPCI). Numbers for 1931 from SPCI, *Annuaire*, 1931, 31; numbers for 1937 from Indochine, *Adresses, annuaire complet*. See also Union des planteurs de caoutchouc en Indochine, *Annuaire*.

86. Chollet, *Planteurs en Indochine*, 31–38. These relationships reflected on a small-scale trade that had once taken place between the Khmer and the Chinese courts. Loubet, *Monographie de la province*; Chandler, *A History of Cambodia*, 102–3. Even before the beginning of latex production, there had been attempts to grow export crops, such as cotton and maize, both on large and small scales. For example, in 1921, Adrien Hallet arranged for the transfer of 18,000 ha of land from the Industrial Cotton Society of the Hallet-Rivaud group to form the plantation of Chup. See Boucheret, "Les plantations d'hévéas," 57.

87. This section draws on Aso, "Rubber and Race."

88. NAC 4369, Aptitude au travail des différentes races qui habitent au Cambodge, 1923, 5.

89. NAVN2 2921, Troubles Phnongs à Budop (Thu Dau Mot), 1915, 1917, 1919, 1921.

90. See NAVN2 VA.2 041, (1) Rapport de l'administrateur de Thu Dau Mot sur le statut des Moïs de la région de Hon Quan et sur les différents survenus entre la Société des plantations d'hevéas de Xatrach et l'administration, 1912, Letter from the administration of Thủ Dầu Một, April 17, 1912, about the "Moïs" of Hớn Quản. N. 16, Procès verbal de la 26e Réunion du Comité 1er avril 1912, 220.

91. Ly, *Le régime des concessions domaniales*, 192–96.

92. NAC 5547, Demande de concession d'un terrain sis à Snoul (Kratié) formulée par la Société des Plantations de Kratié, 1924–33. See also NAC 11885, Certificat médical de Bui Quang Chieu concernant la situation du paludisme à Kratie, 1919; Tirouvanziam and Phleng, "Le paludisme à Kratié."

93. NAVN2 IIB.54/013, (1) Domaines, ouverture à la colonisation de Phu Rieng, demandes de concessions, cartes de régions, rapports, 1927 (région Moï) Letter from Goucoch to Director of Michelin, 13 juin 1927. This plantation was the setting for Trần's *The Red Earth.*

94. See Morère, *Le dialogue interrompu.* See NAVN2 III.59/N68, Rapport à GGI, Krautheimer, Goucoch, 1er avril 1933. ANOM FM EE/II/1762/Krautheime. Nominated for the "croix de chevalier à cause de mérite agricole," July 14, 1914, for promoting rubber plantations.

95. ANOM GGI 65478, Notes mensuelles sur les événements politiques—octobre 1932. *Saigon*, 15 janvier 1934. GGI *Arrêté* 17 mai 1929. For more on the Interprovincial Conference, see Jennings, *Imperial Heights*, 108–9.

96. Amicale des Anciens Planteurs d'Hévéa, "L'œuvre d'une Française en Cochinchine," 48.

97. Fortunel, "L'Etat, les paysanneries," 158.

98. Bonneuil, "Science and State Building," 268. See also Bonneuil, *Des savants pour l'empire*, and Brockway, *Science and Colonial Expansion*.

Chapter Two

1. Caresche, "Visite de la station agricole d'Ong Yem," 13. See also Achard, "Rapport sur les champs d'essais."

2. Caresche, "Visite de la station agricole d'Ong Yem," 14. See also Robequain, *The Economic Development*, 202. For agricultural experiments until 1897, see Société des études indochinoises, *Essais agricoles et industriels*.

3. For theorizing about Science, Technology, and Society in Asia, see Anderson, "How Far Can East Asian STS Go?" and Abraham, "Contradictory Spaces."

4. NAVN1 AFC 460, Letter, MDC to GGI, 31 août 1897.

5. For exploration in rubber, see Jumelle, *Les plantes à caoutchouc*. For the problem of incorporating "local" latex knowledge into "global" science, see Coll., "Nomenclature des principales lianes à caoutchouc."

6. Robequain, *The Economic Development*, 201.

7. Grove, *Green Imperialism*; NAVN2 IA.4/082, (6) Report, M. Carle, 31 décembre 1907. See also Carle, *L'hévéa brasiliensis*.

8. Chambre de commerce de Marseille, *Compte rendu*, 158.

9. NAVN1 Hà Đông 3472, Circulaire 20, RST, 22 mars 1900.

10. NAVN1 AFC 456, Report, M. Mieville, novembre 1908. See also NAVN1 AFC 456, Report, M. Jacques, n.d.

11. NAVN1 AFC 456, Letter, MDC to GGI, 30 août 1906; Note, n.d., on propagation of *ficus* by cuttings; Letter, Head of the Agricultural Service in Tonkin to AFC, 14 septembre 1906; Minute, AFC, 23 octobre 1906; Letter, Head of the Agricultural Service in Tonkin to AFC, 26 octobre 1906. For an understanding of agricultural concessions in Tonkin and northern Annam, see Tạ, *Đồn Điền*, and Tạ, *Việc Nhượng Đất*.

12. NAVN1 AFC 456, Letter, RST to AFC, 6 avril 1905.

13. Kohler, *Landscapes & Labscapes*.

14. Basalla, "The Spread of Western Science."

15. NAVN1 AFC 443, MDC to GGI, 24 novembre 1910.

16. NAVN2 IA.4/N4, (8) Report, Service agricoles de la Cochinchine, 29 juin 1908, and Robequain, *The Economic Development*, 239.

17. NAVN1 AFC 443, Supplement to Ministerial Circular 7.

18. NAVN1 AFC 443, Letter, RSC to GGI, 5 avril 1911, and Letter, Service agricoles et commercial locaux au Cambodge to RSC, 20 mars 1911.

19. NAVN1 GGI 7774, Letter, Institut scientifique de l'Indochine, 29 octobre 1919; NAVN1 GGI 7440, Lettre, Directeur de l'Institut scientifique de l'Indochine à Directeur de la Service économique, 29 décembre 1921.

20. ANOM GGI 64329, Rapports mensuels sur la politique, l'économie, les arrondisements etc. 1908. For the political reports, note the letter from Outrey, Goucoch, to GGI, May 29, 1908, concerning the unrest in Annam. While Cochinchina

seemed peaceful, mention is made of Gilbert Chiểu and his work on *Nông Cổ Mín Đàm,* the agricultural journal of Canavaggio and Jeantet. For more on Chiểu, see Brocheux, "1908: Le chassé-croisé G. Chiểu-P. Văn Trường"; Peycam, *The Birth;* Tai, *Radicalism.*

21. IRCI, *Cahiers,* 1942, 1944; L. Enderlin, "Industrie indochinoise." For more on this agreement, see Bauer, *The Rubber Industry.*

22. Aso, "The Scientist, the Governor, and the Planter"; ANOM Agefom, Carton 201, Dossier 143, Indochine, Institut français du caoutchouc (IFC), 1936–1939. See also Bureau du caoutchouc de l'Indochine française, *Réglementation.*

23. NAVN1 AFC 446, Girard, Report of the Head of the Agricultural Service in Cochinchina on rubber plants. See also Đặng, *100 Năm Cao Su,* 41–44. Belland is credited by French planters as being the first to grow *hevea* on a sizable level in Cochinchina. For currency information, see Bassino and Nakagawa, "Exchange Rates and Exchange Rate Policies."

24. NAVN1 AFC 446, "Les caoutchoucs dans les colonies françaises," *La Dépêche coloniale,* 12 juillet 1905. There were many in this period that thought rubber had "a great future." For example, NAVN1 AFC 451, Letter, Josselme à Goucoch, 30 avril 1897. Viz., Eberhardt and Dubard, *L'arbre à caouthcouc.*

25. See APCI, *Annales,* 1916, for a list of plantations.

26. APCI, *Annales,* 1911, 57–58.

27. Ibid., 11.

28. Ibid. See also NAVN1 RST 78050, Order 1740, GGI, 7 juin 1910, and NAVN1 AFC 460.

29. NAVN1 AFC 460, Lettre, N. Patouillard à GGI, 29 mai 1911, 184. The tipping point may have been this letter from N. Patouillard, the president of the Mycological Society in France. For more on this debate, see Aso, "The Scientist, the Governor, and the Planter."

30. See Dean, *Brazil,* and Grandin, *Fordlandia.* For more on coffee diseases, see Mc-Cook, "Global Rust Belt."

31. Petch, *The Diseases and Pests,* 2.

32. Petch, *The Physiology & Diseases;* quote from Petch, *The Diseases and Pests,* 1.

33. Chevalier, "Situation de la production," 78; Chevalier, "Rapport sur la cinquième Exposition internationale," 340–42.

34. Steinmann, *De ziekten en plagen van Hevea brasiliensis in Nederlandsch-Indië.* See preface of Sharples, *Diseases and Pests,* and Wildeman, "Plantations et maladies de l'hevea," 19.

35. Sharples, *Diseases and Pests,* 4.

36. Barlow, *The Natural Rubber Industry;* see chapter 4, "The Technologies of Production," 112–59. Weeding regimes could have other environmental consequences, and industrial agriculture and transnational capital had to deal with issues such as erosion. "Actes officiels: circulaire au sujet de l'obligation pour les acquéreurs à titre onéreux de terrains domaniaux de borner leurs propriétés avec des matériaux de longue durée" and "Renseignements à l'usage des Planteurs de caoutchouc (suite)," *APCI* 11 (1911):

79, 81. See also Blaikie, *Political Economy of Soil Erosion*. More recently see Anderson, *Eroding the Commons*, and Showers, *Imperial Gullies*.

37. "Renseignements à l'usage des planteurs de caoutchouc, Visite de M. H. N. Ridley," *APCI* 21 (1911): 613; Henry, *Terres rouges et terres noires basaltiques*, 73.

38. For more on weeding, see Knight, "A Precocious Appetite," and Moon, *Technology and Ethical Idealism*. Quote from *Hévéa Brasiliensis* (Rapport de mission de M. G. Vernet), 5e partie (suite), 147. For Vernet's passing, see SPCI, "Annonce du décès."

39. For a detailed analysis of clean weeding, see Bonneuil, "Mettre en ordre," chapter 5.

40. Caty, "Visite de la Plantation de Canque," 21. See also Caty, "L'amélioration des plantes" and Scott, *Seeing Like a State*.

41. Sharples, *Diseases and Pests*, x, 454.

42. Coolidge and Roosevelt, *Three Kingdoms*, 14. See also Mitman, *Reel Nature*.

43. Cochinchine française, *Rapports au Conseil colonial*, 1895, 1896, 1897, 1898, 1900–1901, 1902, 1903, 1904. No. 17, Procès verbal de la 27e/28e Réunion du Comité mai-juin 1912, 267, Primes à la destruction des fauves. No. 18, Procès verbal de la 29e Réunion du Comité juillet 1912, 301, Primes à la destruction des fauves.

44. NAVN2 IB.24/247, Chasse: Règlementation, décret du 10 mars 1925; Rapport des administrateurs: Bien Hoa, Baria, Thudaumot, directeur du Jardin Botanique, Service forestier, 1924–26.

45. ANOM INDONF, Carton 225, Dossier 1838. Folder 1933, Réclamation de M. Moutet . . . au sujet des coolies de caoutchouc, Plantation Gia Nhan. See also Boucheret, "Les plantations d'hévéas," 425.

46. For protective value of cattle, see Roubaud, "La différenciation des races zootropiques"; Roubaud, "À propos des races zoophiles"; and Mesnard and Toumanoff, "Note préliminaire sur l'anthropophilie."

47. *Bulletin de Société des études indochinoise*, 1910–11. See "Congrès colonial de Paris, 1910," including commissions on agriculture, etc. "Monographie de Thu Dau Mot," 22, 24–25. Includes information on Ông Yêm, Xa Trạch plantation, and Stieng. Background on Thủ Dầu Một from *BSEI*. Also see Morice and Tips, *People and Wildlife*.

48. Anon., *Bien Hoa en 1930*, 173; NAVN2 IIB.55/304 Baria, Divers, Colonisation, 1913–26, correspondance avec planteurs, etc., 1900–23, 1923. For more on epizootics, see NAVN1 116 AFC Rapports économiques des années 1898–1909 sur la situation économique dans la province Tay Ninh, Administrateur de Province à M Lt Gouv, (168) juillet/août, septembre/octobre: Mentions human and animal epidemics and the movement of Cambodians, Laotians, and Cham across the border, 87–89. 1904 juillet/août, 5 septembre: Epizootie, 93. For discussion of veterinary medicine, see Annick Guénel and Sylvia Klingberg, "Veterinary Science and Cattle Breeding in Colonial Indochina." For comparative recent work, see Brown and Gilfoyle, *Healing the Herds*.

49. NAVN2 IA.3/195, Includes several brochures and advertisements for agricultural implements, tents, etc. *APCI*, no. 27, avril 1913. NAVN1 RST 39791, Questionnaire sur le marché des tracteurs agricoles et forestiers, 1920–1921. Industrie et commerce,

Maison Renault à Billancourt, questionnaire sur le marché des tracteurs agricoles et forestiers, 1921. *APCI*, no. 14, 3 avril 1911, Renseignements à l'usage des planteurs de caoutchouc (suite), 241.

50. Barlow, *Natural Rubber Industry*. For an overview of colonization, see Boué, "La colonisation européenne"; and Pasquier, *La colonisation des terres incultes*.

51. APCI, *Annales*, 1913, 6.

52. Voon, *Western Rubber Planting*.

53. APCI, *Annales*, 1913, 15.

54. Barlow, *Natural Rubber Industry*, chapter 4; Grandin, *Fordlandia*; Dean, *Brazil*.

55. Bray, Coclanis, Fields-Black, and Schaefer, *Rice*.

56. APCI, *Annales*, 1913.

57. Henry, *Terres rouges et terres noires basaltiques*, 7. Henry also wrote about agricultural economics more generally, *Économie agricole*. For a contextualization of the study of tropical soils, see Chatelin, "Genèse, mutation et éclatement des paradigmes."

58. Henry, *Terres rouges et terres noires basaltiques*, 10.

59. Ibid., 17.

60. Ibid., 33. See also Castagnol, *Problèmes de sol*, chapter 6.

61. Henry, *Terres rouges et terres noires basaltiques*, 36.

62. Ibid., 44 bis. For earlier discussions of desiccation, see Grove, *Green Imperialism*.

63. Gourou, *Land Utilization*, 492.

64. Quotes from report of Henry, 23 mars 1925, in Carton, "L'œuvre de l'institut des recherches," 116.

65. Tilley, *Africa as a Living Laboratory*. For later work on soils, see Du Pasquier, *Les problèmes d'utilisation*.

66. Kleinen, "Tropicality and Topicality," 334.

67. Anon., *Scandales de l'agriculture*. For more on Henry, see Coll., *Hommes et Destins*.

68. Cucherousset, "Nos banques agricoles."

69. Lettre de l'Association 'colonies et sciences', *Bulletin du syndicat des planteurs de caoutchouc de l'Indochine* 90 (1926): 363–64. See also letter from ASC asking for financial contributions in "Rapport sur le fonctionnement de l'Association *Colonies-Sciences* en 1926," *Bulletin du syndicat des planteurs de caoutchouc de l'Indochine* 100 (1927): 95. For a comparison of agriculture in Indochina and in France, see Chevalier, *L'organisation de l'agriculture coloniale*.

70. Efforts to gain equality for native and French forestry agents had continually been rebuffed. See ESASI, *Bulletin de l'Association amicale des anciens élèves de l'École supérieure d'agriculture et de sylviculture de l'Indochine*, no. 7, 1938: 184.

71. Nguyễn and Đỗ, *Cahier des vœux*, 22.

72. Chollet, *Planteurs en Indochine*, 149–53, quote from page 153.

73. See Conklin, *A Mission to Civilize*. For Việt Nam, see Ha, "From 'Nos Ancêtres, Les Gaulois' "; Kelly, *Franco-Vietnamese Schools*; Trinh, *L'école française*; Bezançon, *Une colonisation éducatrice*. For the popularization of Western knowledge at the beginning of the twentieth century, see Affidi, "Vulgarisation du savoir."

74. ANOM, Rapport sur l'Enseignement, Fontaine, Hanoi 1 août 1904, p. 9, Section "Collèges Indigènes." Bureaux included Travaux Publics, le Cadastre, les Postes et Télégraphes, les Douanes et Régies le Commerce et l'Industrie.

75. For the reforms of 1908, see Gantès and Nguyễn, *Vietnam le moment moderniste.* For *mise en valeur*, see Conklin, *A Mission to Civilize*. See also Sarraut, *La mise en valeur.* For imperial Britain see Hodge, *Triumph of the Expert.*

76. There had been Vietnamese who had obtained degrees in agricultural engineering in France, such as Bùi Quang Chiêu and Lâm Ngoc Chân (Monpellier), but these engineers were few. Caresche, "Visite de la station agricole d'Ong Yem."

77. The French, as one historian has quipped, tend to implement social, political, and economic programs by setting up schools. Artz, *Technical Education in France,* 220–30. For an early summary of agricultural schools, see Pham, *Vietnamese Peasants,* 15–19. For similar questions in the Dutch East Indies, see Moon, *Technology and Ethical Idealism,* 35–41.

78. Li, *The Will to Improve,* 4–5. Li convincingly argues that government officials, from the Dutch East Indies to the New Order, repeatedly invoked this term to intervene, often violently, into Indonesians' lives. For trusteeship in Việt Nam, see Bradley, *Imagining Vietnam and America.*

79. See complaints of low enrollment in ANOM Agefom, Carton 243, Dossier 325, Lettre, Directeur du Service d'Education à GGI, Hanoi, 5 décembre 1923. This was an enrollment of mediators rather than elite. See Moon, *Technology and Ethical Idealism,* 54–69.

80. NAVN1 GGI 7492, Création à l'Ecole pratique d'Agriculture de Ben Cat (Thu Dau Mot) d'une section d'ouvriers spécialistes agricoles en faveur des pupilles de la Société de protection de l'enfance en Indochine, 1923–24. For the founding of the penal agricultural colony, see "La colonie pénitentiaire d'Ong-yem," *La quinzaine coloniale: organe de l'Union coloniale française* 9, no. 17 (1905): 571–72.

81. Kelly, *Franco-Vietnamese Schools,* and Ha, "From 'Nos Ancêtres, Les Gaulois,'" 109–10.

82. Of the others, twenty-nine were working for the agricultural department, thirteen for the forestry department, fourteen as farmers, and one miscellaneous. Pham writes that Bên Cát closed and reopened due to low attendance. Pham, *Vietnamese Peasants,* 18.

83. Names: "Ecole pratique d'agriculture," "Ferme-Ecole de Ben Cat," and "Ecole pratique d'agriculture et de Sylviculture de Ben Cat." *Khoa Học Phổ Thông,* no. 42, 16 mai 1936, 605, and no. 43, 1 juin 1936, 634, "Trường Canh Nông Bên Cát," Nguyễn Háo Ca. In 1936, Bến Cat and Xano, a school opened the previous year near Cần Thơ to teach rice farming, were set to join. For their second year, students could choose to study either rice at Xano or rubber, *cao-su*, and forestry, *kiểm lâm*, at Bến Cat. See also NAVN2 IB.24/271, Conseil colonial: Livre vert, Services Agricole, 13p.

84. ESASI, *Bulletin de l'Association amicale des anciens élèves de l'École supérieure d'agriculture et de sylviculture de l'Indochine,* no. 7 (1938): 180–84. See also *L'Effort indochinois,* no. 91, 15 juillet 1938, Đại letter dedicated to M. le Recteur Bertrand.

85. ANOM Agefom, Carton 243, Dossier 325, En Indo-Chine: Création d'une école spéciale d'agriculture et de sylviculture, 1. Cambodians and Laotians who were admitted but did not meet all the educational requirements were offered supplementary courses. All student expenses were covered by the general budget. See Marr, *Vietnamese Tradition on Trial*, and Tai, *Radicalism*. See also *L'Ecole Spéciale d'Agriculture et de Sylviculture Jules Brévié, Organisation et Programme* (Hanoi: IDEO, 1938); *Arrêté* du 15 août 1938 modifié par ceux du 21 octobre 1938 et 23 janvier 1939.

86. See *L'Effort indochinois*, no. 91, 15 juillet 1938, 3; and no. 95, 12 août 1938, 3.

87. ESASI, *Bulletin de l'Association amicale des anciens élèves de l'École supérieure d'agriculture et de sylviculture de l'Indochine*, no. 7, 1938: 178.

88. ANOM Agefom, Carton 243, Dossier 325, *La Verité, Agence indochinois*, août 1938, Autour de la création de l'école spéciale d'agriculture et de sylviculture en Indochine. See also Amicale mutuelle des employés indigènes de l'agriculture en Cochinchine, au Cambodge et dans le sud Annam (A.M.E.I.D.A.C.). *Bulletin de renseignements*.

89. Nguyễn, "The Vietnamese Confucian Literati," 236. See also Durand, "Note de technologie vietnamienne"; Huard and Durand, "Science au Viet-Nam." For an idea of why Japan might have been a useful model, see Walker, "Meiji Modernization." See also Marr, *Vietnamese Tradition on Trial*, and Peycam, *The Birth*. Citing numerous texts ranging from Buddhist prayer texts to racy detective novels, Shawn McHale makes the convincing case that Vietnamese were thinking about more than just revolution and nation in his *Print and Power*. See Phan, *La Fraternité*, cited in Brocheux, "Chassé-Croisé," 208. For an effort to translate scientific terms from the late 1930s, see Trân, "Tieng Annam."

90. Woodside, "The Development of Social Organizations," 49. The first article was written by Nguyễn Thượng Huyền, "Mấy Điều Mới Phát-Minh về Nghề Nông Pho," 31–33, and the second was by Đỗ-Văn-Minh, a teachers' training college student, titled "Nông-Phố Học," 569–73. See also Lan, *Notions d'agrologie/Địa Học Yếu Lược*; Lan, *Notions de botanique/Thực Vật Học Yếu Lược*; Nguyễn, *Những Sự Kỳ-Quan*; Nguyễn, *Sách Làm Ruộng*.

91. For more on mediators, see Raj, *Relocating Modern Science*; Elman, *On Their Own Terms*; and Au, *Mixed Medicines*.

92. For some examples of their work see NAVN2 IA.8/033, (2) Enseignement: Ấu Học Nông Phố (Premières lectures agricoles de P. Braemer); NAVN2 IA.8/032, (8) Enseignment: Canh Nông Chưởng Pháp (précis agriculture) par Nguyễn Thanh Chưởng, 1911–12.

93. *Khoa Học Phổ Thông*, no. 1, 1 septembre 1934: 1–2.

94. The journal was published in Sài Gòn, and its editor was Trân Văn Đôn, who, after beginning his career as a bureaucrat, received his medical degree in France partly because of his contributions to the war effort during World War I. After submitting his thesis in 1920, he returned to Sài Gòn to work as a doctor and started a political party.

95. A few articles were translations from French authors, but most were written by Vietnamese agricultural engineers and other scientists. From the mundane agricultural

topics covered, one can guess that the intended audience included rural farmers. But the journal also published more philosophical pieces, such as "What Is Farming?," with quotes from J. J. Rousseau on the importance of labor.

96. *Khoa Học Phổ Thông*, no. 53, 1er novembre 1936; no. 54, 16 novembre 1936; no. 61, 1 mars 1937: "Khí Hậu với Sinh Vật Trong Một Xứ," Vũ Đức Hiện.

97. *Khoa Học Phổ Thông*, "Ra Dời," no. 1, 1er septembre 1934: 2; *Khoa Học Phổ Thông*, "Agriculture française–Agriculture annamite," Nguyễn Háo Ca in Supplément en français, no. 73, 1er septembre 1937: 37.

98. *Khoa Học Phổ Thông*, nos. 13–16, 1er mars 1935. Exact title was Khâm sứ Trung-Kỳ.

99. See, for example, Bray, *Science and Civilisation in China* and the works of Georges Métailié.

100. Trần, "Tiển Đồ Khoa-Học."

101. Some of these authors, not surprisingly, had political views, and Phạm Ngọc Thạch, later a famous medical doctor in the DRV, contributed articles.

102. Barlow, *Natural Rubber Industry*, 220, 242.

103. Ibid., 243.

104. See Murray, " 'White Gold,' " 59, where Murray terms plantations "European operations *par excellence*."

105. Nam Co, "Su bien doi cua huong thon tu xua den nay," *Nam Phong*, no. 76 (juin 1923): 327–28, as cited in Pham, *Vietnamese Peasants*, 143–46. See also Chevalier, "Economie rurale."

106. As Suzanne Moon so poetically points out in the case of Indonesia, "Small might be beautiful; at times, however, it has also been colonial." Moon, *Technology and Ethical Idealism*, 150. For trains in Indochina, see del Testa, "Paint the Trains Red."

107. ANOM FM Guernut 87, Report of admininstrateur de Thu Dau Mot, M. Wolf, chef de la province, 1936, délégation de Hon Quan, 5. Former plantation workers and their children were one possible source of everyday rubber knowledge. For instance, according to one local official, most of the inhabitants of Thanh Son village, Tan-Minh canton, deep in red-earth territory, were former "coolies" on the surrounding plantations. Yet, instead of owning their own trees, the men in this village of 197 people continued to work for plantations for a salary of 0.35 piaster per day without food. ANOM FM Guernut 87, M. Nguyên Tấn Biên, Instituteur de 7ème classe à Giadinh, à l'école primaire de Phu-Hoà-Dông gives an example from Gia Định of the income of a poor family.

108. In 1927, a former student of ESASI subsequently moved to Sài Gòn and requested employment on a plantation. Not only agricultural graduates were seeking positions, as was shown by the request of a recent graduate from l'Ecole de Médecine de l'Indochine, who hoped to earn 300 piasters per month plus furnished housing while working on a plantation. See *Cochinchine agricole*, no. 7, July 1927; no. 9, September 1929; and no. 2, February 1929, respectively, for the 32-year-old Phạm-Tuân, Đào-Văn-Thác, and Tran-Tran-Lai. There was also turnover in the plantations as two years later, someone who had worked since 1926 on the Courtenay plantation in Baria as a secretary, warehouseman, and su (*surveillant*, overseer) sought a new position after his job was cut.

109. In addition to the examples discussed above, the government passed laws intended to promote smallholders. For example, Goucoch Circulaire du 1 décembre 1925; GGI *arrêté* du 19 septembre 1926 cited in *Cochinchine agricole*, no. 1, 1927: 4–5.

110. See Bonneuil, "Mettre en Ordre," 286 bis: Of 575 native-owned operations, 520 were 0–40 hectares, 43 were 40–100 hectares, 12 were 100–500 hectares, and nothing was above.

111. Ibid., 289. In terms of numbers, fewer than ten European-owned plantations (and conglomerates) produced most of the rubber, with these mega-plantations employing by far the greatest number of workers from Tonkin and Annam.

112. Moon, *Technology and Ethical Idealism*; Goss, "Decent Colonialism?" and Goss, *The Floracrats*.

113. Pyenson, *Civilizing Mission*; Anon., *Scandales*, 37–40; Bonneuil and Petitjean, "Les chemins de la création de l'ORSTOM."

114. *Saigon*, 20 octobre 1933, no. 141, and 26 avril 1935, p. 1 "Sanh sản rồi, bán cho ai đây? Nam-kỳ sẻ sanh sản thêm ba món quí: bắp, đào lộn hột và đậu nành, Nhưng phải nhờ ở sức của chánh-phũm phòng thương-mãi và phòng canh-nông . . ." [Things are ready, but whom to sell to? Cochinchina will be ready to sell three more valuable foodstuffs: corn, cashew nut, and soybean, But it must rely on the power of the government, the chamber of commerce and agriculture.]

115. Bonneuil, "Mettre en Ordre," 289.

116. Instead, these farmers grew mostly tobacco and fruit. See records from ANOM FM Guernut 87.

117. See APCI, *Annuaire*, 1937.

118. By contrast, the "coolies" (the derogatory term for an unskilled worker) working on his plantations were likely to earn a little less than 150 piasters per year, based on standard wage rates for the area, a tiny amount even if housing and some rice were provided. ANOM FM Guernut 87, report of Admininstrateur de Thu Dau Mot, Circonscriptions de Chef-lieu et de Ben Cat, 26 avril 1938, Lê van Truyện, 19, a landowner could be called well off with 3,400 piasters in gross income.

119. See ANOM FM Guernut 87, report of Admininstrateur de Thu Dau Mot. Annual budget of rich local planter, 18. A more modest, and more typical, example of a Vietnamese rubber grower also comes from the province of Gia Định, close to population centers, such as Sài Gòn, and on gray earth soils. This family of Vo Van Vê, the head of a canton, owned thirty hectares on which he produced five tons of rubber, which brought in 5,000 piasters. In his rubber growing, he incurred the following costs: plantation maintenance, 1,000 piasters; labor and harrowing, 100 piasters; taxes and land tax, 500 piasters; lighting and heating, 100 piasters; cattle, 200 piasters; fertilizer, 100 piasters; and acid and crates and transport, 600 piasters. Judging by these numbers, clearly this family did most of their own growing with the help of a few workers, most likely about ten at any one time. Both Madame Nguyễn Đức Nhuận, owner of *Phụ Nữ Tân Văn*, and Nguyễn Văn Của, the Vietnamese publisher with a contract from the colonial government, and thus incredibly wealthy, owned plantations in Biên Hòa.

120. Peycam has discussed, for example, how a one-sou increase in coffee prices collectively instituted by some Chinese shop owners in August 1919 met with an outcry from the Vietnamese controlled press. Although this activism failed to garner active mass support, it represented one of the early attempts to do so. Peycam, *The Birth*.

121. NAVN1 RST 70 124, au sujet de la croisière touristique organisées par Ho Van Lang à Saigon, 5.12.1937, 4.1.1938 (5); Peycam, *The Birth*.

122. Creation of indigenous plantations: ANOM GGI 2239, Agriculture caoutchouc.

123. Boucheret, "Les plantations d'héveas," and Aso, "The Scientist, the Governor, and the Planter."

124. This technique, which consisted of matching three parts of the tree—the roots, the trunk, and the foliage—creating what today might be called a GMO (genetically modified organism), had been developed in the Dutch East Indies. The difference in yield could be quite substantial. For example, between the fifth and the twelfth years a hectare planted with seedlings on the best-quality red earth soil could yield 700 kilograms of latex per year. By contrast, between the fifth and the twelfth years a hectare planted with grafted trees on the best-quality red earth soil could yield 1200 kilograms of latex per year. ANOM FM Guernut 26, BAREME, 4.

125. Li notes that programmers enact "programs to move populations from one place to another, better to provide for their needs; programs to rationalize the use of land, dividing farm from forest; programs to educate and modernize," all of which "are implicated in contemporary sites of struggle." Li, *The Will to Improve*, 1.

126. Ibid., 7.

127. Biggs, "Breaking from the Colonial Mold." Michael Dove has written a series of articles on smallholder latex production among the Dayaks in Borneo in which he shows the importance of preexisting patterns of trade and agriculture and explores human-plant metaphors in the local language. Dove, "Smallholder Rubber and Swidden Agriculture"; Dove, "Political vs. Techno-Economic Factors"; and Dove, "Living Rubber, Dead Land." Dove cites an anonymous 1925 poem that illustrates the uptake of rubber among the Dayaks of Borneo. "Now all is changed great peace and quiet / The sharp-edged sword becomes the tapper's knife. / The carved shield becomes a swing / Wherein is wrapped in clothes the babe whose future lies / In the price of rubber tapped in a ring." In Dove, "Transition from Native Forest Rubbers," 392. Even though latex production had taken place in Việt Nam before the introduction of *hevea*, it seems there was no Vietnamese word before *cao su* for latex; the word *nhựa két* was used for gutta-percha.

128. ANOM FM Guernut 87, report of Administrateur de Thu Dau Mot. Expanding Vietnamese production was mentioned, for example, in a summary of the desires presented to the Guernut commission by the village of An-Tây-Thôn, which was also in Thủ Dầu Một province, west of RC 13, south of Bên Cát (near the Củ Chi tunnels). There were some spin-off industries from rubber that were bringing changes to farmers and artisans in the countryside. For example, the administrator of Thủ Dầu Một noted that the local ceramics industry was producing bowls for rubber collecting. For ceramics, see p. 17; for forges that made knives, probably for tapping, see p. 19; for carts, see p. 13. See also "Que le caoutchouc soit planté comme auparavant," "Vœu contre la

restriction du caoutchouc," and "Vœux présentés par les habitants du village de AN-
TAY-THON (sic) le 11-11-37." In "Autoriser les planteurs à replanter à replanter [sic] les
arbres à caoutchouc," some potential Vietnamese planters said that the 1934 international agreement restricting rubber production was limiting access to rubber planting
at a time when more and more Vietnamese wished to plant.

129. *L'Effort indochinois*, no. 91, 15 juillet 1938. Đại discussed two main problems that
Indochina faced: (1) industrialization and (2) the overpopulation of certain parts
of Indochina. Đại discussed what he knew (i.e., industrial agriculture). Up to this point
the government's efforts at colonization had largely failed. One reason was that the
government was sending those with essentially no training to the countryside. Đại
noted approvingly the example of the British empire, where the government sent
those with appropriate training to Canada and India.

130. ANOM INDO GGI SE1642, Belisaire; Lâm Văn Vàng article in *Khoa Học Phổ
Thông*; ANOM SE2240, Caoutchouc. For rubber in the late 1930s, see Vaxelaire, *Le
caoutchouc*.

131. See Cucherousset, *Le Tonkin*.

Chapter Three

1. This composite picture is drawn from many sources, including Trần, *The Red
Earth*, Schultz, *Dans la griffe*, and in particular Kalikiti, "Rubber Plantations." For original contracts, see among others NAVN1 RND 3277, M11 series. Contrats de travail des
coolies de Nam Dinh engagés par la Société civile d'Etudes de Cultures de la Cochinchine et du Cambodge à destination de la Cochinchine de juillet et novembre 1926.

2. For material on labor regulations for recruits, see NAVN2 IIA.46/223, Main
d'œuvre, dossiers relatifs à la règlementation de la main d'œuvre indigène et javanaise en Indochine: affaires diverses concernant la main d'œuvre des plantations
caoutchouc en Cochinchine, 1905, 1910–18. Officially, recruiters were responsible for
feeding workers while they made the journey from north to south, but in practice
these recruits often received inadequate food rations. For malaria rates in migrants see
Bordes and Mesnard, "Notes sur le taux d'infection palustre."

3. For an overview from the French perspective, see ANOM Agefom, Carton 190,
Dossier 106; Bos, "Le développement et l'avenir." See also NAVN2 IA.8/272(2), Rap-
port sur le fonctionnement de l'Assistance médicale en Cochinchine, Exposition de
Marseille, 25.

4. For discussion of an "epidemic" of malaria, see Grall, "Paludisme 'Épidémié.'"
Guénel, "Malaria Control"; Guénel, "Malaria, Colonial Economics and Migrations";
and Guénel and Guerin, "'L'ennemi, c'est le moustique.'"

5. See Packard, *White Plague*, inter alia, 5; Wilder, "Colonial Ethnology and Political
Rationality in French West Africa," 337, 354.

6. For older forms of protest, see NAVN2 IIB.54/112, (2) Domaines, dossier de BIF,
déplacement de la gare de Trang Bom, protestation des coolies Tonkin de BIF contre
le non paiement de leurs salaires dus, 1912–14.

7. Amrith, *Decolonizing International Health*, 28.

8. Packard, *The Making of a Tropical Disease*, 35 and n. 33, 262; Morin, *Entretiens sur le paludisme*, xxx. See also Morin and Bader, "Recherches sur quelques facteurs physico-chimiques," 1933 and 1934.

9. NAVN1 AFC 566, Ordonnance de la 5e Année de Tu-Duc (1852) et de la 17e Année de Tu-Duc (1864); Thompson, *Vietnamese Traditional Medicine*. Given the recent use of the plant artemisia to treat malaria, folk medicine might well have been effective. Marr, "Vietnamese Attitudes," 177. See Au, *Mixed Medicines*, for the generation of resistance to colonial medicine. For an early account of Nguyễn court medicine, see Duvigneau, *Le Thai-Y-Vien*.

10. Gouvernement générale de l'Indochine, *L'Indochine française*, 316, for numbers from 1876. For missionary care, see Daughton, *An Empire Divided*, and Marr, "Vietnamese Attitudes," 179.

11. Candé, *De la mortalité des Européens*; Monnais, *Médecine et colonisation*, 45, 47. Cholera, along with plague and smallpox, received most of the colonial medical attention.

12. For writing about malaria on the coasts, see Simond, "Paludisme." See also Monnais, *Médecine et colonisation*, 80; Kérandel, "Riziculture et distribution géographique." For ideas about views of malaria among lowlanders and uplanders, see Morin, *Entretiens sur le paludisme*, xxx. For understandings of *falciparum*, see Brau, "Paludisme à P. falciparum."

13. Candé, *De la mortalité des Européens*, 38. For health conditions in the Mekong Delta during the nineteenth century, see Tai, *Millenarianism*; Borel, "Anophèles et paludisme"; Gouvernement générale de l'Indochine, *L'Indochine française*, 317; Kérandel, "Riziculture et distribution géographique."

14. Guénel, "The Creation of the First Overseas Pasteur Institute," 22. See also Bernard, *La vie et l'œuvre*. For a critical perspective on the research of Calmette for the alcohol industry in Việt Nam, see Sasges, "Contraband, Capital, and the Colonial State." For an account of life at the Pasteur Institute in Nha Trang from the perspective of a woman, see Hoenstadt, *Three Years in Vietnam*.

15. See discussions in the *BSPCI*.

16. For 1896 law, for example, see NAVN2 IB.23/124, IIA.46/224, and IIA.46/311. For a look at *mise en valeur* through the lens of rubber and rice, see Bone, "Rice, Rubber, and Development Policies."

17. NAVN1 AFC 158, Lettre, Goucoch Rodier à GGI, 21 juin 1907, au sujet de la colonisation européenne. NAVN2 IIA.46/311, Lettre, Outrey à Cremazy et Thiollier, 14 septembre 1907, au sujet des coolies tonkinois. See the discussion of informal and formal labor recruitment processes set up between Tonkin and Annam and Cochinchina in Kalikiti, "Rubber Plantations," and Boucheret, "Les plantations d'hévéas." Archival files contain clippings of Chinese press attacks on plantation conditions. A few plantations, such as Lộc Ninh, directed by Paul Cibot, employed several Javanese to tap trees. Cibot also translated rubber knowledge into French, e.g. Stanley Arden, *L'hévéa brasiliensis dans la péninsule malaise* (Paris: Augustin Challamel, 1904).

18. See ANOM FP 36APC/1, Cibot papers. Later, in 1927, the director of Bù Đốp plantation stated that death rates on the nearby Lộc Ninh had been higher than even Bù Đốp's astronomical rates that year.

19. See annual reports from 1911 in NAVN2 IA.8/113 and IA.8/103 and provincial material from 1906 in IA.8/077. Cochinchine française, *Rapports au CC*, 1909, xx, xxvi. The year 1910 witnessed an epidemic of cholera, which struck the Mekong Delta especially hard near the end of the rainy season in November.

20. NAVN2 IA.8/103, Rapport pour le mois de février et mars. See also Trần, "Cong Trình Khảo Cứu" and "Note sur le paludisme contracté."

21. Cochinchine française, *Rapports au CC*, 1915, 249.

22. Cochinchine française, *Rapports au CC*, 1910, xxl; Allain, "Paludisme et quinine d'État"; Hermant, "Fonctionnement du service," 231–33. Skepticism among villagers is understandable, given that injections could cause tetanus, Gaide and Dorolle, "Sur des cas de tétanos."

23. *BSPCI*, no. 88 (1926): 340–41; Gouvernement général de l' Indochine, *L'Indochine française*; and *BSPCI*, no. 36 (1921), and no. 41 (1922): 613, about free quinine. For Yersin's work on quinine, see Yersin and Lambert, *Essais d'acclimatation*.

24. For an overview of quinine, see Morin, "Le paludisme et sa prophylaxie," 400. See also Société des Nations, *Enquête sur les besoins en quinine*.

25. NAVN2 IB.24/039; see also IA.10/126, Rapport 7325, l'ingénieur en chef, 30 juillet 1915, au sujet de l'emprunt de 90 millions, compte rendu de l'état d'avancement des travaux au 1er juillet. Quote from Anon., *Bien Hoa en 1930*, 219. For malaria at Gia Rây, see Borel, "Enquête malariologique."

26. See discussions about difficulties of recruiting labor articulated in *BSPCI*.

27. NAVN2 IIA.46/223, Main d'œuvre, dossiers relatifs à la règlementation de la main d'œuvre indigène et javanaise en Indochine: affaires diverses concernant la main d'œuvre des plantations caoutchouc en Cochinchine, 1905, 1910–18.

28. NAVN2 IIA.46/223, Lettre de L'Inspecteur des Affaires politiques et administratives de la Cochinchine, Quesnel, à Goucoch, 11 avril 1918, au sujet de la main d'œuvre agricole.

29. Thompson, *Labor Problems*.

30. NAVN2 IIA.46/223, Main d'œuvre, travaux de la commission instituée pour le projet de réglementation de la main d'œuvre indigène, procès-verbaux de 1918. See also *BSPCI*, no. 10 (1918).

31. NAVN2 IIA.46/224, e.g., Lettre, Crémazy à Goucoch, 22 juillet 1913 et Lettre, Cibot à Crémazy, 9 juillet 1913; NAVN2 IIA.46/223, Main d'œuvre, travaux de la commission instituée pour le projet de réglementation de la main d'œuvre indigène, 2e partie, procès-verbaux 10, 23 février 1918.

32. NAVN2 IIA.46/223, Main d'œuvre, travaux de la commission instituée pour le projet de réglementation de la main d'œuvre indigène, 2e partie, procès-verbaux 10, 23 février 1918.

33. *BSPCI*, no. 10 (1918). The local laws for tenant and sharecropping farmers were to be kept in place. Javanese migrant workers were covered by *arrêtés* of 1910 and 1913.

Other "Asiatics" were to be regulated by the 1918 *arrêté*. NAVN2 IIA.46/223, Main d'œuvre, travaux de la commission instituée pour le projet de réglementation de la main d'œuvre indigène, 2e partie, procès-verbaux 10, 23 février 1918.

34. NAVN2 IIA.46/223, Main d'œuvre, travaux de la commission instituée pour le projet de réglementation de la main d'œuvre indigène, 2e partie, procès-verbaux 11, 7 mars 1918. See also *BSPCI*, no. 10 (1918), Article 11, and 61, point 4.

35. See Brocheux and Hémery, *Indochina*; *BSPCI*, no. 6 (1918), "Note sur la main-d'œuvre"; NAVN2 IIB.56/029, "Rapport du 11 septembre 1928 du médecin chargé de l'assistance médicale de la province de Bienhoa au sujet des résultats de l'enquête faite à la plantation Michelin à Phu Rieng."

36. Nam, *Chi Pheo and Other Stories*. See also Hardy, *Red Hills*; *BSPCI*, no. 122 (1929): 344–46; Tran, *Ties That Bind*, 49–60.

37. ANOM INDO GGI 16360, Admission dans les hôpitaux de la Cochinchine des indigènes originaires du Tonkin de l'Annam et du Cambodge, 1912.

38. NAVN2 IA.8/224. We must be skeptical of claims that Lực had been injured before being recruited, given how brutal the trip could be. See also *arrêtés* of December 29, 1905, free medical care for natives in NAVN2 A.8/155(23).

39. NAVN2 IIA.46/306, dossiers concernant la main d'œuvre agricole asiatique en Cochinchine, statuts, etc., 1922–25. See letters including Administrator of Thudaumot to labor inspector, June 1923; Administrator of Bienhoa to labor inspector, May 15, 1923; Administrator of Tayninh to labor inspector, May 18, 1923.

40. NAVN2 IIA.46/306, dossiers concernant la main d'œuvre agricole asiatique en Cochinchine, statuts, etc., 1922–25, Administrator of Baria to labor inspector, May 18, 1923. For a planter account of planter-worker relations, see Duval and Chébaut, *L'appel de la rizière*.

41. NAVN2 IIA.46/306, dossiers concernant la main d'œuvre agricole asiatique en Cochinchine, statuts, etc., 1922–25, Procès verbaux des visites.

42. Delamarre, *L'émigration et l'immigration*, 28, chart after 34; Boucheret, "Les plantations d'hévéas," 213–15, in particular fig. 32, 214.

43. SPCI, *Annuaire*, 1926, 59. For malaria in Hà Tiên see Montel, "Un essai de quinine préventive."

44. NAVN2 IIA.46/306, dossiers concernant la main d'œuvre agricole asiatique en Cochinchine, statuts, etc., 1922–25. Nguyen Dinh Tri, délègue administrateur du chef-lieu, à l'administrateur de Hatien, 29 juillet 1924.

45. Trần, *The Red Earth*, 26.

46. NAVN2 IIB.56/029, Travail, rapports et Procès-verbal de la visite des plantations de l'inspecteur du travail, 1927–28; see Bù Đốp visits on August 22, 1927, March 23, 1928, and May 18, 1928.

47. Cochinchine française, *Rapports au CC*, 1922, 275; 1921, 181. See Murray, " 'White Gold,' " and Brocheux, "Le prolétariat des plantations."

48. Morin argued that despite gross underreporting, over the last fifteen years malaria was the number one cause of hospitalizations (12 percent) and number two in

immediate cause of death (10 percent, second to cholera). Morin, "Le paludisme et sa prophylaxie," 391. For more on Morin, see chapter 4.

49. Monnais, *Médecine et colonisation*, 80, Tableau 5: Calendrier des épidémies secondaires de la péninsule indochinoise (1897–1931). Furthermore, as Monnais points out, malaria works to weaken the immune system, and there were doubtless many hospitalizations and deaths attributed to other diseases that had malaria as a contributing factor.

50. Borel, "La constitution du sol." For work on Southeast Asia, see Walton et al., "Population Structure and Population History"; Walton et al., "Genetic Population Structure"; Manguin et al., "Scar Markers," 51. For work on Việt Nam, see Van Bortel et al., "Population Genetic Structure"; Trung et al., "Malaria Transmission"; Trung et al., "Behavioural Heterogeneity."

51. Another one of the paradoxes Morin refers to in "Le paludisme et sa prophylaxie."

52. Brocheux, "Le prolétariat des plantations," 265.

53. The most common regime in the late 1920s appears to have been 0.75g, 0.50g, 0.25g, three times a week. Brocheux, "Le prolétariat des plantations," 218, n. 862, notes the managers' belief that workers didn't like to take quinine, and on page 261 he argues that quinine alone was not enough to keep workers healthy—a reasonable conclusion. See also Assali, "Intolérance absolue à la quinine."

54. Phạm Ngộc Hồng, interview transcript, 9.

55. Perrow, *Normal Accidents*, 14; and Walker, *Toxic Archipelago*, 17, for a discussion of Perrow's ideas.

56. League of Nations, *Report of the Intergovernmental Conference*, especially chapter 4, 65–85; Navarre, *Manuel d'hygiène coloniale*, 222.

57. Oberländer, "The Rise of Western 'Scientific Medicine' in Japan"; Carpenter, *Beriberi, White Rice, and Vitamin B*; Thịnh, *Etude sur l'étiologie du béribéri*, 41.

58. NAVN2 IIB.56/029, Procès-verbal de la visite effectuée le 15 mai 1928, sur la Plantation de Xatrach; McIntyre, "Eating the Nation," 297, n. 22. See also Delaye, "Colonial Co-Operation," 470. See also Bernard, *Les Instituts Pasteur d'Indochine*.

59. For Snoul, see Chollet, *Planteurs en Indochine française*, 85–92. For foraging, see ANOM INDONF, Carton 225, Dossier 1838, especially Folder 1933, Réclamation de M. Moutet au sujet des coolies de caoutchouc, Gia Nhan plantation. A letter from a plantation manager and five workers complained about work conditions on the plantations, saying that when rice was not enough, the workers went into the forest for food. For high prices, see Boucheret, "Les plantations d'hévéas," 425.

60. Delamarre, *L' émigration et l'immigration*, 20, 34, and after. See also Boucheret, "Les plantations d'hévéas," 441.

61. *BSPCI*, no. 87 (1926), 307. For more on Guillerm and fish sauce, see McIntyre, "Eating the Nation," 171, 330. *Nước mắm*, or fish sauce, contributed 7 percent of average daily protein needs. See also Guillerm, *Rapport à la commission*; Guillerm and Morin, "Essai de détermination statistique"; and later reprint, Morin, "La ration alimentaire ouvrière." See also "Résumé de la partie non politique du discours de M. le Gouverneur de la Cochinchine," *L'Eveil économique de l'Indochine* 10, no. 481 (1926): 4.

62. Delamarre, *L'émigration et l'immigration*. This section is also based on the meeting minutes of the labor inspectorate after 1927 that are held in NAVN2. See also NAC 8218, Rapport sur la situation matérielle et morale des émigrés tonkinois . . . 1927; NAC 1654, Rapports du médecin du service mobile d'hygiène, 1929, Le Nestour.

63. Cucherousset, "La question de la main-d'œuvre"; Monet, *Les jauniers*. This work in turn inspired writers such as Schultz, *Dans la griffe*, and Viollis, *Indochine S.O.S.*

64. Boucheret, "Les plantations d'hévéas," 223, and Schultz, *Dans la griffe*.

65. McIntyre, "Eating the Nation," 287, n. 16, citing a 1933 article from Nguyễn Văn Kiệm, Hardy, et al. Morin called for 750–800 grams in Morin, "La ration alimentaire ouvrière."

66. *BSPCI*, no. 101 (1927); *BSPCI*, no. 109 (1928). For prison conditions, see Zinoman, *Colonial Bastille*. NAVN2 IIB.56/029, Rapports et procès-verbaux des visites des plantations de l'inspecteur du travail, 1927–28. In the mid-1920s, the Xuân Lộc plantation bought twenty-five tons of rice per month from Binhtây and the Société des distilleries in Cholon and husked it on site.

67. Kalikiti, "Rubber Plantations," 15. He cites *Revue économique*, no. 20, décembre 1927, 310.

68. During work as a nurse's assistant, Trần Tử Bình observed how many of the medical staff were not interested in providing real care, most often following the orders of management. For health work undertaken at Suzannah, see Borel, "Paludisme en Cochinchine" and in Indochina, Grima, *La lutte antipalustre*.

69. Đỗ Hữu Chuẩn and Nguyễn Thị Thứ, interview transcript, 22. Lê, *Đất Đỏ Miền Đông*, 42.

70. NAVN2 IIB.54/112, (2) Domaines, dossier de BIF, déplacement de la gare de Trang Bom, protestation des coolies Tonkin de BIF contre le non paiement de leurs salaires dus, 1912–14. For bricks, Đỗ Hữu Chuẩn and Nguyễn Thị Thứ, interview transcript, 21. Several articles appear in the 1930s in *Khoa Học Phổ Thông* and other science journals discussing mushrooms. See ads for doctors in *BSPCI*, e.g. no. 102 (1927).

71. Gouvernement général de l'Indochine, "Réglementation sur la protection."

72. NAVN2 IIB.56/029, Procès-verbal de la visite de Phú Riềng. See also Trần, *The Red Earth*, 38, 44–45. See also Boucheret, "Les plantations d'hévéas," 265–82, and NAVN2 D.1/391, Procès verbaux des visites des plantations. For an excellent treatment of Michelin's plantations, see Panthou and Tran, *Les plantations Michelin au Viet Nam*.

73. *BSPCI*, no. 100 (1927): 13. *Paillotes* were an economic phenomenon and not a cultural one. There is no evidence whether planters were ignorant or hypocritical, but they were wrong in any case. Also, paillotes were more than racial separation and represented a question of temporary versus permanent, at least at the beginning of colonization. On plantations, temporary huts were reserved for manual laborers, who were Vietnamese, Khmer, and so forth. See also Homberg, *Les grands produits coloniaux*.

74. Boucheret, "Les plantations d'hévéas," and Le Comité de l'Indochine, "Mémoire adressé par le Comité de l'Indochine au ministre des colonies touchant le régime de la main-d'œuvre indigène en Indochine," 1929. With the world economic depression of the 1930s, planters complained even more loudly about production costs, Comité de l'

Indochine, *La crise mondiale.* Boucheret shows parallels with developing work laws in France, though the colonial context affected implementation. Boucheret, "Les plantations d'hévéas," 986, 1001. See also Marcille, "L'extension à l'Indochine." For a critique of such arguments, see Murray, "'White Gold,'" and Brocheux, "Le prolétariat des plantations."

75. From Delamarre, *L'émigration et l'immigration,* 16. For the killing of the labor recruiter, see Vann, "White Blood on Rue Hue." See also Note pour le ministère des Colonies, 5 juin 1930, in Boucheret, "Les plantations d'hévéas," 1039–43, and ANOM FM SLOTFOM III, Carton 39, Assassinat à Hanoi le 9 février par Léon Văn Sanh, de M. Bazin, directeur de l'office général de recrutement de main-d'œuvre Indochinoise.

76. Delamarre, *L'émigration et l'immigration,* 22.

77. "Đời Lao-Động, Tai Nạn Trong Khị Làm Việc," *Saigon,* 3 juin 1935, 8.

78. NAVN2 IIA.45/224, (1) "Enquête au sujet des incidents des 23 et 24 Mai à Dâu-tiêng," 14 juin 1937; Bernal, "The Nghe-Tinh Soviet Movement."

79. ANOM FM Guernut 24, Notes: Esquivillon, "La législation du travail libre, Cochinchine" 1937, 20p.–Notes: 12, 13, 16, 17. See also Esquivillon, "Régime de la main-d'œuvre engagée, Cochinchine," 1937, 21p et graphiques. This report focuses on the contract workers of the rubber plantations. For quote, see NAVN2 IIA.45/194. See also Bunout, "La main-d'œuvre et la législation du travail."

80. Godart, *Rapport,* 46. For more on birds in Vietnamese culture, see Nguyễn, "Les Vietnamiens et les oiseaux." For wishes presented to Godart, see Pham, Le, and Nguyen, *Vœux présentés par la commission permanente.*

81. ANOM FM Guernut 103, D7: Cochinchine. Vœux divers, villages, fonctionnaires, cultivateurs, etc. . . . sous-dossiers numbers 1 à 100. No. 40, Voeux du Syndicat des planteurs; ANOM GGI 65505, Généralités. Lettre, Goucoch à GGI 15 juin 1937, au sujet du maintien de l'ordre, 10. In the political report from December 1937, Pagès wrote about "minorities ethniques," while still using older terms such as "moïs."

82. Quote from Charles Fourniau in introduction to Godart, *Rapport,* 31. Image is published in Aso, "Profits or People?," 44.

83. "Việc nước nhà," *Saigon,* 1 mai 1935, 2.

84. Thompson, *Labor Problems,* 206–9; NAVN2 IIA.45/224, (1) "Rapport sur l'organisation du travail et conditions d'existence de la main-d'œuvre de la Plantation Michelin à Dâutiêng," 19 juin 1937, 7. See also Verney, "Le nécessaire compromis colonial."

85. NAVN2 IIA.45/224, (1) "Rapport sur l'organisation," 10–11.

86. Ibid., 15–19.

87. Nguyễn-Marshall, *In Search of Moral Authority.* For Lạc Thiện, see IMTSSA 167, Affaires diverses-Documentation sanitaire, correspondance et texte de la conférence sur "l'Organisation médico-sociale en Indochine et particulièrement au Vietnam" (1949). For Ánh Sáng, see Herbelin, "Des HBM au Việt Nam," and ANOM FM 1AFFPOL/3242, Assistance et œuvres sociales en Indochine: Rapport de l'Inspection générale de l'hygiène et de la santé publique. ANOM INDONF, Carton 272, Dossier 2404, 2406, *L'Effort indochinois,* 26 novembre 1937. ANOM GGI 65445, Sureté reports,

Cambodge, 1921. See also *La Jeune indochine*, cited by *Tribune indochinoise*, no. 221, 10 février 1928. ANOM GGI 65482, Notes mensuelles sur les événements politiques–1938. See also NAVN2 IIA.45/275, (3) 4 Avril, Saigon, Diffamation par voie de presse, SIPH contre Nguyễn Văn Mai, gérant du journal *Dân Mới*. Mai challenged the decision on March 28, 1939, and on April 4, the Court confirmed its previous decision without qualification. For more, see NAVN2 IIA.45/293.

88. Lê, *Đất Đỏ Miền Đông*, 42. See also Brocheux, "Le prolétariat des plantations"; Murray, " 'White Gold' "; Kalikiti, "Rubber Plantations"; and Boucheret, "Les plantations d'hévéas."

89. McCook, "Managing Monocultures."

90. For studies on the important convergence of health and ecology in the U.S. context, see Mitman, Murphy, and Sellers, *Landscapes of Exposure*.

Chapter Four

1. For Gourou on architecture, see Herbelin, *Architectures du Vietnam colonial*, 314.

2. Claval, "Colonial Experience," 301.

3. Quote from Bruneau, "From a Centred to a Decentred Tropicality," 310. For more on Gourou, see Bowd and Clayton, "Tropicality, Orientalism, and French Colonialism"; Bréelle, "The Regional Discourse of French Geography"; Bruneau, "Pierre Gourou (1900–1999)"; Kleinen, "Tropicality and Topicality."

4. Bowd and Clayton, "French Tropical Geographies," 280. See Gouvernement général de l'Indochine, "Contrat de la colonie"; Morin, "Sur l'organisation du service antipaludique." For public works and malaria, see Pouyanne, *Historique succinct*.

5. Tilley, *Africa as a Living Laboratory*, 24; Sears, "Ecology."

6. Cooper and Stoler, "Between Metropole and Colony," 9, 10. Paul Kramer, too, has emphasized the role empire played in the formation of race, arguing that race and empire worked to construct each other. In Kramer's view, race was "a dynamic, contextual, contested, and contingent field of power" enabling, but also at times impeding, empire, and the colonies were locations where concepts of race were applied and reworked. Kramer, *The Blood of Government*, 2.

7. Saada, "Race and Sociological Reason," 381–82. See also Firpo, *The Uprooted*. The question of métis was a so-called imperial problem, and there have been analyses of the question of race mixing in the American, French, British, and Dutch empires. For example, see Blanckaert, "Of Monstrous Métis?" "The politics of empire and colonialism," writes Nancy Stepan, "were other sources of new reflections on, and scientific study of, embodied difference, the politics of women and their rights another." Stepan, "Race, Gender, Science and Citizenship," 30–31. For general histories of scientific racism and race science, see Barkan, *Retreat of Scientific Racism*, and Stepan, *The Idea of Race in Science*. For acclimatization in Cochinchina, see Morice, "Quelques mots sur l'acclimatement," and Osborne, *Nature, the Exotic*. For post-1954 typological thinking in Việt Nam, see Pelley, " 'Barbarians' and 'Younger Brothers.' "

8. NAC 4369, Aptitude au travail des différentes races qui habitent au Cambodge, 1923, 4.

9. Chollet, *Planteurs en Indochine*, 145; Kalikiti, "Rubber Plantations," 239.

10. Brocheux, "Le prolétariat des plantations," 68; NAVN2 IIB.56/029, Rapports et procès-verbaux des visites des plantations de l'inspecteur du travail, 1927–28, visite de la Plantation de Cam-Tien, 14 janvier 1928. See also Bourgois, *Ethnicity at Work*.

11. NAC 36063. Rapport de l'inspecteur du travail sur l'affaire d'au-chhok/plantation de Stung-Trang, 3. For the perceived racial difference between those from Tonkin and those from Annam, see Chollet, *Planteurs en Indochine*, 102. For more on racial conflict and communist activity, see Trân, *The Red Earth*. For the intersection of cultural identity, worker protest, and class formation, see Tran, *Ties That Bind*, 49–60.

12. Gourou, *Les pays tropicaux*, quoted in Farinaud and Choumara, *Infestation palustre et démographie*, 3.

13. ANOM FP 100APOM/170, Congrès d'agriculture coloniale, "Programme relatif à la réforme de l'hygiène et de l'Assistance médicale indigène en Indo-Chine."

14. Đỗ Hữu Chuẩn and Nguyễn Thị Thứ, interview transcript, 27. For Bernard quote, see his "Notions générales sur le paludisme," 141. NAVN2 IA.7/236, (7) Procès-verbal de la Réunion du Comité local d'hygiène de la Cochinchine, 21 janvier 1919, 5. That same year, Bernard published a treatise on preventing malaria on agricultural and forestry exploits in the south as Gouvernement de la Cochinchine, *Notions générales sur le paludisme*. For Girard's work on rubber, see Chevalier and Girard, *L'hévéa en Indochine*; Girard, *L'hévéa en Cochinchine*.

15. Gouvernement de la Cochinchine, *Notions générales sur le paludisme*; NAVN2 IA.7/236, (7) Lettre, 21 janvier 1919, Gaide à Goucoch, au sujet de la prophylaxie anti-paludéenne. See NAVN2 IA.7/236, (7) for planters' fights with the forestry department. See also Gaide, *L'assistance médicale et la protection*. For an overview of the Pasteur Institute to 1931, see Bordes, *Le paludisme en Indochine*.

16. Loubet, *Province de la province*, 19–28; Méry, *History of the Mountain People*, 122. Guénel and Guerin, "L'ennemi, c'est le moustique" and Guérin, *Paysans de la forêt*; Farinaud, "Le paludisme chez les Phnongs."

17. For comparison, see Packard, *White Plague, Black Labor*. For implications of Montagnard health for southern Indochina, see Mesnard and Bordes, "L'infection paludéenne chez les moïs."

18. Méry, *History of the Mountain People*. An *arrêté* creating reservations in Cochinchina was passed in 1937. For more on reservations, see NAVN2 IIA.45/194, Dr F. D. Clément, rapport, Délégué administrateur de Nui-Bara, Médecin chef de l'assistance médicale des Nui Bara et Budop délégation du Song-Be, Bien Hoa, "Des moyens envisagés pour sauvegarder l'integrité physique et l'originalité éthnique des Moïs"; and report, administrator of Hớn Quản, November 11, 1937, in which he wrote of the "question politique de race à suivre à l'égard des minorités ethniques de l'hinterland indochinois."

19. Borel, "Résultats d'une enquête épidémiologique"; Farinaud and Mesnard, "Recherches sur le réservoir," 723.

20. NAVN2 III.59/N130, Rapport annuel, Institut Pasteur, Saigon, 1937, 11–19, 41, 43–46. Includes summary of results of study of Montagnards and malaria. Farinaud and Mesnard, "Recherches sur le réservoir," 726. See also Borel, "Contribution à l'étude" ; Bordes and Mesnard, "L'importance du réservoir." NAVN2 S.4/26, Chống bệnh rét rừng, 1943, Dr. P. Delbove and G. Wormser, "Difficultés de la colonisation en zone d'hyperendemicité palustre de l'Indochine méridionale." For a layperson's perspective on the health of uplanders, see, for example, the Beaucarnot diary translation project by David del Testa.

21. Anderson, *Colonial Pathologies.* See also Anderson, "Excremental Colonialism"; Anderson, "Immunities of Empire"; and Arnold, "'An Ancient Race Outworn.'"

22. Daston, "The Moral Economy of Science," 4. Daston's approach brackets the political in a way that Fassin finds unsatisfying and he returns to the kinds of moral economies explored in the works of E. P. Thompson that were more directly relevant for studies of political and social authority. Fassin, "Les économies morales revisitées." The concept of moral economies has been developed for the Southeast Asian context in Scott, *The Moral Economy*; Popkin, *The Rational Peasant*; and Brocheux, "Moral Economy." For a recent discussion of the Scott-Popkin (non) debate, see Montesano, "War Comes to Long An."

23. Paucot and Le, *Notions d'hygiène*, 34–35. See also Simond, "Hygiène de l'Indochine."

24. Borel specifically refers to an article by P. Bussy, "Etude agricole des terres de la Cochinchine," *Bulletin agricole de l'Institut scientifique de Saigon* 2, no. 1 (1920): 1–11. Borel, "La constitution du sol." Although Borel did not refer to Vietnamese cultural understandings of the upland forests as regions of *nước độc*, other researchers did. See also Présentation d'ouvrage, "Les moustiques de la Cochinchine."

25. In 1930, the results of his studies were compiled and posthumously published as Borel, *Les moustiques.* Outside of Việt Nam, the literature on malaria is immense and already in the 1970s too voluminous for a United Nations bibliography. For the Malaysian context, see Manderson, *Sickness and the State*; and Liew, "Making Health Public." For recent work on malaria and agriculture, see Kitron, "Malaria, Agriculture, and Development"; Stapleton, "Internationalism and Nationalism." Scott has recently argued that malaria formed part of a strategy of state avoidance by upland peoples in *Art of Not Being Governed.*

26. For more on Morin, see http://data.bnf.fr/13024048/henry_gabriel_sully _morin/, last accessed January 1, 2018; and for general context, see Osborne, *The Emergence of Tropical Medicine.* For Morin's summary of malaria statistics on plantations, see NAVN2 VIA.8/305 (10). For a comparison with other French colonies doctors in Asia, see Ackerman, "The Intellectual Odyssey."

27. Anon., "Institut Pasteur de Saïgon. Rapport annuel sur le fonctionnement des services en 1928," 14 p., 1929; Morin, *Entretiens sur le paludisme*, 141–42.

28. For other studies of Kon Tum, see Lieurade, *Essai de démographie.* See Opinel and Gachelin, "Theories of Genetics."

29. See Jennings, *Imperial Heights* and Morin, *Entretiens sur le paludisme.*

30. Morin and Robin, *Essai sur la prévention pratique*, 1. See also Morin and Martin, "Possibilités d'utilisation pratique."

31. Morin, *Entretiens sur le paludisme*, 248.

32. On synthetics, see inter alia Blanc, "Essais de traitement."

33. Adam, "De l'écologie agricole," 1–3, 11.

34. Ibid., 11.

35. See Lecomte, Humbert, and Gagnepain, *Flore générale de l'Indo-Chine*, Biodiversity Heritage Library, http://biodiversitylibrary.org/search.aspx?SearchTerm=l%27indo -chine&SearchCat=, last accessed April 29, 2013. For more on Carton, see Institut National d'Agronomie Coloniale, *Anciens élèves de l'Institut national d'agronomie coloniale* (Paris). For more on Carton's theoretical work in ecology, see Carton, "Importance des facteurs écologiques" and Carton, "Considération sur l'action".

36. In 1921, Girolamo Azzi held the first chair in Italy in agricultural ecology. In reviewing Azzi's research, the Royal Academy of the Lynxes (Reale Accademia dei Lincei) noted how the science of plant ecology "could take on an eminently practical character when it considered the plant grown by man for his needs." In Azzi, *Ecologie agricole*, 5–6. See also Carton, "L'Ecologie." For an overview of Carton's understanding of meteorology, see *Feuille mensuelle de Renseignements de l'IGA*, 6–8. See also Henri Lecourt, "La météorologie agricole en Indochine," *L'Eveil économique de l'Indochine* 12, no. 566 (1928): 6–9. This summary depicts climate as the average plus the extremes of weather (i.e., a statistical phenomenon) and notes the importance of large plantations for extending the meteorological network in Indochina.

37. Carton, "L'œuvre de l'institut," 119. Carton considered rubber in "Le caoutchouc en Indochine."

38. Tilley, *Africa as a Living Laboratory*, 134; Carton, *Cours de climatologie et d'ecologie*, 1, 3.

39. Carton, "Le climat et l'homme," 161. For conclusion, see 172–73. On page 174, Carton discusses how the same species could be an active transmitter or inoffensive, depending on "microclimate," as a comparison between A. malcultatus in Malaya and Indochina showed.

40. See Jennings, *Imperial Heights*, 40–41 and 50–51 for "geoclimatic determinism"; Carton, "Le climat et l'homme," 106, quoting Marchoux, "Afrique Occidentale Française," 23; and Carton, "Le climat et l'homme," 108.

41. Bradley, *Imagining Vietnam*, 51–57.

42. Bayly, "French Anthropology," 583, 596–97, 600.

43. Huxley, Haddon, and Carr-Saunders, *We Europeans*, 7. Despite the "Statement on Race" published by the United Nations Educational, Scientific, and Cultural Organization (UNESCO) in the 1950s, Stepan has argued that the 1940s may have marked the end of race science but not the end of race thinking and typologies. Stepan, *The Idea of Race*, 189. See also Saada, "Race and Sociological Reason."

44. Vũ, "Le problème des Eurasiens en Indochine." See Institut indochinois de l'étude de l'homme, *Compte Rendu des séances de l'année 1943*, inter alia, 451, "Recherches sur le pied des Annamites," Đỗ Xuân Hợp. The foot was used as an object to show

evolution. Kleinen, "Tropicality and Topicality," 343; Huard, "L'Extrême-Asie"; Huard and Bigot, "Introduction à l'étude des Eurasiens." See also Lochard, "Science pure." In Latin America, the unattractiveness of hereditary determinism (especially of the racist kind prominent in the United States and Britain) and an environmentalist and sanitary tradition in medicine resulted in an emphasis on the possibility of medical intervention, "constructive miscegenation," and the creation of a "purified, unified, and homogenized" cosmic race. Stepan, *The Hour*, 73, 106, 138. See also Stern, "From Mestizophilia to Biotypology," and Stern, *Eugenic Nation*.

45. For race and French tropical medicine, see Osborne, *The Emergence of Tropical Medicine*.

46. Arnold, "Tropical Governance," 17, 20. See also Neill, *Networks in Tropical Medicine*.

47. Morin, "Le paludisme et sa prophylaxie," 410. See also, e.g. Farinaud, "Développement de la lutte."

48. Guénel, "The Conference on Rural Hygiene"; Bynum, "Malaria in Inter-war British India." See also ANOM FM Guernut 22, Dossier Bb, Société des Nations, Organisation d'hygiène, Conférence d'hygiène rurale des Pays d'Orient (Bandoeng, 3–13 août 1937). Rapport sur l'organisation des services sanitaires et médicaux (Point I de l'ordre du jour) présenté par le Dr. P. M. Dorolle, Médecin de 1ère classe de l'Assistance médicale indigène en Indochine. Hermant (Dr. P) Programme d'organisation des services d'assistance et d'hygiène en Indochine, Paris, 1938. League of Nations, *Intergovernmental Conference*, 32. In 1936 the Pasteur Institute printed a volume of articles on the science of malaria in Indochina.

49. Pasteur Institute Archives, Canet letter, "Considérations sur l'organisation de la médecine préventive et curative dans les grandes plantations de caoutchouc en Indochine." Canet was one of only a few European doctors to work outside Sài Gòn after 1945. For published work, see Canet, "Note sur une petite épidémie"; Canet, "Organisation de l'hygiène"; Canet, "Prophylaxie collective"; and Schneider, Canet, and Dupoux, "Traitement curatif du paludisme." See also Chollet, "Travaux d'assainissement." As Eric Panthou pointed out in his communication for HOMSEA 2014, the number of plantation hospitals in French Indochina lagged far behind their counterparts in the British empire in both date of construction and number.

50. Pasteur Institute Archives, Canet letter, "Considérations sur l'organisation de la médecine préventive et curative dans les grandes plantations de caoutchouc en Indochine." See also Mingot and Canet, *L'hévéaculture en Indochine*, 48. For other reports, see ANOM FM Guernut 24, extract from Dr. Cambaudon, "Le service médical du groupe des plantations de Mimot," *Revue coloniale de médecine et chirurgie* (1937): 1–5, and NAC 1600, Plantation Kantroy, 1928–32, which discusses Cambaudon's work for Mimot.

51. Mingot and Canet, *L'hévéaculture en Indochine*, 54–55. See also Thompson, *French Indo-China*, for a good summary of French ideas about "native psychology" during colonialism.

52. Morin, *Entretiens sur le paludisme*, 246.

53. Mingot and Canet, *L'hévéaculture en Indochine*, 55, 57. For a comparison of villages constructed on rubber plantations run by Henry Ford, see Grandin, *Fordlandia*. See also ANOM FM Guernut 88, Cochinchine: réponses à l'enquête n°2 sur l'habitat (classement par provinces).

54. WHO, Centralized files, First and Second Generations (1946–1955), 453-4-5, Dr H.G. S. Morin "de l'Institut Pasteur de Dalat" on rural malaria control demonstration projects in the Far East, 1947, "Rural Malaria Control Demonstration Projects in the Far East," 2. Morin, *Entretiens sur le paludisme*, 128.

55. Guénel and Guerin, " 'L'ennemi, c'est le moustique.' " See sketch in NAVN4 RSA 3932. For malaria and troop health, see Bouvier and Rivoalen, *Troupes du groupe de l'Indochine*.

56. NAVN4 3722, Etudes des mesures prophylactiques contre le paludisme par le Dr. Moreau, 1937–38. For regional ties based on malaria, see Moreau, "Notes sur un voyage d'études malariologiques."

57. NAVN4 3722, Etudes des mesures prophylactiques contre le paludisme par le Dr. Moreau, 1937–38.

58. Ibid. See also Morin and Robin, *Indications pratiques pour le drainage*.

59. NAVN4 3722, Etudes des mesures prophylactiques contre le paludisme par le Dr. Moreau, 1937–38. For artemisia, see Packard, *Making of a Tropical Disease*, 114, and Tran et al., "A Controlled Trial." For more on traditional medicine, see Thompson, *Vietnamese Traditional Medicine*. The connection between economic well-being and malaria was widely discussed, e.g. Marchoux, "Influence du bien–être."

60. Arnold, "Tropical Governance." For meetings in Indochina, see Far Eastern Association of Tropical Medicine, *Comptes-rendus*, 1914, 1939.

61. Boucheret, "Les plantations d'hévéas," 223, and Schultz, *Dans la griffe*. For maternal medicine, see chapter 5, Au, *Mixed Medicines*. Quote from Homberg, *La France des cinq parties*, 6.

62. See Hunt, *A Colonial Lexicon*, and Au, *Mixed Medicines*, 101–3. On degeneration, see Pick, *Faces of Degeneration*; Schneider, *Quality and Quantity*, 56–63, 84, 147. See also Schneider, "The Eugenics Movement in France."

63. NAVN4 3723, Elaboration d'une notice de prophylaxie contre paludisme, études des Dr. Moreau, Antoine, Dang Van Du sur le drainage et le projet antimalérien, 1937–38. For an important discussion of nationalism and the medical profession in Asia, see Lo, *Doctors within Borders*. In a letter from April 12, 1938, the RSA argued that simply translating the work of Moreau on malaria wouldn't be effective because it was too complicated for popular audiences. For appropriate roles of women and children, see Marr, *Vietnamese Tradition on Trial*.

64. "Bệnh Sốt Rét," *Khoa Học Phổ Thông* (1938), 2. Other Vietnamese medical doctors were interested in malaria, e.g. Le, "Le paludisme en Cochinchine"; Nguyễn, *Note sur les précautions à prendre*.

65. NAVN4 3723, Elaboration d'une notice de prophylaxie contre paludisme, études des Dr. Moreau, Antoine, Dang Van Du sur le drainage et le projet antimalérien, 1937–38.

66. Bùi, "Le médecin en face." Quotes from Morin and from Le Nestour on pp. 33–34.

67. Canet, "Note sur l'alimentation," 60. For more on plantations during World War II, see ANOM FM 1AFFECO/59, Réalisations sociales dans les plantations de caoutchouc d'Indochine, 1942.

68. For a reassessment of the oeuvre of Robequain and fellow geographer Pierre Gourou, see the special issue of the *Singapore Journal of Tropical Geography* 26, no. 3 (2005). See also Bréelle, "The Regional Discourse of French Geography."

69. Robequain, *The Economic Development*, 181–82.

70. ANOM FM SLOTFOM III, Carton 117. Tracts divers en langue française; Correspondance 1930–35. Contains communist propaganda tracts denouncing certain aspects of "imperialist" rule in Indochina, including a few articles on rubber and plantations. ANOM GGI 65490, Notices mensuelles sur les événement politiques (1940). June 28, GD, Tân-Phú-Trung and Tân-An-Hội. See timeline in Trần, *Địa Chí Đồng Nai*. Another Lê Đình Cự, perhaps the same, organized a labor union in 1956. NAVN2 PTT DICH 4466; Trịnh, *Phong Trào Lao-Động Việt-Nam*; Trần, Nguyễn, and Trần, *Phong Trào Công Nhân Cao Su Bình Sơn*, 24, 25; Kalikiti, "Rubber Plantations," 25.

71. Nghiêm, "Điều Tra Nhỏ: Những Tiểu Đồn Điền"; Nghiêm, "Điều-Tra Nhỏ: Những Tá Điền"; and Nghiêm, "Nạn Dân Đói." For an analysis of the journal, see Brocheux, "La revue 'Thanh Nghị,'" 317–31. Thanh Nghị drew its name from a movement in China from the late nineteenth century that went by the name *qingyi*. This word means approximately "informed opinion."

72. Recognizing the limits of this policy, Nghiêm concluded, the government issued a new policy on March 20, 1936, concerning the "colonies de peuplement." Nghiêm, "Điều-Tra Nhỏ: Những Tá Điền," 265, 266. See Brocheux, "La revue 'Thanh Nghị.'"

73. Nghiêm, "Nạn Dân Đói," 225–27. See esp. n. 107, on famine, and n. 62, on cafe and smallholder colonization. For more on famine, see Gunn, *Rice Wars*, 1.

74. For rubber during World War II, see Haumant, "Industrie indochinoise." Labor shortages were still a problem: ANOM GGI SE 2241 Cultures industrielles de caoutchouc. n. 2441 Inconvénients resultants, pour le recrutement de la main d'oeuvre nécessaire aux plantations de caoutchouc, de la levée d'un contingent supplémentaire de militaries indigènes en Cochinchine (vœu formulé par le Syndicat des planteurs de caoutchouc à Saigon, 1938). See also NAVN2 III.59/N130, (1) Hôpitaux, asiles, infirmeries, divers, au sujet de la poste médical de Trang Bang, 1942. Discusses improvements in the Tây Ninh post that would "rendre plus efficace notre action sur cette population dont l'état sanitaire parait très médiocre." Đỗ Hữu Chuẩn and Nguyễn Thị Thứ, interview transcript, 19.

Chapter Five

1. See Boucheret, "Les plantations d'hévéas," 585. Tertrais, *La piastre et le fusil*, 373, graphique 16 et tableau 23, "évolution de la production indochinoise de caoutchouc," from *Bulletin statistique de l'Indochine*; *Le Monde*; Conseil économique, with continuing

growth for South Vietnam until the 1960s. In 1958, almost 63,000 tons of rubber was produced in the RVN according to the records of the Vietnamese Institute for Rubber Research (IRCV), or 72,000 tons according to the Agricultural Economics and Statistics Services. Numbers for the late 1940s can be found in IRCI, *Rapport annuel, 1948–1951*.

2. For a recent treatment of war memories and landscapes, see Pholsena and Tappe, *Interactions with a Violent Past*. For more on *đoàn thể*, see Wehrle, *Between a River and a Mountain*, 31, and Woodside, *Community and Revolution*.

3. Bowd and Clayton, "French Tropical Geographies," 274.

4. Brocheux, "The Economy of War."

5. Chollet, *Planteurs en Indochine*, 173. For the 1944–45 famine, see Gunn, *Rice Wars*, and Huỳnh et al., *Lịch Sử Phong Trào*, 207. For more on 1945, see Marr, *Vietnam 1945*, and Marr, *Vietnam*. For an international perspective, see Lawrence, *Assuming the Burden*, and Goscha, *Historical Dictionary*.

6. Wormser, *Des quelques facteurs*, 18. See also Malye, "Les problèmes d'Indochine."

7. For figures, see Boucheret, "Les plantations d'hévéas," 588; Huỳnh et al., *Lịch Sử Phong Trào*, 228, 253; ANOM Haut-Commissariat de France pour l'Indochine, Conseiller aux affaires sociales 75, 104, 105.

8. Marr, *Vietnam*, 453–54.

9. Huỳnh et al., *Lịch Sử Phong Trào*, 256, "người thợ không bằng một cây cao su." For a comparison with the Philippines, see Bankoff, " 'The Tree as the Enemy of Man' ".

10. Huỳnh et al., *Lịch Sử Phong Trào*, 237; ANOM Haut-Commissariat de France pour l'Indochine, Conseiller aux affaires sociales 75: Main d'œuvre pour plantations de caoutchouc, travailleurs du nord, terres rouges. See report on communist organization of rubber workers, including information on *Cao Su Chiến* (Guerre du caoutchouc), a monthly starting in January 1950 and an "organe de propagande de l'Union syndicale des plantations d'hévéas du Nam-Bo et du Cambodge." Ann Stoler has shown that during the Indonesia revolution after World War II, the interests of Javanese workers on the estates of North Sumatra and those of *laskar*, or popular militias, did not always coincide. Stoler, *Capitalism and Confrontation*.

11. Điệp, *Máu Trắng, Máu Đào*, 133; Brocheux, "The Economy of War," 315; Huỳnh et al., *Lịch Sử Phong Trào*, 236.

12. Statistics from January 1950 meeting, Huỳnh et al., *Lịch Sử Phong Trào*, 231; Trần, *Địa Chí Đồng Nai*, 33; Vũ, *Công Nhân Cao Su*, 71; Đặng, *100 Năm Cao Su*, 206–7, 10.

13. Lawrence, *Assuming the Burden*.

14. Huỳnh et al., *Lịch Sử Phong Trào*, 216. Circulation numbers are hard to judge, and the 4,000 number given here is most likely an estimate. For journals, see Đồng Nai 300 Nam, *Gia Định Thành Thông Chí*, Chương VI, Biên Niên 1858–1954, and "Lịch Sử Đảng Biên Hòa. My thanks to a Council on International Education Exchange student for asking the fascinating question of how rubber workers got their news. Huỳnh et al., *Lịch Sử Phong Trào*, 232–33.

15. For a discussion of the subtle difference in meaning between *cần lao* and *lao động*, see Taylor, *A History of the Vietnamese*, 556. Huỳnh et al., *Lịch Sử Phong Trào*, 232–33.

16. Đặng, *100 Năm Cao Su*, 218–19. Đặng mentions the 1/19 aircraft. For a French perspective on rubber to the mid-1940s, see Bouvier, *Le caoutchouc* and Camus, *L'oeuvre humaine et sociale*.

17. Huỳnh et al., *Lịch Sử Phong Trào*, 251, 258. For an example in Malaya, see Han, *And the Rain My Drink*. Đặng, *100 Năm Cao Su*, 212–13, 234.

18. Huỳnh et al., *Lịch Sử Phong Trào*, xxx.

19. NAVN3 Ủy Ban Kháng Chiến Hành Chính Nam Bộ, 1945–1955, HS 449, "French Plantations." The report argued that the economic strength of the French in southern Vietnam came from rice and rubber plantations, but while the rice plantations had suffered extensive damage since the beginning of the war, the rubber plantations had been harder to sabotage. For sabotage methods, see Vũ, *Công Nhân Cao Su*, 72–73. This method was called "khoanh cỡ chó," or "ceinturage à mort," in the report "French Plantations." See also Huỳnh et al., *Lịch Sử Phong Trào*, 241, 267.

20. Tate, *The RGA History*, 540, n. 9, Malaysian communists also slashed trees. Vũ, *Công Nhân Cao Su*, 73. See also Trần, *Địa Chí Đồng Nai*, 33; Huỳnh et al., *Lịch Sử Phong Trào*, 238–39.

21. Vũ, *Công Nhân Cao Su*, 69. This claim was reprinted in Trần, *Địa Chí Đồng Nai*, and Huỳnh et al., *Lịch Sử Phong Trào*, 241, 253. See also NAVN3 Ủy Ban Kháng Chiến Hành Chính, HS 37, 86, and Ủy Ban Kháng Chiến Hành Chính Nam Bộ, 1945–1955, HS 449, "French Plantations."

22. Vũ, *Công Nhân Cao Su*.

23. Huỳnh et al., *Lịch Sử Phong Trào*, 240, 260–61. See Goscha, *Vietnam*, 104.

24. Huỳnh et al., *Lịch Sử Phong Trào*, 271, 272.

25. See, for example, Đặng, *100 Năm Cao Su*, 223–27. See also Huỳnh et al., *Lịch Sử Phong Trào*, 269, 270, for more about scientific and effective destruction of rubber (phá hoại cao su "một cách khoa học và có hiệu quả hơn"), see May 1948 report. Vũ, *Công Nhân Cao Su*, 77. Huỳnh et al., *Lịch Sử Phong Trào*, 247, claimed that all sections of the Trade Union had party cells.

26. Huỳnh et al., *Lịch Sử Phong Trào*, 254. NAVN3 2973 BNL Tập tài liệu về hội nghị cao su năm 1948 do Bộ Canh Nông tổ chức ngày 20 đến 21-01-1949. "Ta chưa tổ chức phá-hoại cao-su được ở Cao Miên."

27. NAVN3 PTT 2012, Công văn, báo cáo của BCN, Liên đoàn Cao su Thủ Dầu Một, Ban Cao su Bà Rịa, Biên Hoà, Liên hiệp Công đoàn Nam Bộ về tổ chức, phá hoại cơ sở cao su ở Nam Bộ năm 1949, 1950, 1952. See also NAVN3 BNL 3022, Báo cáo của Liên hiệp Công đoàn Nam Bộ về tình hình phá hoại cao su của địch từ tháng 7 đến tháng 10 năm 1949. For 2/NB, signed June 22, 1949, see Huỳnh et al., *Lịch Sử Phong Trào*, 264–65.

28. Hồ, *Lịch Sử Liên Trung Đoàn*, 237; Goscha, *Vietnam*, 110–13; "Nguyễn Bình" in Goscha, *Historical Dictionary*.

29. Huỳnh et al., *Lịch Sử Phong Trào*, 270, 271. There were one hundred people in a "đại đội cao su." Tate, *The RGA History*, 539, n. 1, 530. Like the Việt Minh, the Malaysian Communist Party looked to disrupt plantations and ransom captured European personnel.

30. NAVN3 BNL 2973, Tập tài liệu về hội nghị cao su năm 1948 do Bộ Canh Nông tổ chức ngày 20 đến 21-01-1949. The burning of dry rubber is more difficult to determine, since the numbers for January through October are aggregated into one category. Hồ, *Lịch Sử Liên Trung Đoàn 301-310*, 202. See Đảng Cộng Sản Việt Nam, ed., *Văn Kiện Đảng*, vol. 10 (1949), 309–13. Điện của Thường vụ Trung ương ngày 17 tháng 11 năm 1949 gửi đồng chí Lê Duẩn và xứ ủy Nam Bộ về nhiệm vụ của Xứ ủy.

31. Huỳnh et al., *Lịch Sử Phong Trào*, 295–97. The strategy of preserving plantations for eventual Việt Minh takeover continued into the late 1950s and 1960s, as recounted in Michon, *Indochina Memoir*, 46. Trần, *Địa Chí Đồng Nai*.

32. Huỳnh et al., *Lịch Sử Phong Trào*, 283–86. Thanh Sơn was possibly short for Nguyễn Thanh Sơn, a member of the Committee of the South responsible for organizing anticolonial activity in Cambodia during the Resistance War against the French. See Lamant, "Les partis politiques," 74. For bộ đội quote, see Boucheret, "Les plantations d'hévéas," 582; see also 571, 572, and 592.

33. NAVN2 Phủ Thủ Hiến Nam Việt Lo, 2–27.

34. Ibid.

35. Lawrence, *Assuming the Burden*, 193; NAVN2 Phủ Thủ Hiến Nam Việt M10-37, labor inspectorate report, October 20, 1948.

36. NAVN2 Phủ Thủ Hiến Nam Việt M10-37, labor inspectorate report, October 20, 1948. This report vastly underreported the numbers of deaths, illnesses, and injuries.

37. On Bảo Đại, see Lawrence, *Assuming the Burden* and "Recasting Vietnam." Ordinance no. 10 of August 6, 1950. Điệp, *Máu Trắng, Máu Đào*, 119.

38. See, for example, NAVN2 PTT Quốc Gia Việt Nam 3109, 21 septembre 1949, Haut-Commissariat de France pour l'Indochine, L. Pignon to Bao Dai, au sujet de "personnes déplacées du Nord-Vietnam"; "Contre le recrutement de main-d'œuvre parmi les détenus viêtnamiens pour les plantations de Cochinchine," *Agence France-Presse*, 7 décembre 1948, citing article in *Công Luân* by Nguyễn Mạnh Hà, former economic minister of the Hồ Chí Minh government. See also NAVN2 PTT Quốc Gia Việt Nam 2995 and ANOM Haut-Commissariat de France pour l'Indochine, Conseiller aux affaires sociales 74.

39. NAVN2 Phủ Thủ Hiến Nam Việt M10-37, M11-43, Recrutement (mộ phu) de main-d'œuvre à Quang Nam pour la Cie de Caoutchouc du Mékong. See letter from October 29, 1949, from GCV to governor of South Vietnam, describing situation.

40. Mus, *Viêt-Nam*, 108–9.

41. NAVN2 PTT Quốc Gia Việt Nam 3109, for April 7, September 10 letters, concerning recruiting methods and SPTR request for workers and for Prince Bửu Lộc and Xuân letters. The exact phrase used was "personnes déplacées à l'occasion des opérations militaires." For more on Nguyễn Văn Xuân, see Goscha, *Historical Dictionary*.

42. NAVN2 PTT Quốc Gia Việt Nam 2994 for Mai Ngọc Thiệu letter.

43. NAVN2 PTT Quốc Gia Việt Nam 2995 for Nguyễn Hữu Trí and Trần Văn Hữu letters; Goscha, *Historical Dictionary*, 324, 462; Tate, *The RGA History*, 555, by Tunku in Malaya in 1955.

44. NAVN2 PTT Quốc Gia Việt Nam 2904 for Đăng Hữu Chí letter. The July 31 letter addressed to the prime minister recounted the meeting of July 24, attended by delegates from three regions of Việt Nam—north, center, and south (but not from Cambodia)—and the delegate from the Planters' Union.

45. ANOM, Haut-Commissariat de France pour l'Indochine, Conseiller aux affaires sociales 96, Travail, divers–2. Phạm Văn Ban. NAVN2 r-685, P22, Plainte d'un coolie de la plantation de Bình Ba contre son directeur qui avait exercé sur la personne de sa femme, 1950. Five piasters figure from Trần, *The Red Earth*, 41–42. Điệp, *Máu Trắng, Máu Đào*, 133; Vũ, *Công Nhân Cao Su*, 71; Huỳnh et al., *Lịch Sử Phong Trào*, 274–76.

46. Tate, *The RGA History*, 556, n. 8; Richard Stubbs, "Counter-Insurgency and the Economic Factor," no. 19, Singapore ISEAS, 1974; Wilder, "Colonial Ethnology and Political Rationality," 337, 354. Contextualizing the "vivant" (the living) within the "vécu" (the lived) experience. See, for example, Canguilhem, "Le Concept et La Vie." While Canguilhem privileges the concept of "vivant," following Fassin in "Les économies morales," I wish to hold both in tension.

47. NAVN2 PTH NV M10-18: Délégation de M. Wolf pour participer à la commission. Letter, April 15, 1946, inspecteur du travail, R. Schneyder, wrote to commissaire de la République en Cochinchine.

48. NAVN2 Phủ Thủ Hiến Nam Việt M10-45, Salaire minimum à allouer aux travailleurs des exploitations agricoles du Sud-Viêtnam (Lương tối thiểu của công nhân cao su, 1949). Secrétariat d'Etat au travail et action sociale, Ngô Quốc Côn, à président du gouvernement provisoire du Vietnam, 1er février 1949, 3, 5. See also, Trần Văn Huê, Ministre du travail et action sociale à président du gouvernement provisoire du Sud Vietnam, 22 février 1949.

49. Huỳnh et al., *Lịch Sử Phong Trào*, 263, 264. See also NAVN2 Phủ Thủ Hiến Nam Việt M10-37, for salary of unskilled worker, 3 piasters; tappers, skilled workers, 3.50 piasters.

50. Huỳnh et al., *Lịch Sử Phong Trào*, 257; NAVN3 BNL 3661; NAVN3 Ủy Ban Kháng Chiến Hành Chính HS 936, 41. For workers in Malaysia, see Barlow, "Changes in the Economic Position of Workers."

51. NAVN2 PTT Quốc Gia Việt Nam 2994 for Mai Ngọc Thiệu letter; NAVN2 PTT Quốc Gia Việt Nam 2904 for Đăng Hữu Chí letter.

52. NAVN2 PTT Quốc Gia Việt Nam 2994 for SPTR report. Its numbers came from *Việt-Nam Kinh-Tế Tập-San (Bulletin économique du Vietnam)* 1, no. 2 (1950)—It seems from the article "Enquête sur les budgets familiaux viêtnamiens," 65.

53. See *Việt-Nam Kinh-Tế Tập-San (Bulletin économique du Viet-Nam)*.

54. Wehrle, *Between a River and a Mountain*, 39, 40, 49, 66–74; "Việt-Nam tại Hội-Nghị Quốc-Tế Cao-Su (Le Viêtnam à la conférence internationale du caoutchouc)," *Việt-Nam Kinh-Tế Tập-San (Bulletin économique du Viet-Nam)* 5, no. 5 (1954): 296–303; Boucheret, "Les plantations d'hévéas," 586–95.

55. *Việt-Nam Kinh-Tế Tập-San (Bulletin économique du Viet-Nam)* 2, no. 8 (1951): 317; 1, no. 1 (1950): 14–15; and 3, no. 3 (1952): 118, for a month-by-month list of prices from

January 1951 to February 1952. For Singapore prices, see Coates, *The Commerce in Rubber,* 320–23.

56. "Tiền Công Thợ Trong Các Đồn Điền Cao-Su tại Việt-Nam (Salaires ouvriers dans les plantations de Caoutchouc au Việtnam)," *Việt-Nam Kinh-Tế Tập-San (Bulletin économique du Viet-Nam)* 5, no. 6 (1954): 373. In December 1954, according to the *Bulletin,* worker salaries had risen to 32.66 piasters, 31.50 piasters, 45.62 piasters, and 83.48 piasters, respectively, while by December 1955, the numbers stood at 40.54 piasters, 34.29 piasters, 48.50 piasters, 85.04 piasters, again respectively. "Tiền Công Thợ trong các Đồn Điền Cao-Su tại Việt-Nam (Salaires ouvriers dans les plantations de caoutchouc au Việtnam)," *Việt-Nam Kinh-Tế Tập-San (Bulletin économique du Viet-Nam)* 6, no. 5 (1955): 233; 7, no. 3 (1956): 131; "Tiền Công Thợ tại Việt-Nam (Salaires ouvriers au Việtnam)," 5, no. 8 (1954): 499–502.

57. NAVN2 S.0/13, Phúc trình hàng năm của Bộ Y tế 1949.

58. Fan, "Redrawing the Map," 534. Like Maoist political doctrine, the DRV "exalted everyday epistemology, and projected a utopian vision of scientific and political modernity." See also Brocheux, *Une histoire économique,* 189. This reality was captured in interviews of Hồ Đắc Vy, a chemical engineer who had worked in the Viện Khảo Cứu Nông Học in the free zone in Thanh Hóa until he returned to Hà Nội in September 1952. For agriculture in communist China, see Schmalzer, *Red Revolution, Green Revolution.*

59. Bonneuil, "Mise en valeur." See also Verney, "Nécessaire compromis colonial." IRCI, *Cahiers,* 1942: 8, 9, 14; Enderlin, "Industrie indochinoise."

60. *Bulletin du syndicat des planteurs de caoutchouc de l'Indochine,* no. 251 (1941): 43–51, Procès-verbal de l'assemblée générale ordinaire du 9 avril 1941. See also NAVN1 RST 75239-12, Rapport annuels sur le fonctionnement de la station expérimentale agricole et forestière de Phu Ho de 1930 (53); NAVN2 N.31/55, commission d'examen de sortie de la section forestière de l'école d'agriculture de Bencat, 1943–45. On the IRCI's board there were three representatives from the Planters' Syndicate and two from the Planters' Union, including Roger du Pasquier, who had studied under Auguste Chevalier. Bonneuil, "Auguste Chevalier." See Coll., "La conférence internationale du caoutchouc," 10.

61. IRCI, "Procès-verbal de la deuxième réunion du comité technique consultatif du 8 Octobre 1941."

62. Boucheret, "Les plantations d'hévéas," 296, 388–89; Enderlin, "Industrie indochinoise"; IRCI, *Cahiers,* 1942. For a sketch of the planned Lai Khê facilities, see Aso, "The Scientist, the Governor, and the Planter," 250.

63. Ordinance no. 10 of August 6, 1950; Pasquier, *Problèmes d'utilisation des terres,* 54.

64. IRCI, *Rapport annuel,* 1958.

65. See Boucheret, "Les plantations d'hévéas," 566, and Brocheux, *Une histoire économique,* 172, for exchange rate. See also Tertrais, *La piastre et le fusil.* In 1946, 1 U.S. dollar=119,3 francs; in 1953–54 = 349,95 francs; 1 piaster=10 francs before 1945; 1 piaster=17 francs from 1945 until 1953, then 1 piaster=10 francs again. There was 4,834,000 piasters of revenue in 1952, up from 552,000 piasters in 1949.

66. Boucheret, "Les plantations d'hévéas." Đặng cites the memoirs of Guy Simon, *Choroniques de Cochinchine 1951–1956*, as evidence in Đặng, *100 Năm Cao Su*, 215–16.

67. ANOM FM Agefom 200/139, Folder B, Législation commerciale, gommes et résines de 1919 au 1er mai 1952. "Creation of the first Vietnamese Rubber Company" exporting raw or semiprocessed rubber happened in 1951; ANOM Agefom 190/106, Folder B, loose: Articles on the diseases of rubber trees. "Travaux d'amélioration de l'hévéaculture en Indochine"; Main d'œuvre de 1930 à 1943 Folder A: Le caoutchouc en Indochine, généralités, 1950–53. "Le caoutchouc en 1951"–Covering the period of plantations in transition; "Barème '*hévéa*'"; "La réorganisation . . . ." Agefom 202/144: three pamphlets: "En Indochine, sur les plantations de caoutchouc" (mai 1952); "IFC/IRCI" (juin 1952); "IFC/IRCI" (mai 1949). ANOM FM Agefom 200, "Le caoutchouc et la hausse des prix en Indochine, extrait du journal intérieur d'une grande plantation," *Revue générale du caoutchouc* 25, no. 12 (1948): 495. M. Bocquet, "Visage du Vietnam vu par un français," Institut français du caoutchouc, 1950. See his work in rubber, e.g., "Homogénéisation".

68. NAVN3 PTT 2012, Công văn, báo cáo của BCN, Liên đoàn Cao su Thủ Dầu Một, Ban Cao su Bà Rịa, Biên Hoà, Liên hiệp Công đoàn Nam bộ về tổ chức, phá hoại cơ sở cao su ở Nam bộ năm 1949, 1950, 1952. Letter August 5, 1952. *Vietnamese Studies* supported by internal documents show the desire for agricultural and rubber expertise during the 1960s.

69. NAVN3 PTT 2012, Công văn, báo cáo của BCN, Liên đoàn Cao su Thủ Dầu Một, Ban Cao su Bà Rịa, Biên Hoà, Liên hiệp Công đoàn Nam bộ về tổ chức, phá hoại cơ sở cao su ở Nam bộ năm 1949, 1950, 1952. Letter August 14, 1952, from Nguyễn Xuân Cung, the Minister of Agriculture, sent to Government Economic Board. Side note: Đặng wrote "Too slow!" (Chậm quá), since this letter was a response to request from April 14, 1952.

70. IMTSSA 167, Affaires diverses. Documentation sanitaire: correspondance et texte de la conférence sur 'l'organisation médico–sociale en Indochine et particulièrement au Vietnam', 1949, 19.

71. Ibid. For teaching on rubber in France, see George, *Le caoutchouc et les industries*.

72. Bowd and Clayton, "Tropicality," 310.

73. The details of Mus's life are drawn from Chandler, "Paul Mus." Other articles in this issue of the *Journal of Vietnamese Studies* have been useful for contextualizing *Viêt-Nam*, in particular Larcher-Goscha, "Ambushed by History." Mus, *Viêt-Nam*, 107–8, 112. Harry Miller writing in the *Straits Times* in 1952 as quoted in Coates, *The Commerce in Rubber*, 358. Miller wrote specifically about Quản Lợi as "a self-contained town, with a central market, magnificent hospital, schools, playgrounds, temples and churches. It was a revelation."

74. A similar situation for mine workers is described in Packard, *White Plague, Black Labor*. Mus, *Viêt-Nam*, 103–6, 108–9.

75. Mus, *Viêt-Nam*, 110–12.

76. For an analysis of the failure of colonizing schemes that considers the individual perspectives of migrants, see Hardy, *Red Hills*, and McElwee, *Forests Are Gold*. Many planters and government officials were simply focused on profit and had no concern for Vietnamese society, an issue I discuss in "Profits or People?"

77. Mus, *Viêt-Nam*, 108. Mus was also most likely ironic when he wrote "all intention of honesty," but whether he took the planters at their word doesn't make a difference here.

78. Ibid., 186.

79. Ibid., 113.

80. NAVN2 Phủ Thủ Hiến Nam Việt D7-359, Letter, May 17, 1952, Tình trạng công nhân ở các đồn điền cao su, 1952. (Pacification/Plantations) Secret Letter, May 17, 1952, governor of South Vietnam to the interior minister, Sài Gòn. Nguyễn, *Tradition and Revolution*.

81. See Trân, "Thiên-Đường Chăng?," *Liên Lạc*, November 1952. For a critique of this propaganda, see Điệp, *Máu Trắng, Máu Đào*, 111–16. For planters' propaganda efforts in France, see Anon., "En Indochine, sur les plantations de caoutchouc."

82. Huỳnh et al., *Lịch Sử Phong Trào*, 237. Nghi was also secretary of the South Vietnam rubber labor union, 1962–75. Similar commando tactics were used by anticommunist forces in the Philippines during the 1950s. See also White, *Speaking with Vampires*. See Bùi, "Cọp Ba Móng." See also Trân, *Địa Chí Đồng Nai*, Diệt cọp Ba Móng ở Chiến khu Đ. The tiger was apparently killed on February 11, 1950. Also see Lê, *Đất Đỏ Miền Đông*, 79. For more on tigers, see Boomgaard, *Frontiers of Fear*.

83. NAVN2 Phủ Thủ Hiến Nam Việt D7-359, M10-37. Duiker, *Communist Road to Power*, 43, 131–32. Simply staying alive was considered an accomplishment. Đặng, *100 Năm Cao Su*, 234. According to the *New York Times*, the economic impact of the loss of a half million rubber trees was the most important consequence. "Rubber Plantations Hit: Typhoon May Have Killed Half Million Trees," *New York Times*, January 6, 1953.

84. Motte, *De l'autre côté de l'eau*, 89; Osborne, *Tropical Medicine*. IMTSSA 167, Affaires diverses. Documentation sanitaire: rapport de la première conférence, "La pathologie spéciale à l'Indochine," 1947, 3.

85. Service de Santé, *Guide d'Hygiène*, ed. Forces armée en Extrême-Orient (Hà Nội: IDEO, n.d.). For a reproduction of the cartoon, see Aso, "Patriotic Hygiene," 437. See also Forces Terrestres en Extrême-Orient, *Le service de santé*.

86. NAVN2 VIA.8/305 (10). Morin hinted at French culpability for Vietnamese deaths when he wrote that the explosion in malaria in Cochinchina during the 1920s was related to plantations. In 1954, he created a demonstration center in Yaoundé for the use of DDT. In the 1930s, he discussed the social aspects of malaria prevention in Morin, "Vers la prophylaxie sociale." See also his correspondence with the WHO in Centralized files, First and Second Generations (1946–1955), 453-4-5, Dr. H. G. S. Morin "de l'Institut Pasteur de Dalat" on rural malaria control demonstration projects in the Far East, 1947. Noël Bernard, too, commented on the end of empire in *De l'empire colonial*.

87. E.g. Điệp, *Máu Trắng, Máu Đào*, 101–2.

Chapter Six

1. Phạm Ngộc Hồng, interview transcript, 10–11.

2. Nguyễn, *Địa Ngục Cao-Su.* Many Vietnamese I talked to recited poems like the one above that emphasize the hard lives of rubber plantation workers under the French and it seems that such poems form part of the school curriculum. I discussed this phrase with Hồng.

3. Phạm Ngộc Hồng, interview transcript, 11.

4. Ibid., 3.

5. Ibid., 6.

6. For community development and U.S. foreign policy, see Immerwahr, *Thinking Small,* and Cullather, *The Hungry World,* and for small technology in Indonesia, see Moon, *Technology and Ethical Idealism.*

7. NAVN2 Sưu Tập tài Liệu Ảnh 582, "La Société Plantations des Terres Rouges," *Vietnam Presse,* July 29, 1958, Morning edition; "Les Plantations d'Hévéas de la Société Michelin au Viet-Nam," *Vietnam Presse,* August 2, 1958, Evening edition; IRCV, *Rapport Annuel,* 1959. For the attack on the Dầu Tiếng plantation, see National Archives and Records Administration Record Group (afterwards NARA RG) 59, Box 4588-4590, Department of State, General Records, 1955–59, 851g.00/1-655–851g.00/8-2558.

8. See Keith, *Catholic Vietnam,* and Hansen, "The Virgin Heads South."

9. For Bảy Viễn, see NAVN2 PTT DICH 10923, TDBCPNP T99-59, and PTT DICH 10830. Other "enemies" such as Bảo Đại and Trần Văn Soái also had plantations confiscated. See PTT DICH 12620, Tiền lời đồn điền tịch thu của Bảo Dai khuyếch trương trồng cao su, 1959.

10. See NAVN2 PTT VNCH 29209, Giải quyết đình công và xích mích tại các sở, đồn điền cao su, 1954–1955. See also Võ, *Phong Trào Công Nhân,* and Nguyễn, *Tư Bản Pháp và Vấn Đề Cao Su.* NAVN2 TDBCPNP R2-128, Về việc công nhân đồn điền cao su Xa Cam xin mở lớp dạy thanh niên thất học, 1955. See NAVN2 PTT VNCH 29209, Giải quyết đình công và xích mích tại các sở, đồn điền cao su, 1954–1955. Letter from Nguyễn Tăng Nguyên, Tổng Trưởng BLĐ và Thanh Niên to the PM and the other ministers, December 10, 1954.

11. NAVN2 PTT VNCH 29209, Letter from Nguyễn Tăng Nguyên, Tổng Trưởng BLĐ và Thanh Niên to the PM and the other ministers, December 10, 1954, and letter from Nguyễn Tăng Nguyên, Tổng Trưởng BLĐ và Thanh Niên to the PM, January 6, 1955. For Diệm's request, see letter, Ngô Đình Diệm, to Minister of Labor and Youth, February 10, 1955. For report, see letter from Nguyễn Tăng Nguyên, Tổng Trưởng BLĐ và Thanh Niên to the PM, February 24, 1955. For the June 26, 1953, law, see letter, Trần Văn Lam, Đại Biểu Chánh Phủ tại Nam Việt, to the province chiefs, in Biên Hòa, Bà Rịa, Thủ Dầu Một, Tây Ninh, and Gia Định, February 23, 1955.

12. NAVN2 PTT VNCH 29209, Giải quyết đình công và xích mích tại các Sở, đồn điền cao su, 1954–1955. See letter from Nguyễn Tăng Nguyên, Tổng Trưởng BLĐ và Thanh Niên to the PM, February 24, 1955, and letter from Nguyễn Tăng Nguyên, Tổng Trưởng BLĐ và Thanh Niên to the PM, March 1, 1955. Letter from Nguyễn

Tăng Nguyên, Tổng Trưởng BLĐ và Thanh Niên to the PM, March 23, 1955. Letter from Ngô Đình Diệm to Nguyễn Tăng Nguyên, April 2, 1955. NAVN2 PTT DICH 16502, Nha tổng giàm đốc Cảnh sát và Công an Về việc công nhận các đồn điền cao su dự Định tranh chấp 1957.

13. NAVN2 PTT VNCH 29209, For Việt Minh control of workforce, see letter from Nguyễn Tăng Nguyên, Tổng Trưởng BLĐ và Thanh Niên to the PM, January 6, 1955. See also letter from Nguyễn Tăng Nguyên, Tổng Trưởng BLĐ và Thanh Niên to the PM, February 24, 1955, and letter from Nguyễn Tăng Nguyên, Tổng Trưởng BLĐ và Thanh Niên to the PM, March 1, 1955.

14. Interview with Nam Pham, Phạm Văn Vỹ's son. NAVN2 PTT VNCH 29209, For proposal for stronger noncommunist organizations, see letter, Trần Văn Lam, Đại Biểu Chánh Phủ tại Nam Việt, to the PM, February 28, 1955.

15. For a summary of Bửu's early life, see Wehrle, *Between a River and a Mountain*, 33–35. Bửu's break with the Việt Minh paralleled the American Federation of Labor internationalists' vindication as anticommunists. Ibid., 41.

16. Tran, *Ties That Bind*, 81–83.

17. ANOM HCI CS 74, Travail Haut-Donnai PMSI—A discussion of Montagnard labor ; Loi, n. 46, 11 avril 1946, suppression du travail forcé dans les territoires Outremer, en Indochine, n. 39369, 8 février 1947. For labor movements in Africa, see, e.g., Cooper, "The Dialectics of Decolonization."

18. Lê, *Đất Đỏ Miền Đông.*

19. NARA RG 59, Box 4588-4590, Department of State, General Records, 1955–59, 851g.00/1-655–851g.00/8-2558, Box 4588: Foreign service despatch, July 6, 1956, Some Changing Aspects of the Labor Situation. From Gardner E. Palmer, Counselor of Embassy, for Ambassador.

20. NAVN2 PTT DICH 16496, Giải quyết yêu sách của công-nhân cạo mủ ở Xa-Trạch, 1957, Phần cây cao và thời gian nghỉ trưa của công-nhân cạo mủ ở Xa-Trạch, 1957. Circulaire 100, Chatel, Xatrach, May 7, 1957. Letter Nguyễn Văn Toán, Tỉnh Trưởng, Tỉnh Bình Long, to Thanh Tra Lao Động và An Ninh Xã Hội Nam Phần, July 22, 1957. Letter Nguyễn Văn Toán, Tỉnh Trưởng, Tỉnh Bình Long, to Thanh Tra Lao Động và An Ninh Xã Hội Nam Phần, July 31, 1957. Letter, Bộ Trưởng Lao Động, Huỳnh Hữu Nghĩa, to Vice President, Economic Minister, BVT PTT, August 9, 1957. Letter from Bộ Trưởng Bộ Kinh Tế, August 20, 1957, agreed with assessment of Bộ Trưởng Lao Động. NARA RG 59, Box 4588-4590, Department of State, General Records, 1955–59, 851g.00/1-655–851g.00/8-2558: Foreign service despatch, January 22, 1958, Weekly Economic Review January 14–20, 1958. From William Gileane, Second Secretary of Embassy, for Ambassador.

21. NAVN2 PTT DICH 16510, Về việc hội cao su người Pháp phàn nàn về bài báo đăng trong Việt Nam Thông tấn xã August 31, 1957. NAVN2 PTT DICH 16502, Nha tổng giàm đốc Cảnh sát và Công an Về việc công nhận các đồn điền cao su dự đinh tranh chấp 1957. NAVN2 PTT DICH 16509, Về việc nghiệp đoàn các nhà trồng cao su trình bày mới lo sợ về hậu quả những cuộc phiến động xã hội 1957. Letter, R. Desplanques, Michelin, to Inspecteur provincial du Travail de Binh Duong, et Chef de la Province de Binh Duong, August 16, 1962.

22. NAVN2 PTT DICH 16510, Về việc hội cao su người Pháp phàn nàn về bài báo đăng trong Vietnam Thông tấn xã August 31, 1957. Letter, Bộ Trưởng Lao Động, Huỳnh Hữu Nghĩa, to VP, Economy Minister, September 26, 1957. Letter, Bộ Trưởng Lao Động, Huỳnh Hữu Nghĩa, to Président du Syndicat des planteurs de caoutchouc, 26 septembre 1957.

23. NAVN2 PTT DICH 17385, Phiếu trình của ông tỉnh trưởng tỉnh Bình Long Về việc đình công tại hai đồn điền cao su Xa Cát và Lộc Ninh, 1960. Letter, Trân Hữu Quyền, TLĐLC, to other labor unions, rubber workers, January 5, 1960. Letter, Nguyễn Văn Toán, Tỉnh Trưởng, Tỉnh Bình Long, to Bộ Trưởng Bộ Nội Vụ, January 11, 1960.

24. NAVN2 PTT DICH 13474, lễ ký kết cộng đồng khế ước lao động cao su 12.3.1960 tại Bộ Lao Động.

25. NARA RG 469, Records of U.S. Foreign Assistance Agencies, 1942–1963, Box 3 Vietnam Desk Subject Files, compiled 1956–1962, P 89, Foreign service despatch, 6-10-60, Collective Agreement Between Rubber Plantation Workers and Owners, "Convention Collective de l'Hévéaculture pour le Vietnam."

26. NARA RG 469, Records of U.S. Foreign Assistance Agencies, 1942–1963, Box 3 Vietnam Desk Subject Files, compiled 1956–1962, P 89, Official Journal of the RVN, No. 25, 6/4/60, Department of Labor. Decree No. 66-BLD-LD/ND, 5/19/60 approving the Collective Convention of Rubber Plantations.

27. NARA RG 469, Records of U.S. Foreign Assistance Agencies, 1942–1963, Box 3 Vietnam Desk Subject Files, compiled 1956–1962, P 89, Memorandum of Conversation with Desplanques, 6/22/61. NAVN2 PTT DICH 17643, Phản đối vụ Nguyễn Văn Vỹ bị sa thải, 1961. Letter, Nguyễn Văn Toán, Tỉnh Trưởng, Tỉnh Bình Long, to Bộ Trưởng Đặc Nhiệm Phối Hợp An Ninh, Bộ Nội Vụ, BLĐ, July 27, 1961. Letter, Trần Kim Tuyến, PTT office, August 11, 1961. Letter, Nguyễn Lê Giang, Inspecteur générale du travail et de la securité sociale to M. Inspécteur générale de Caoutchoucs d'Extrême-Orient, 2 Sept 1961. See NAVN2 PTT VNCH 29209, Giải quyết đình công và xích mích tại các sở, đồn điền cao su, 1954–1955. See also Võ, *Phong Trào Công Nhân*, and Nguyễn, *Tư Bản Pháp và Vấn Đề Cao Su*, 98–99, and Appendix II. Đặng, *100 Năm Cao Su*, 214. NAVN2 PTT DICH 17889, BLĐ vụ tranh chấp tại đồn điền cao su, 1961–62. Letter, Nguyễn Thoại, Nha Tổng Thanh Tra Lao Động to President's office, ca. December 28, 1961. Letter, Nguyễn Lê Giang, Đổng Lý Văn Phòng Lao Động to Đổng Lý Văn Phòng Bộ Nội Vụ, February 1, 1962, Về việc tranh chấp tại IRCV, 59–61

28. See Catton, *Diem's Final Failure*; Chapman, *Cauldron of Resistance*; Jacobs, *America's Miracle Man*; and Miller, *Misalliance*.

29. Drabble, *An Economic History of Malaysia*, 5. See also Miller, *Misalliance*, 95–108. For economic histories of Singapore and Indonesia, see Huff, *The Economic Growth of Singapore*, and Booth, *The Indonesian Economy*. For South Vietnam, see Đặng, *Kinh Tế Miền Nam*. NARA RG 59, Box 4588-4590, Department of State, General Records, 1955–59, 851g.00/1-655–851g.00/8-2558 Box 4588, President Diệm's Speech on the Government's Economic Policy Delivered at Tuy Hòa on September 17, 1955.

30. NARA RG 59, Box 4588-4590, Department of State, General Records, 1955–59, 851g.00/1-655–851g.00/8-2558 Box 4588, Foreign service despatch, 9-19-55, Economic Policy Discussions. From Gardner E. Palmer, Counselor of Embassy, for Ambassador and American Aid and Economic Recovery of Vietnam, August 20, 1955, translation of paper handed to Senator Mike Mansfield at meeting with President and Ministers. NAVN2 TDBCPNP D30-237, Letter, Cảnh sát công an Nam phần to Bộ Nội Vụ, December 28, 1956.

31. NAVN2 PTT DICH 10000, Letter, Trần Văn Minh, Thiếu Tướng Đệ Nhất Quân Khu, to PTT, July 21, 1956.

32. Miller, *Misalliance*, 132–36; NAVN2 TDBCPNP D30-237, PTT DICH 4466, and TDBCPNP D5-356.

33. NAVN2 PTT DICH 5394, vụ cướp đồn điền cao su Minh Thạnh, Bình Long, January 4, 1958. PTT DICH 5902, Tòa Đại Biểu Chính Phủ tại Miền Đông Nam Phần Về việc điều tra VC đổi vùng trong các đồn điền cao su 1959.

34. NAVN2 PTT DICH 7136, Về việc diệt trừ nạn trộm mủ cao su 1960–1961. Meeting, Toà Hành Chánh Bình Dương, August 2, 1960. Letter, Huỳnh Văn Điểm, GĐ Kế Hoạch, to PTT, September 1, 1960.

35. NAVN2 PTT DICH 7895, biện pháp ngăn chặn việc lấy trộm mủ cao su tại đồn điền, 1962. Letter, Nguyễn Thành, Cảnh sát và Công an, to Bộ Nội Vụ, May 25, 1962. Letter, Bùi Văn Lương, Bộ Nội Vụ, to Tỉnh Trưởng Biên Hòa, May 4, 1962. Letter, Bùi Văn Lương, Bộ Nội Vụ, to Trưởng Tư Pháp, June 11, 1962. Letter, Nguyễn Hữu Thi, Ty Lao Động Tây Ninh, to Tỉnh Trưởng Lao Động Tây Ninh, May 5, 1962. Letter, Bùi Văn Lương, Bộ Nội Vụ, to Tỉnh Trưởng Miền Đông, June 4, 1962. PTT DICH 8460, nạn trộm mủ cao su và cạo mủ lén tại Lai Khê, 1963. Report to Bình Long province chief, on theft from Xa Cát, May 22, 1963. Mentions Nguyễn Văn Xin, who stole forty tons of rubber, was caught, and released by the court.

36. NAVN2 PTT DICH 7896, An ninh tại đồn điền cao su 1962. Letter, Nguyễn Lê Giang, Đổng Lý Văn Phòng Lao Động to Đổng Lý Văn Phòng Bộ Đặc Nhiệm An Ninh etc., January 5, 1962. Note, anon, nd, Summary, VC pushing workers to demand raises, from 30 January 1962. Letter, Huỳnh Hữu Nghĩa, BLĐ, to Bộ Đặc Nhiệm An Ninh, March 8, 1962. Letter, Nguyễn Lê Giang, Đổng Lý Văn Phòng Lao Động to Đổng Lý Văn Phòng Bộ Nội Vụ, August 18, 1962. Letter, Nguyễn Lê Giang, Đổng Lý Văn Phòng Lao Động to Đổng Lý Văn Phòng Bộ Nội Vụ, August 31, 1962. Note de service from P. D'Ornano, Ste Hévéas Tây Ninh, August 31, 1962. Letter, Nguyễn Văn Tam, Tỉnh Trưởng Lao Động Bình Dương, to Thanh Tra Lao Động, August 23, 1962. Telegram, R. DesPlanque, Michelin, Plantations A, Dầu Tiếng and Thuận Lợi, August 24, 1962. Letter, Nguyễn Văn Tam, Tỉnh Trưởng Lao Động Bình Dương, to Thiếu Tá Tỉnh Trưởng Bình Dương, September 21, 1962.

37. Brocheux, *Une histoire économique*, 224. The importance of the French came from "L'inspiration principale concerna la voie à suivre, les methodes et les savoir-faire pour poursuivre la modernisation." Brocheux mentions the work of Nguyễn Công Tiểu published in *Khoa Học Tạp Chí*. He also cites *Khoa Học Phổ Thông*. Brocheux

argues that the family space and a sort of home economy helped bring about the transition to *đổi mới*. See also Schmalzer, *Red Revolution, Green Revolution*.

38. Trịnh, "Contribution à la mise en pratique"; Đặng and Đăng, *Bệnh Sốt Rét*, 34, 193, 203–8, 217–30, 285–87.

39. Đặng and Đăng, *Bệnh Sốt Rét*, 34.

40. Ibid., xvi.

41. Prakash, *Another Reason*; Sở Quân Dân Y, *Bịnh Sốt Rét*, 30. The authors discuss both Western and Eastern medical approaches to the disease and argue that only independence, unification, democracy, peace, and socialism would enable the complete elimination of malaria.

42. As quoted in Đặng et al., "Studies of the epidemiology of malaria in northern Vietnam, Volume 1," 293.

43. For more on Lysenko and Soviet malaria prevention research, see Bruce-Chwatt, "Malaria Research and Eradication in the USSR."

44. The DRV argued that its techniques were more successful than those of the WHO. Bộ Y Tế, *Bệnh Sốt Rét*; Đặng, *Những Kỷ Niệm*, 44, 48.

45. Bộ Y Tế, *Bệnh Sốt Rét*, 39, 59, 62–64; Đặng, *Những Kỷ Niệm*, 50–51; Sở Quân Dân Y, *Bịnh Sốt Rét*, 27–28.

46. Malarney, "Germ Theory," 111, 115; Đặng, *Những Kỷ Niệm*, 38; Trương, Chanoff, and Doan, *A Vietcong Memoir*, 160, 162.

47. NAVN2 PTT D1CH 15867.

48. Although Miller cautions that the Ngô family did not simply copy models from Malaya, this caution is more apt for counterinsurgency. Miller, *Misalliance*, 232–33. For Việt Nam-Malaya contact, see NAVN2 PTT DICH 15867.

49. IRCV, *Rapport annuel 1960–61–62*, 93. See also NAVN2 PTT DICH 12619, trồng tỉa và khai thác cao su 1958–59. PTT DICH 10827, tình hình hoạt động đồn điền cao su Việt Nam 1957. PTT DICH 10000, Letter, Nguyễn Hữu Dung, ĐLVP Bộ Canh Nông, to ĐLVP PTT, July 11, 1956.

50. Đặng, *100 Năm Cao Su*, 79–84.

51. NARA RG 469, Records of U.S. Foreign Assistance Agencies, 1942–1963, Box 33, Sheldon Tsu, "Outline of the Importance of Rubber and the Situation in Vietnam," April 1957, 1–10. See also NAVN2 PTT DICH 10829, BCN, BKT, TNGĐKH, hoạt động của đồn điền cao su, 1957, Letter, ĐLVP BKT, ca. March 1957.

52. NARA RG 469, Records of U.S. Foreign Assistance Agencies, 1942–1963, Box 33, Sheldon Tsu, "Outline of the Importance of Rubber and the Situation in Vietnam," April 1957, 1–10.

53. NAVN2 PTT DICH 12619, trồng tỉa và khai thác cao su 1958–59.

54. Ibid. NĐ 287 KT.

55. Mùi, "Avenir des plantations de caoutchouc appartenant à des vietnamiens," in Association culturelle populaire, *Le problème de l'hévéaculture au Viet-Nam*. "Nghiệp-đoàn khai-thác cao-su Việt-Nam" and "Đại Hội Nghiệp-đoàn K.T.C.S.V.N.," *Tuần San Phòng Thương Mại Sài Gòn*, August 22, 1958, 4–5, 33. For a more complete evaluation, see NAVN2 PTT DICH 12843, thành lập "quỹ khuếch trương trồng tỉa cao su," 1959.

Also PTT DICH 12853, Nhân xét dân biểu Quốc hội về vay tiền thực hiện chương trình "quỹ khuếch trương trồng tỉa cao su" 1958–59. And PTT DICH 12854, Vay tiền "quỹ khuếch trương trồng tỉa cao su" 1959. Yêu Cầu Sửa Đổi Nội Dung Bản Dự Thảo Cho Các Nhà Trồng Tỉa Cao Su Vay."

56. NAVN2 PTT DICH 11946, đoàn thể, cá nhân cao su xin vay tiền của Quỹ Khuếch Trương trồng tỉa cao su, 1958. PTT DICH 13472, khuếch trương trồng tỉa cao su Việt Nam 1960.

57. Rapport de visite de la plantation de Mme Võ Thị Sen (Gia Trập), IRCV, visite effectuée le 7 juillet 1959. Also McIndoe, *A Preliminary Survey*, Appendix 1, 9. See also NAVN2 PTT DICH 12843, thành lập "quỹ khuếch trương trồng tỉa cao su" 1959. From NAVN3 PTT 2012, 5. NAVN2 PTT DICH 11737, đầu tư công ty cao su Michelin, kế hoạch khuếch trương ngành cao su tiến tới tự túc về tài chính, 1958. NAVN2 PTT DICH 18953, Viện nghiên cứu cao su Việt Nam khai thác đồn điền cao su đồn điền 1955-60. "Vietnamese Rubber Plantations Increase," *The Nation*, Sài Gòn, July 15, 1959.

58. NAVN2 PTT DICH 13472, khuếch trương trồng tỉa cao su Việt Nam 1960. Note from Banque Nationale du Vietnam, Nguyễn Hữu Hạnh, to Ngô Đình Diệm, March 9, 1960. Note from Banque Nationale du Vietnam, Nguyễn Hữu Hạnh, to Ngô Đình Diệm, September 9, 1960. For more on Nguyễn Đình Quát, see NAVN2 PTT DICH 22401; PTT DICH 10000.

59. NAVN2 TDBCPNP D30-237. See also PTT DICH 22401 and PTT DICH 10000. Letter, Nguyễn Hữu Dung, ĐLVP Bộ Canh Nông, to ĐLVP PTT, July 11, 1956. PTT DICH 16509: Về việc nghiệp đoàn các nhà trồng cao su trình bày mới lo sợ về hậu quả những cuộc phiến động xã hội 1957. PTT DICH 16510, "Nghiệp-đoàn khai-thác cao-su Việt-Nam" and "Đại Hội Nghiệp-đoàn K.T.C.S.V.N." *Tuần San Phòng Thương Mại Sài Gòn*, August 22, 1958, 33.

60. Association culturelle populaire, *Le problème de l'hévéaculture*.

61. Ibid.

62. Ibid., 51.

63. See report by François Schmitz, Xuân Lộc, April 10, 1960, in NARA Commodities: Rubber Vietnam, 1957–1961.

64. NAVN2 PTT DICH 14317, Đề nghị của Ông Nguyễn Chí Mai về vấn đề cao su Việt Nam 1961. Đặng, 100 Năm Cao Su, 121–33; McIndoe, *A Preliminary Survey*, Appendix 1.

65. Miller, *Misalliance*, 171–84. NAVN2 PTUDDNV 568, Bảo vệ an ninh 3 địa điểm dinh điền ở đồn điền cao su Dầu tiếng, 1963. Letter Đại Úy Trần Đình Minh, sĩ quan kiểm soát các khu trù mật, to Tổng Thư Ký Bộ Cải Tiến Nông Thôn, Sài Gòn, January 2, 1963. Letter, Bộ Trưởng Nội Vụ, meeting about road running along Tây Ninh, Bình Dương, Gia Định, and Long An, January 21, 1963. Letter, Trung Tá Nguyễn Cao, to Đại Tá, February 19, 1963. Letter, Đại Tá Nguyễn Văn Thiệu to TU Dinh Điền và Nông-Vụ, March 13, 1963. Letter, Trung Tá Nguyễn Cao, TU Trường Dinh Điền và Nông-Vụ, to Bộ Trưởng Cải tiến Nông Thôn, April 8, 1963. PTT DICH 12634, Công nhân đồn điền cao su và Hoa kiều tại Cambodge xin hồi hương đi dinh điền 1959. PTT DICH 10830.

66. Tan, "'Swiddens, Resettlements, Sedentarizations, and Villages,'" 210–11, 233. See also Scott, *Seeing Like a State*. NAVN2 PTT DICH 13477, Tài liệu viện khảo cứu cao su về thể thức thành lập một đồn điền cao su 1000 ha . . . . 1957–1960. PTT DICH 13476, Về việc trồng cao su tại địa diểm dinh điền 1959–1960. PTT DICH 10828, Trồng cây cao su 1957.

67. NAVN2 PTT DICH 11946, đoàn thể, cá nhân cao su xin vay tiền của Quỹ Khuếch Trương trồng tỉa cao su, 1958. Wehrle, *Between a River and a Mountain*; Miller, *Misalliance*, 67–71, 164–65. ANOM Fonds privés, Monguillot (1946–54), 56 PA/13–Caoutchouc, Santé. Indochine–Affairs Economique–Caoutchouc.

68. NAVN2 PTT DICH 13472, khuếch trương trồng tỉa cao su Việt Nam 1960. Le Secretaire d'etat au travail et a la Reforme Agraire, Huỳnh Hữu Nghĩa, à Directeur Général du B.I.T., Geneve, June 13, 1960. Letter, anon, nd, ca. July 2, 1960. Report on Hội Đồng Liên Bộ, Nguyễn Ngọc Thơ, July 7, 1960. General director of planning, Huỳnh Văn Đểm, to BVT BLĐ, September 19, 1960.

69. NAVN2 PTT DICH 13464, See also PTT DICH 18953. Lễ Khai mạc phiên họp đầu tiên của ủy ban tư vấn cao su 14.4.1960.

70. NAVN2 PTT DICH 13464, Ph. J. P. Richard, L'Ecole supérieure d'application de l'agriculture tropicale, Commissaire Générale au Développement Agricole, L'Avenir des "Small holdings" en Hévéaculture au Vietnam, presented March, 2, 3.

71. NAVN2 PTT DICH 13464, Ph. J. P. Richard, L'Ecole supérieure d'application de l'agriculture tropicale, Commissaire Générale au Développement Agricole, L'Avenir des "Small holdings" en Hévéaculture au Vietnam, presented March, 13.

72. NAVN2 PTUDDNV 6299, Về việc thành lập và phát triển công ty cao su 1963. Letter, Philippe J. P. Richard to Commissaire Général du Développement agricole et d'agriculture, January 3, 1963, 7. PTUDDNV 6856, Về chương trình cao su 1960. Philippe J. P. Richard, Expert d'Agriculture au Commissariat Gén. à M. le Commissaire Général au Développement Agricole, objet, transmission de mon rapport de mission, Saigon, 24 février 1960. NAVN2 PTT DICH 15202, thành lập khu dinh điền Trung Nguyên, công nhân 1961–62. NAVN2 PTUDDNV 6112, 1-6. This file is from the office of agricultural reform from 1962. Philippe Richard à M. le Commissaire Générale au Développement Agricole, objet, transmission de mon rapport de mission, Mars 1962. NAVN2 PTUDDNV 6114, Thành lập Văn phòng Hợp tác cao su 1962.

73. NAVN2 PTT DICH 13404, Tuyển mộ công nhân đồn điền cao su để phát triển cộng đồng và làm khu trù mật 1960. PTT DICH 15869: thiết lập các dinh điền gần đồn điền cho công nhân cao su 1962–63.

74. NARA RG 472, Records of the U.S. Forces in Southeast Asia, 1950–1976; A1 461, Box 3. Brocheux, *Une histoire économique*, 172–76, 1960, 1 U.S. dollar = 60 dong and 4.9 francs, source R. L. Bidwell, Currency Conversion tables, London, 1970. Similar trend elsewhere on terres rouges. NAVN2 PTT DICH 15869, thiết lập các dinh điền gần đồn điền cho công nhân cao su 1962–63.

75. NAVN2 PTT DICH 15910, Cao Đài Tây Ninh chiếm canh đất Quốc gia trồng cao su ở rừng cấm Cẩm Giang, 1962–63. PTT VNCH 22822, công ty đồn điền cao su Vietnam

và các cá nhân xin trưng khẩn đất quốc gia tại Phước Tuy 1962–1966. PTT VNCH 22827, Công chức, quân nhân xin khẩn đất trồng cao su tại Phước Long 1960–1966. 30 November 1960, letter from general secretary of president, Nguyen Thanh Cung, to the minister of land and agrarian reform about land grants to officials. Arrêté du GGI, 28-3-1929, complété par arrêté du 12-5-1939 pourtant réglementation du régime des concessions gratuites de terres domaniales en Indochine. NARA RG 469, Records of U.S. Foreign Assistance Agencies, 1942–1963; Box 3 Vietnam Desk Subject Files, compiled 1956–1962, P 89, Foreign service despatch, Security Conditions on Rubber Plantations, 1-9-61, Gardiner. PTT VNCH 21888, Toà giám mục Sài Gòn xin chỉnh trang và xây dựng tại vườn cao su Phú Tho Hoà Gia Định 1963–1964. PTUDDNV 6411, Về việc trồng cao su tại địa điểm Biên Hòa, 1963.

76. NAVN2 PTT VNCH 22827, Công chức, quân nhân xin khẩn đất trồng cao su tại Phước Long 1960–1966. Letter, July 2, 1966, from Ủy viên Canh Nông to Ô Phụ tá tại Phủ chủ Tịch Ủy Ban Hành Pháp Trung ương.

77. NAVN2 PTT DICH 15202, thành lập khu dinh điền Trung Nguyên, công nhân 1961–62. NAVN2 PTUDDNV 6299, Về việc thành lập và phát triển công ty cao su 1963. Letter, Philippe J. P. Richard to Commissaire Générale au Développement Agricole et d'Agriculture, January 3, 1963. Mentions following letter: Thuyết trình giải thích của Bộ trưởng cải tiến nông thôn trước Quốc hội, December 28, 1962.

78. NAVN2 PTT DICH 19233, tin VTX Về việc trồng sản xuất, xuất khẩu cao su Việt Nam 1961–62. PTT DICH 19234, tin VTX đưa tin các khóa huấn luyện phương pháp trồng tỉa cao su và cung cấp các loại cây ăn trái, cây kỹ nghệ cho đồng bào các Dinh Điền, 1961–62. PTUDDNV 6360, Về việc cấp đất trồng cao su cho di dân địa điểm Chur-Su (Thăng Trị) khu Đắc Lắc I vùng Cao Nguyên Trung Phần n. 1963. NAVN2 PTUDDNV 6112, 1-6. This file is from the office of agricultural reform from 1962. "Tu don dien cao-su den dinh dien cao nguyen," 1962.

79. NAVN2 PTUDDNV 6298, Về việc canh tác trồng cây cao su theo chương trình dinh điền 1963. "Le développement de l'hévéaculture vietnamienne avec les nouveaux concessionnaires." Also in PTUDDNV 6300, kỹ thuật trồng cao su, bông vải, thầu dầu, cây gai, Nha kỹ thuật 1963. Đặng, 100 Năm Cao Su, 121–33. See also McIndoe, *A Preliminary Survey*, Appendix 1.

80. Kumar, *Indigo Plantations*. See also Smith-Howard, *Pure and Modern Milk*. Fabre, "Les plantations de caoutchouc," 396–97.

81. Tully, *Devil's Milk*. NARA RG 469, Records of U.S. Foreign Assistance Agencies, 1942–1963, Box 3 Vietnam Desk Subject Files, compiled 1956–1962, P 89, International Cooperation Administration (ICA), U.S. Assistance to Rubber Production Programs. ICATO Circ FA 3, 10-14-61.

82. Natural Rubber Research Conference, 9/26–10/1/1960. NAVN2 PTT DICH 12621. PTT DICH 10829: BCN, BKT, TNGĐKH, hoạt động của đồn điền cao su, 1957.

83. NAVN2 TDBCPNP D5-356, Việt Minh assassination of Jacques Gaxotte, vice director of IRCI, July 17, 1956.

84. Nguyễn, *Rapport stagiaire*; IRCV Reports; Anon., "A l'IRCC, travaux de recherches sur l'hévéa."

85. IRCV, *Rapport annuel*, 1958, 1–9; 1959, 147–74; and 1962, 1–2, 93–99.

86. IRCV, *Rapport annuel*, 1958, 7, 8, 72–76; IRCV, *Rapport annuel*, 1959, 61–71; Taysum, "A Possible Basis for Phyto-Immunity."

87. NAVN2 PTT DICH 18953, Viện nghiên cứu cao su Việt Nam khai thác đồn điền cao su đồn điền 1955–60, Bouthillon. "L'Institut de recherches sur le caoutchouc au Viet-Nam," *Vietnam Presse*, November 28, 1959, Morning edition. See Padirac from BNF on rubber research between 1936 and 1960s, e.g. from Patrice Compagnon.

88. See, for example, Heng-Lon, "Industry of Rubber," *Kambuja* 1, no. 6 (1965), 6p; "Le rendement à l'hectare du latex vietnamien est supérieur à celui des autres producteurs," *Journal d'Extrême-Orient*, July 10, 1957; "Le Viet-Nam pourrait, à bref délai, fabriquer de 500 000 à 1 million de pneus de bicyclette et réaliser ainsi une économie annuelle de plusieurs dizaines de millions de piastres en devises étrangères," *Journal d'Extrême-Orient*, July 11, 1957. NARA RG 469, Records of U.S. Foreign Assistance Agencies, 1942–1963, Box 3 Vietnam Desk Subject Files, compiled 1956–1962, report by François Schmitz, Xuân Lộc, April 10, 1960.

89. "Le caoutchouc, richesse 'numéro un' du Viet nam, comment et pourquoi l'exploitation du latex sur place est possible et rentable," *Journal d'Extrême-Orient*, July 9, 1957; IRCV, *Rapport annuel*, 1958, 8.

90. NAVN2 PTT DICH 8460, nạn trộm mủ cao su và cạo mủ lén tại Lai Khê, 1963. Letter, EDC Baptiste, Vice Président IRCV, to Président IRCV, July 3, 1963.

Chapter Seven

1. Phan, *The Republic of Vietnam's Environment and People*, iii.

2. NAVN2 Toà Đại Biểu Chánh Phủ tại Nam Việt N9-99, Don dien cao su, Kiem soat rung ru.

3. See Hecht, Morrison, and Padoch, *The Social Lives of Forests*; Bowd and Clayton, "French Tropical Geographies," 282. For visibility, see NARA RG 472, Records of the U.S. Forces in Southeast Asia, 1950–1976, Box 42.

4. For smallholders in Indonesia, see Barlow, Colin, Muharminto, "The Rubber Smallholder Economy."

5. Nguyễn, *Tư Bản Pháp*, 75, 81, 98–99, Appendix II. This information comes from Võ, *Phong Trào Công Nhân Miền Nam*. For more on the labor movement, see Trịnh, *Phong Trào Lao-Động*, 50. See also Lê, *Đất Đỏ Miền Đông*.

6. NAVN2 PTT Việt Nam Cộng Hòa 15654, Nội Vụ, Về việc bảo vệ an ninh đồn điền cao su tại vùng 3 chiến thuật 1963–1966. Lá thư, Nguyễn Văn Sáng, Tổng Thanh Tra Lao Động và An Ninh Xã Hội cho Bộ Lao Động, November 25, 1963. General report on situation, April 30, 1964. Lá thư, Nguyễn Xuân Sinh, Trưởng Ty Thanh Tra Lao Động cho Tổng Thanh Tra Lao Động và An Ninh Xã Hội, Long Khánh, November 22, 1963. NAVN2 PTT Việt Nam Cộng Hòa 21760, Lá thư, Bộ Lao Động, Đàm Sỹ Hiến cho

Thủ Tướng, Sài Gòn, July 2, 1964. Bãi bỏ thuế đặt biệt cao su và giảm thuế các nhà trồng tỉa cao su để ngăn chặn các cuộc đình công lớn, 1964.

7. Fabre, "Les plantations de caoutchouc." See chapters 4 and 5 in Michon, *Indochina Memoir*. Đặng, *100 Năm Cao Su*, 218.

8. NAVN2 PTT Việt Nam Cộng Hòa 15319, đồn điền cao su của Pháp kiều trong vùng VC kiểm soát đã đóng thuế cho VC, 1965. Report on ransom and taxes paid by plantations to VC, July 27, 1965. Note from PTT, June 7, 1965. Letter, Phạm Văn Liễu, Director of National Police, to Interior Commissioner, August 9, 1965. NAVN2 PTT Việt Nam Cộng Hòa 15654, PTT, Bộ Nội Vụ, Về việc bảo vệ an ninh đồn điền cao su tại vùng 3 chiến thuật 1963–1966. April 13, 1964, Summary of information from Trần Gia Huấn about the kidnapping of Jean Chateau, director of Quản Lợi, SPTR. Handwritten letter from Trần Gia Huấn, February 1, 1964. Lá thư, Trần Thanh Ben, Tổng Giám Đốc Cảnh Sát Quốc Gia cho PTT, May 8, 1964. Lá thư, Trung Tướng Trần Ngọc Tám cho Bộ Nội Vụ, April 29, 1964.

9. Santiago, *Ecology of Oil*. NAVN2 PTT Việt Nam Cộng Hòa 22361, Ông Dương Minh Nghĩa xin sang bộ hai đồn điền cao su mua của Ông Leo Servain tại Tây Ninh 1964–1965. NAVN2 PTT Việt Nam Cộng Hòa 15654, PTT, Bộ Nội Vụ, Về việc bảo vệ an ninh đồn điền cao su tại vùng 3 chiến thuật 1963–1966. Lá thư, Nội Vụ cho Thủ Tướng, May 28, 1964. NAVN2 PTT Việt Nam Cộng Hòa 21760, Lá thư, Bộ Lao Động, Đàm Sỹ Hiến cho Thủ Tướng, Sài Gòn, July 2, 1964, về việc Pháp kiều chủ nhân đồn điền cao su lo ngại về tin đồn tài sản của họ sẽ bị quốc hữu hóa đã xúi dục công nhân tranh chấp để cản trở việc quốc hữu hóa, 1964.

10. NAVN2 PTT Việt Nam Cộng Hòa 31643, VC gây áp lực cho cuộc tranh chấp, đình công của công nhân đồn điền cao su, 1964–1975. Tổng Trưởng Bộ Lao Động, Đàm Sỹ Hiến to Tổng Trưởng Bộ Nội Vụ, October 1, 1964.

11. NAVN2 PTT Việt Nam Cộng Hòa 31643, VC gây áp lực cho cuộc tranh chấp, đình công của công nhân đồn điền cao su, 1964–1975. Lá thư, Tổng Trưởng Bộ Lao Động, Đàm Sỹ Hiến cho Tổng Trưởng Bộ Nội Vụ, October 1, 1964. Lá thư, Nguyễn Lưu Viên, Tổng Trưởng Bộ Nội Vụ cho Tỉnh Trưởng Miền Đông, October 12, 1964. Lá thư, Tổng Trưởng Bộ Lao Động, Đàm Sỹ Hiến cho Tổng Trưởng Bộ Nội Vụ, October 14, 1964. Lá thư, Phạm Văn Vy, Cao-Văn-Lung, Nguyễn Hữu-Linh, Hội Đồng Quản Trị Nghiệp-đoàn Công-Nhân đồn điền Bình Dương cho Thủ Tướng, Bộ Trưởng Bộ Lao Động, November 11, 1964.

12. NARA RG 472, Records of the U.S. Forces in Southeast Asia, 1950–1976, Box 3, p. 4.

13. NAVN2 PTT Việt Nam Cộng Hòa 29460, Giải quyết những yêu sách của công nhân đồn điền cao su 1965. Lá thư, Trần Thanh Hiệp, Bộ Trưởng Bộ Lao Động cho Thủ Tướng, June 10, 1965. Lá thư, Đàm Quang San, Đổng Lý Văn Phòng Bộ Lao Động cho Đổng Lý Văn Phòng Thủ Tướng, June 12, 1965. Về việc điều chỉnh thuế xuất cảng cao su, Lê Đức Hợi, Đổng Lý Văn Phòng Thủ Tướng, May 6,1965. Tờ trình hội đồng chánh phủ về thuế xuất cảng cao su, Trần Văn Kiện, Tổng Trưởng Tài Chánh, March 13, 1965. Lá thư, Tu Ngọc Đình, Đổng Lý Văn Phòng Tài Chánh cho Đổng Lý

Văn Phòng Bộ Lao Động, June 15, 1965, về việc thuế suất cảng cao-su. Lá thư, Đại Tá Dương Hồng Tuân, Đổng Lý Văn Phòng Phủ Tổng Thống, Ủy Ban Hành Pháp Trung Ương, June 30, 1965. Lá thư, Lê Hoàng Chương, chủ tịch Văn Phòng Liên Đoàn Đồn Điền, November 5, 1965, về việc đình-công đang đe-doạ các đồn điền. NARA RG59 Box 1300, Central Foreign Policy Files 1967–69.

14. Binh, *The Red Earth*; Nguyễn, *No Other Road to Take.*

15. Điệp, *Máu Trắng, Máu Đào.* See Phụng's humorous take on Vietnamese women marrying Western men in Vũ, *Kỹ Nghệ Lấy Tây.*

16. Binh, *The Red Earth*, 33.

17. Điệp, *Máu Trắng, Máu Đào.*

18. NARA RG 472, Records of the U.S. Forces in Southeast Asia, 1950–1976, Box 3, TARA report, p. 3. Trương, Chanoff, and Van, *A Vietcong Memoir*, 160, 162.

19. Military Assistance Command, Vietnam, MACJ, VC use of Rubber, tea, and cinnamon plantations, June 27, 1966, accessed online through Texas Tech Archives.

20. VC document on Production of Bến Củi Rubber Plantation 1967, English translation, captured in 1966, Texas Tech Archives. NARA RG 472, Records of the U.S. Forces in Southeast Asia, 1950–1976, P 97, Box 2, 1971, "Conditions at the Binh Son Rubber Plantation."

21. Military Assistance Command, Vietnam, MACJ, VC use of Rubber, tea, and cinnamon plantations, June 27, 1966. VC document on Production of Bến Củi Rubber Plantation 1967, English translation, captured in 1966, p. 10. NARA RG 472, Records of the U.S. Forces in Southeast Asia, 1950–1976, P 97, Box 2, 1971, "Attitudes of Rubber Plantation Workers in Binh Long Province," March 18, 1971. NARA RG 472, Records of the U.S. Forces in Southeast Asia, 1950–1976, P 97, Box 2, 1971, "Attitudes of the An Loc Rubber Plantation Workers, Long Khanh," March 12, 1971, p. 5.

22. Military Assistance Command, Vietnam, MACJ, VC use of Rubber, tea, and cinnamon plantations, June 27, 1966.

23. VC document on Production of Bến Củi Rubber Plantation 1967, English translation, captured in 1966, Texas Tech Archives, pp. 7, 11. Minutes from Lai Khê Party Committee, November 6–17, 1966, p. 3.

24. For 1939–45, see Everett Guy Holt, *Report on Indochina Rubber Industry and Siamese Rubber Production Outlook* (Washington, DC: U.S. Department of Commerce, 1946). See also Finlay, *Growing American Rubber*, and Garfield, *In Search.*

25. NAVN2 PTT Việt Nam Cộng Hòa 15654, PTT, Nội Vụ, Về việc bảo vệ an ninh đồn điền cao su tại vùng 3 chiến thuật 1963–1966. Lá thư, đồn điền cao su Lộc Ninh, Bộ Quốc phòng, August 2, 1966. See also NAVN2 PTT Việt Nam Cộng Hòa 21760, Lá thư, Bộ Lao Động, Đàm Sỹ Hiến cho Thủ Tướng, Sài Gòn, July 2, 1964, về việc Pháp kiều chủ nhân đồn điền cao su lo ngại về tin đồn tài sản của họ sẽ bị quốc hữu hóa đã xúi dục công nhân tranh chấp để cản trở việc quốc hữu hóa, 1964.

26. Bowd and Clayton, "French Tropical Geographies," 282. For Thompson remark, see his *No Exit*, 9.

27. NARA RG 472 270/76/21/5-7, Military Assistance Command, Vietnam, J3-05, after action reports, r215, b3, pp. 1–3.

28. For protest banners and VC propaganda, see NARA RG 472, Records of the U.S. Forces in Southeast Asia, 1950–1976, Box 722, 735.

29. NARA RG 472, Records of the U.S. Forces in Southeast Asia, 1950–1976, P 97, Box 2, 1971, "Attitudes of Rubber Plantation Workers in Binh Long Province," March 18, 1971.

30. NARA RG 472, 270/76/21/5-7, Military Assistance Command, Vietnam, J3-05, after action reports, r572, b10. After Action Report from March 29, 1967, pp. 3–11; r587, b11, pp. 1–18.

31. NARA RG 472, 270/76/21/5-7, Military Assistance Command, Vietnam, J3-05, after action reports, r587, b11.

32. NARA RG 472, Records of the U.S. Forces in Southeast Asia, 1950–1976, Box 3.

33. Ibid.

34. Ibid. Also see T. Huu, "OPERATION JUNCTION CITY (22 February–14 April 1967)," *Vietnamese Studies* 57, no. 3 (1980): 75–118. Quote from NARA RG 472, Records of the U.S. Forces in Southeast Asia, 1950–1976, Box 3, TARA report, p. 5.

35. NAVN2 PTT Việt Nam Cộng Hòa 3886, International conference on cao su 1954. Report by Piechaud and Hoa Văn Mùi. Various interviews. For more on the rubber industry in Cambodia during the 1950s and 1960s, see Anon., "L'hévéa au Cambodge"; Anon., "Hévéaculture"; Anon., "A l'IRCC"; Chau, "L'hévéaculture"; Ho, "L'économie de l'hévéaculture."

36. NARA RG 472, Records of the U.S. Forces in Southeast Asia, 1950–1976, P 97, Box 2, 1971, "Attitudes of Phuoc Hoa Rubber Plantation Workers in Binh Duong Province," February 2, 1971, p. 5. NARA RG 472, Records of the U.S. Forces in Southeast Asia, 1950–1976, P 97, Box 2, 1971, "Attitudes of Rubber Plantation Workers in Binh Long Province," March 18, 1971. NARA RG 472, Records of the U.S. Forces in Southeast Asia, 1950–1976, P 97, Box 2, 1971, "Conditions at the Binh Son Rubber Plantation." Report prepared by Biên Hòa Province Advisors, July 12, 1969.

37. NARA RG 472, Records of the U.S. Forces in Southeast Asia, 1950–1976, P 97, Box 2, 1971.

38. Đoàn, *Measures.*

39. NARA RG 472, Records of the U.S. Forces in Southeast Asia, 1950–1976, P 97, Box 2, 1971, "Attitudes of Rubber Plantation Workers in Binh Long Province," March 18, 1971. NARA RG 472, Records of the U.S. Forces in Southeast Asia, 1950–1976, A1 461, Box 3.

40. NARA RG 472, Records of the U.S. Forces in Southeast Asia, 1950–1976, P 97, Box 2, 1971, "Attitudes of the An Loc Rubber Plantation Workers, Long Khanh," March 12, 1971, p. 6; "Attitudes of Ven Ven Rubber Plantation Workers, Tây Ninh Province," February 13, 1971, pp. 4–5; "Attitudes of Rubber Plantation Workers in Phuoc Long, Son Giang," February 11, 1971, p. 2; "Attitudes of Rubber Plantation Workers in Binh Long Province," March 18, 1971; "Attitudes of Phuoc Hoa Rubber Plantation Workers in Binh Duong Province," February 2, 1971, p. 5.

41. NARA RG 472, Records of the U.S. Forces in Southeast Asia, 1950–1976, P 97, Box 2, 1971, "Conditions at the Binh Son Rubber Plantation." Report prepared by Biên Hòa Province Advisors, July 12, 1969, p. 3.

42. Ibid., 5, 6. NARA RG 472, Records of the U.S. Forces in Southeast Asia, 1950–1976, P 97, Box 2, 1971, "Attitudes of Dau Tieng Rubber Plantation Workers, Binh Duong," February 2, 1971, p. 5.

43. NARA RG 472, Records of the U.S. Forces in Southeast Asia, 1950–1976, P 97, Box 2, 1971, "Attitudes of Ven Ven Rubber Plantation Workers, Tây Ninh Province," February 13, 1971, pp. 5, 6; "Opinions of the Workers at the Terre Rouge Rubber Plantation in Binh Long Province," March 22, 1971, p. 4; "Attitudes of the An Loc Rubber Plantation Workers, Long Khanh," March 12, 1971, p. 8; "Attitudes of Phuoc Hoa Rubber Plantation Workers in Binh Duong Province," February 2, 1971.

44. NARA RG 472, Records of the U.S. Forces in Southeast Asia, 1950–1976, P 97, Box 2, 1971, "Attitudes of Phuoc Hoa Rubber Plantation Workers in Binh Duong Province," February 2, 1971; "Attitudes of Rubber Plantation Workers in Phuoc Long, Son Giang," February 11, 1971; "Attitudes of Quan Loi Rubber Plantation Workers, Binh Long," March 11, 1971. NAVN2 PTT Việt Nam Cộng Hòa 6392.

45. NARA RG 472, Records of the U.S. Forces in Southeast Asia, 1950–1976, P 97, Box 2, 1971, "Opinions of the Workers at the Terre Rouge Rubber Plantation in Binh Long Province," March 22, 1971; "Attitudes of the An Loc Rubber Plantation Workers, Long Khanh," March 12, 1971, p. 4.

46. NARA RG 472, Records of the U.S. Forces in Southeast Asia, 1950–1976, P 97, Box 2, 1971, "Attitudes of Rubber Plantation Workers in Binh Long Province," March 18, 1971, pp. 4, 5.

47. NARA RG 472, Records of the U.S. Forces in Southeast Asia, 1950–1976, P 97, Box 2, 1971, "Opinions of the Workers at the Terre Rouge Rubber Plantation in Binh Long Province," March 22, 1971, p. 9. NAVN2 PTT Việt Nam Cộng Hòa 31643, VC gây áp lực cho cuộc tranh chấp, đình công của công nhân đồn điền cao su, 1964–1975. Tổng Trưởng Bộ Lao Động, Đàm Sỹ Hiến to Trung-Tá Tỉnh Trưởng Bình Dương, February 21, 1969. Police report, February 27, 1971, on Michelin strike.

48. NARA RG 472, Records of the U.S. Forces in Southeast Asia, 1950–1976, P 97, Box 2, 1971, "Opinions of the Workers at the Terre Rouge Rubber Plantation in Binh Long Province," March 22, 1971; "Conditions at the Binh Son Rubber Plantation"; "Attitudes of Dau Tieng Rubber Plantation Workers, Binh Duong," February 2, 1971.

49. NAVN2 PTT Việt Nam Cộng Hòa 23804, IRCV xin ứng trước số tiền bồi thường giải toả và tiền thuê mướn nhà do quân lực Hoa Kỳ chiếm dụng tại Trung Tâm thí nghiệm Lai Khê 1969. Lá thư, Tôn Thọ Xương cho Thủ Tướng, Sài Gòn, October 27, 1969. Letter from Crighton Abrams, General U.S. Army, June 1, 1969, to Tôn Thọ Xương, director of Lai Khe rubber Experimental Center. NAVN2 PTT Đệ Nhị Cộng Hòa 2227: xin bồi thường cho viện khảo cứu cao su Lai Khê, 1967–68. Record of Conversation, July 18, 1968, Poliniere, Whalen, and Hirsch. NAVN2 Cơ Quan Phát Triển Quốc Tế Hoa Kỳ (U.S. Agency for International Development) 2795, Thâu hồi VN$ 13.5 million do IRVN mượn của quỹ đặc biệt viện trợ Mỹ 1969–1971.

50. NAVN2 PTT Việt Nam Cộng Hòa 23804: IRCV xin ứng trước số tiền bồi thường giải toả và tiền thuê mướn nhà do quân lực Hoa Kỳ chiếm dụng tại Trung Tâm thí

nghiệm Lai Khê 1969. Letter from Tôn Thọ Xương to the U.S. Ambassador to Vietnam, Sài Gòn, October 22, 1969. NAVN2 Cơ Quan Phát Triển Quốc Tế Hoa Kỳ 694, Quỹ đặc biệt thuộc chương trình Nông vụ phát triển cao su, 1969–70. Contract signed, Tôn Thọ Xương and Jean-Paul Poliniere, January 6, 1971.

51. Record of Conversation, July 18, 1968, Poliniere, Whalen, and Hirsch; Record of Conversation, July 25, 1968, RRIV, Director and G. Wormset, p. 3; McIndoe, *A Preliminary Survey*, 38; Doc IRCV 134, No B1/71JP E, Reasons of RRIV's Development into the Technical Service Institute for Agriculture and Agro-Industries, 10-5-1971 (1st ed. 24-1-71), 2. See also Improved structures for international assistance in developing countries, Polinière, 20-1-71 and No B7/71JP.E, Profile of an expert in the production of corn, peanut, sorghum, and soybean, 7-6-1971. Đoàn and Polinière, *Avenir*; Roudeix, *Rapport*. See also Doc IRCV, No B12/73BY.F, Situation du potentiel hévéicole des petites et moyennes plantations <500ha dans la 3e région militaire au 31-12-70.

52. NAVN2 PTT Việt Nam Cộng Hòa 17597, 7 Pháp kiều thuộc đồn điền cao su viễn đông, CEXO, cao su bắt giữ, 1972.

53. Đoan and Polinière, *Avenir*; Roudeix, "Rapport." See also Doc IRCV, No B12/73BY.F, Situation du potentiel hévéicole des petites et moyennes plantations <500ha dans la 3e région militaire au 31-12-70. NARA RG 472, Records of the U.S. Forces in Southeast Asia, 1950–1976, A1 461, Box 3. NAVN2 Cơ Quan Phát Triển Quốc Tế Hoa Kỳ 855, Nha Nông Nghiep và Thực Phẩm về chương trình phát triển và kỹ nghệ cao su, 1970–71. NAVN2 Cơ Quan Phát Triển Quốc Tế Hoa Kỳ 865, Chương trình phát triển cao su của viện dịch vụ kỹ thuật và kỹ nghệ nông nghiệp, 1971–72. Report, W. C. Tappan, Meeting on Natural Rubber Development Program, November 20, 1972. Report on Rubber Industry by IRCV/Technical Service Institute for Agriculture and Agro-Industries. NAVN2 Cơ Quan Phát Triển Quốc Tế Hoa Kỳ 871, Asian Development Bank, Technical Assistance to the RVN for a Rubber Rehabilitation Project, November 12, 1973. Also letter from National Bank of Viet Nam, September 1. NAVN2 PTT Việt Nam Cộng Hòa 20849, chương trình phát triển liên hiệp quốc về phương pháp đánh giá các đồn điền cao su, 1974.

54. NAVN2 PTT Đệ Nhất Cộng Hòa 10976, SPTR xin phép đốn rừng để trồng thêm cây cao su, 1957. See, for example, forester report in NAVN2 PTT Đệ Nhất Cộng Hòa 10976, SPTR xin phép đốn rừng để trồng thêm cây cao su, 1957.

55. IRCV, *Rapport annuel*, 1958, 4.

56. Ziegler, Fox, and Xu, "The Rubber Juggernaut." For more on the importance of soil and the use of fertilizer for plantations, see NAVN2 PTT Đệ Nhất Cộng Hòa 18953, "Future of 'Small holdings' in Rubber Agriculture in Vietnam," the Minister of Agriculture. U.S. Agency for International Development had been involved in rubber in Việt Nam as early as 1960; see NAVN2 PTT Đệ Nhất Cộng Hòa 13474. See Beinart and Hughes, *Environment and Empire*; Wan Rahaman and Sivakumaran, "Studies of Carbon Sequestration in Rubber," in the Proceedings from the Joint Workshop of the Secretariat of the United Nations Conference on Trade and Development and the International Rubber Study Group on Rubber and the Environemnt (Bali, Indonesia,

1998): 56–65; and Chong Koon Kee, *Difficulties in Planning Water Supply Schemes in West Malaysia* (Singapore: Science Council of Singapore, 1972), 97.

57. Tschirley, "Defoliation in Vietnam." See also National Academy of Sciences, *The Effects of Herbicides.*

58. NARA RG 472, Records of the U.S. Forces in Southeast Asia, 1950–1976, Box 18, Notes: Herbicide damage to rubber, Cambodia; NARA RG 84, Records of the Foreign Service Posts of the Department of State, 1788–ca. 1991, Box 4-8, rubber damage files; Texas Tech Archives, Herbicide Damage to Rubber and Fruit Trees in Cambodia 1969.

59. NAVN2 PTT Việt Nam Cộng Hòa 31643, VC gây áp lực cho cuộc tranh chấp, đình công của công nhân đồn điền cao su, 1964–1975. Letter, Président-Directeur Gén. R. Fontes, Michelin, to Col. Chef de la Province Binh Duong, October 10, 1965. Trương, Chanoff, and Van, *A Vietcong Memoir*, 82. A French planter who lived in the area after 1954 also noted the Việt Cộng people's courts. Michon, *Indochina Memoir*. NARA RG 472, Records of the U.S. Forces in Southeast Asia, 1950–1976, P 97, Box 2, 1971, "Attitudes of Dau Tieng Rubber Plantation Workers, Binh Duong," February 2, 1971, p. 5; NARA RG 472, Records of the U.S. Forces in Southeast Asia, 1950–1976, P 97, Box 2, 1971, "Attitudes of Rubber Plantation Workers in Phuoc Long, Son Giang," February 11, 1971, p. 6.

60. NARA RG 472, Records of the U.S. Forces in Southeast Asia, 1950–1976, P 97, Box 2, 1971, "Attitudes of Quan Loi Rubber Plantation Workers, Binh Long," March 11, 1971. See NARA RG 472, Records of the U.S. Forces in Southeast Asia, 1950–1976, P 97, Box 2, "Attitudes of Ven Ven Rubber Plantation Workers, Tây Ninh Province," February 13, 1971, p. 4. The 1970 decree was No. 66-BLD-LD/ND.

61. NAVN2 PTT Việt Nam Cộng Hòa 30752, Liên đoàn công nhân đồn điền Cao su Việt Nam yêu cầu lãnh đủ phần gạo theo cộng đồng khế ước 1971. Lá thư, Liên đoàn công nhân đồn điền Cao su Việt Nam, Lê Hoàng Chương, Phạm Văn Vỹ cho Thủ Tướng, April 29, 1971. Lá thư, Liên đoàn công nhân đồn điền Cao su Việt Nam, Lê Hoàng Chương cho Thủ Tướng, May 25, 1969. Tổng Trưởng Bộ Lao Động, Đàm Sỹ Hiến cho Văn Phòng Thủ Tướng, June 5, 1971. Lá thư, Trung Tướng Nguyễn Văn Minh cho Thủ Tướng, June 22, 1971. Analysis of Prime Minister's office, June 29, 1971. Phan Văn Chí, Tổng Liên Đoàn Lao Động Việt Nam to Trưởng Ty Lao Động Liên Tỉnh Bình Dương/Hậu Nghĩa, May 29, 1971.

62. NAVN2 PTT Đệ Nhị Cộng Hòa 4679, Về việc hội phụ nữ Việt Nam phụng sự XH xin mua đồn điền cao su BMT 1974; NAVN2 PTT Việt Nam Cộng Hòa 27460.

63. NAVN2 PTT Việt Nam Cộng Hòa 26253, Không chấp thuận đầu tư của công ty Michelin vào các đồn điền Thuận Lợi và Dầu Tiếng, 1973. NAVN2 PTT Việt Nam Công Hòa 28202, Hoạt động của các đồn điền và những khó khăn trong việc bán cao su của Nghiệp-đoàn trồng tỉa cao su Vietnam, 1968–1974. Letter on February 27, 1974, from Hoa Văn Mùi to Prime Minister of RVN. Letter on March 5, 1974, from Trần Quốc Bửu to Prime Minister of RVN. Letter on November 14, 1969, from Hoa Văn Mùi

to Economic minister. Letter on February, 27 1974, from Hoa Văn Mùi to Prime Minister of RVN. Letter on March 5, 1974, from Trần Quốc Bửu to Prime Minister of RVN. Tổng Trưởng Bộ Lao Động, Đàm Sỹ Hiến to Prime Minister, May 8, 1974. Tổng Trưởng Bộ Lao Động, Đàm Sỹ Hiến to Prime Minister, March 27, 1975. NAVN2 PTT Việt Nam Cộng Hòa 25646, cải thiện đời sống công nhân đồn điền và trợ cấp nhân khoán để xuất cảng cao su, 1970–1972. Lá thư, Trần Thiện Khiêm cho Tổng Thống, Sài Gòn, January 30, 1970. Wehrle, *Between a River and a Mountain*.

64. "Phnom Penh Notes Efforts to Develop Natural Rubber Production," Domestic Service in Cambodian 2300 GMT 19 Jun 75 BK, Texas Tech Archives; Lê, *Đất Đỏ Miền Đông*, 230; Phạm Ngọc Hồng, interview transcript, 13. See also Aso, "De plantations coloniales à la production socialiste."

65. Hecht, "Rupture-Talk." See chapter 4, Prakash, *Another Reason*. Đặng, *100 Năm Cao su*.

Conclusion

1. Trần et al., *Trần Tử Bình*. The 2007 book was given to me in 2009 by a former plantation worker and Communist Party member.

2. Post-1975 relations from Đỗ Hữu Chuẩn and Nguyễn Thị Thứ, interview transcript, 17, and Phạm Ngọc Hồng, interview transcript, 7. See Chu, *Cuộc Đời Dài Lắm*, for observations on Communist Party takeover.

3. See McElwee, *Forests Are Gold*, 72–74.

4. See Giebel, *Imagined Ancestries*.

5. Vinh published books on the history of the rubber industry both in Việt Nam and worldwide, including *100 Năm Cao Su*. NAVN3 PTT 2216, Công văn của PTT và BNN về việc Viện Cao su Việt Nam đóng góp vào kinh phí sưu tập giống cao su ở Nam Mỹ năm 1980: (14) Letter, CN144s/O-GĐ và CN158s/O-GĐ, Phó Viện Trưởng, Nguyễn Hữu Chất gửi BNN và BTC, June 20, 1980, về việc xin ngoại tệ góp vào kinh phí sưu tập giống cao su Nam Mỹ; Letter, 3080 V7, PTT gửi BNN, BTC, Bộ Ngoại Giao, Ngân Hàng Nhà Nước, July 21, 1980, về việc đóng góp vào kinh phí sưu tập giống cao su ở Nam Mỹ. See also 135 BNN Tập Quan điểm nhân sự tháng 01.1977 của BNN.

6. Đặng, *Lịch Sử Kinh Tế*, VII Các cơ sở nghiên cứu, 39, 43. Chương 3, Các ngành kinh tế, 41.

7. Graveline, *Des hévéas et des hommes*; Bergot, *Sud lointain*; Orsenna, *L'Exposition coloniale*.

8. Erik Orsenna took part in writing the script. Norindr, *Phantasmatic*, 139, 133–39.

9. Wargnier, *Indochine*. There are many films set on plantations, e.g Fleming, *Red Dust*.

10. V.Q.H., H.Y., and TH.H, "Đầu Tư 18, 73 Triệu USD Trồng Cao Su và Cây Nguyên Liệu tại Lào," *Sài Gòn Giải Phóng*, May 11, 2006; Cohen, "The Post-Opium Scenario"; Manivong and Cramb, "The Adoption of Smallholder Rubber Production"; Mann and D'Aluisio-Guerrieri, "Addicted to Rubber"; Thongmanivong, Yayoi, Khamla, and

Thoumthone, "Agrarian Land Use Transformation"; Ziegler, Fox, and Xu, "The Rubber Juggernaut." See also Baird, "Land, Rubber and People." Quote comes from Đỗ Hữu Chuẩn and Nguyễn Thị Thứ, interview transcript, 14.

11. Food and Agriculture Organization of the United Nations, factsheet.

12. See McElwee, *Forests Are Gold*, chapter 5.

13. Yves Stavridès, "De l'Indochine au Vietnam," *L'Express*, April 26, 2004, 4. For bioterrorism framing, see Lieberei, "South American Leaf Blight."

Bibliography

Interviews

I worked with transcripts of digital recordings made from the following interviews with former rubber workers, except those from June 26, 2007, and January 2, 2008. In addition, I benefitted from informal conversations with several rubber experts and historians.

Nguyễñ Van Sung, June 26, 2007, Aix-en-Provence, France.
Phạm Ngọc Hồng, May 22, 2008, Binh Phước, Việt Nam.
Đỗ Hữu Chuẩn and Nguyễn Thị Thứ, January 2, May 23, and August 16, 2008, Biên Hòa, Việt Nam.

Archives

Cambodia
    National Archives of Cambodia, Phnom Penh.
France
    Académie des sciences d'outre-mer, Paris.
    Archives de l'Institut Pasteur, Paris.
    Archives nationales de France, Paris.
    Archives nationales d'outre-mer/Centre d'archives outre-mer, Aix-en-Provence.
    Bibliothèque nationale de France, Paris.
    Centre de coopération internationale en recherche agronomique pour le développement, Paris.
    Institut de médecine tropicale du service de santé des armées, Marseille.
    Muséum national d'histoire naturelle, Paris.
Singapore
    Institute of Southeast Asian Studies–Yusof Ishak Library.
Switzerland
    World Health Organization Records and Archives, Geneva.
United States
    National Archives and Records Administration, College Park.
    Texas Tech Archives, Lubbock.
Việt Nam
    Ecole française d'Extrême-Orient, Hà Nội, Thành Phố Hồ Chí Minh.
    Lai Khê Research Center, Lai Khê.Trung Tâm Lưu Trữ Quốc Gia I, II, III, IV, Hà Nội, Thành phố Hồ Chí Minh, Đà Lạt.

Thư Viện Khoa Học Tổng hợp, Thành Phố Hồ Chí Minh.
Thư Viện Khoa Học Xã Hội, Hà Nội.
Thư Viện Quốc Gia Việt Nam, Hà Nội.

Primary Sources

Achard, E. L. "Rapport sur les champs d'essais de la Cochinchine, champs d'essais de Ong-Iêm." *Bulletin économique de l'Indochine*, no. 25 (1900): 333–41.

Adam, Jean. "De l'écologie agricole à l'écologie coloniale." *l'Agronomie coloniale* 237 (1937): 1–12.

Allain, J. "Paludisme et quinine d'État en Annam pendant l'année 1912." *Bulletin de la Société de pathologie exotique* 6, no. 10 (1913): 730–44.

Allouard, P. M. *Pratique de la lutte contre les feux de brousse.* Hanoi: IDEO, 1937.

Amicale mutuelle des employés indigènes de l'agriculture en Cochinchine, au Cambodge et dans le sud Annam (A.M.E.I.D.A.C.). *Bulletin de renseignements.* Thudaumot: Imprimerie Nam-Trung, 1939.

Anon. "A l'IRCC, travaux de recherches sur l'hévéa." *Etudes cambodgiennes* 2, no. 5 (1966): 12–16.

———. *Annales du jardin botanique et de la ferme expérimentale des mares.* Saigon: Imprimerie du Gouv. Schroeder, K. et A., 1878.

———. "Bien Hoa en 1930." Unpublished manuscript.

———. "En Indochine, sur les plantations de caoutchouc." Paris: Société d'Éditions techniques coloniales, 1952.

———. "Hévéaculture." *Etudes cambodgiennes* 3, no. 11 (1967): 12–13.

———. *Les scandales de l'agriculture en Indochine.* s.l.: Imprimerie de la presse indochinoise, 1932.

———. "L'hévéa au Cambodge." *Cambodge d'aujourd'hui* 2, nos. 7–8 (1959): 13–26.

———. *L'Indochine forestière.* Saigon, 1942.

———. "Miscellaneous notes, XXIII." *Bulletin of Miscellaneous Information (Royal Botanic Gardens, Kew)*, no. 4 (1906): 121–22.

Assali, J. "Intolérance absolue à la quinine." *Bulletin de la Société médico-chirurgicale de l'Indochine* 4, no. 5 (1926): 221–23.

Association culturelle populaire. *Le problème de l'hévéaculture au Viet-Nam: Exposés et résumés des débats du séminaire organisé par la Section des études économiques et financières de l'Association culturelle populaire.* Saigon: l'Association, 1960.

Association des planteurs de caoutchouc de l'Indochine (APCI). *Les annales des planteurs de caoutchouc de l'Indochine.* Saigon: Imprimerie Commerciale Marcellin Rey, 1911, 1912; C. Ardin, 1916.

Azémar, Henri. "Les Stiengs de Brolam." *Excursions et Reconnaissances*, nos. 27, 28 (1886): 89–146, 215–50.

Azzi, Girolamo. *Ecologie agricole.* Paris: J. B. Baillière et fils, 1954.

Baillaud, Emile. "Le régime forestier dans les colonies françaises." In *Le régime forestier aux colonies*, edited by L'Institut colonial international, 202–15. Brussels, Belgium: L'Institut colonial international.

Baudrit, A. "Monographie d'une rivière cochinchinoise, Le fameux Song-Bé." *Bulletin de la Société des études indochinoises* 11 (1936): 7–42.

Bauer, P. T. *The Rubber Industry, a Study in Competition and Monopoly*. Cambridge, MA: Harvard University Press, 1948.

Bernard, Noël. *De l'empire colonial à l'Union française*. Paris: Flammarion, éd., 1951.

———. *La vie et l'œuvre de Albert Calmette, 1863–1933*. Paris: A. Michel, 1961.

———. *Les Instituts Pasteur d'Indochine: centenaire de Louis Pasteur, 1822–1895*. Saigon: Imprimerie nouvelle, A. Portail, 1922.

———. "Notions générales sur le paludisme et les moyens de le combattre dans les centres agricoles et forestiers de la Cochinchine." *BSPCI* 13 (1919): 133–42.

———. *Yersin: Pionnier—Savant—Explorateur, 1863–1943*. Paris: La Colombe, 1955.

Blanc, F. "Essais de traitement et de prophylaxie du paludisme par la paludrine en Indochine, Juillet-Novembre 1947." *Médecine tropicale: revue du corps de santé colonial* 9, no. 2 (1949): 143–72.

Bocquet, M. "Homogénéisation et standardisation du caoutchouc brut." *Revue générale du caoutchouc* 146 (1938): 326–33.

Bordes, L. A. *Le paludisme en Indochine (historique, épidemiologie etat actuel de la lutte antipalustre)*. Exposition coloniale internationale. Paris. Hanoi: IDEO, 1931.

Bordes, L. A., and J. Mesnard. "L'importance du réservoir de virus authochtone dans la lutte contre le paludisme en Indochine." *Archives des Instituts Pasteur d'Indochine* 12 (1930): 53–64.

———. "Notes sur le taux d'infection palustre de la main d'œuvre importée à son arrivée en Cochinchine." *Bulletin de la Société médico-chirurgicale de l'Indochine* 9, no. 6 (1931): 492–97.

Borel, Emile. "Anophèles et paludisme dans la région de Chaudoc (Cochinchine). Résultats d'une enquête faite du 16 au 21 janvier 1926." *Bulletin de la Société de pathologie exotique* 19 (1926): 806–11.

———. "Contribution à l'étude de la mortalité infantile en Cochinchine; renseignements concernant quelques plantations situées en terres rouges, dans la province de Biên-Hoa." *Bulletin de la Société médico-chirurgicale de l'Indochine* 4, no. 11 (1926): 577–81.

———. "Enquête malariologique à la station d'essai de Giaray (Cochinchine)." *Bulletin de la Société de pathologie exotique* 21 (1928): 312–14.

———. "La constitution du sol et le paludisme en Cochinchine." *Bulletin de la Société de pathologie exotique* 19 (1926): 935–42.

———. *Les moustiques de la Cochinchine et du Sud-Annam*. Collection de la Société de pathologie exotique. Paris: Masson et Cie, Editeurs, 1930.

———. "Paludisme en Cochinchine: Résultats des mesures prophylactiques à la plantation de Suzannah (11 au 13 août 1926)." *Bulletin de la Société de pathologie exotique* 19 (1926): 811–15.

———. "Résultats d'une enquête épidémiologique à Yaback (Annam)." *Bulletin de la Société de pathologie exotique* 19 (1926): 845–52.

Borri, Christoforo, and Samuel Baron. *Views of Seventeenth-Century Vietnam: Christoforo Borri on Cochinchina & Samuel Baron on Tonkin.* Edited by Olga Dror and Keith Weller Taylor. Ithaca, NY: Southeast Asia Program Publications, 2006.

Bos, Maurice. "Le développement et l'avenir des plantations de caoutchouc en Indochine." *Revue generale du caoutchouc* 13, no. 125 (1936): 32–40.

Boudillon, A. *Le régime de la propriété foncière en Indochine.* Paris: E. Larose, 1915.

Boué, Emile. "La colonisation européenne en Indo-Chine (au 31 décembre 1899)." *Bulletin économique de l'Indochine* 3, no. 25 (1900): 325–32.

Bouvier, E.-J., and A.-F.-M. Rivoalen. *Troupes du groupe de l'Indochine: Direction du service de santé: Instruction technique sur la prophylaxie et le traitement du paludisme.* Hanoi: Imprimerie de G. Taupin, 1941.

Bouvier, René. *Le caoutchouc: brillante et dramatique histoire de l'hévéa.* Paris: Flammarion, 1947.

Brau, P. "Paludisme à P. falciparum: Diagnostic et traitement." *Bulletin de la Société de pathologie exotique* 15, no. 9 (1922): 834–42.

Brenier, Henri. *Essai d'atlas statistique de l'Indochine française.* Hanoi-Haiphong: IDEO, 1914.

Bùi, Kiến Tín. "Le médecin en face du problème démographique de l'Indochine." PhD diss., Faculté de Médecine de Paris, 1940.

Bunout, René. "La main-d'œuvre et la législation du travail en Indochine." PhD diss., Université de Bordeaux, 1936.

Bureau du caoutchouc de l'Indochine française. *Réglementation en Indochine française de la production et de l'exportation du caoutchouc.* Saigon: SILI (C. Ardin), 1934.

Camus, J. J. *L'oeuvre humaine et sociale dans les plantations de caoutchouc d'Indochine.* Paris: Société d'éditions techniques coloniales, 1949.

Candé, J. B. *De la mortalité des Européens en Cochinchine depuis la conquête jusqu'a nos jours.* Paris: Challamel, Ainé, Libraire-éduiteur, 1881.

Canet, Jean. "Note sur l'alimentation des coolies de plantations, quelques mesures récentes prises à ce sujet par les grandes exploitations." *Bulletin d'Institut indochinois pour l'étude de l'homme* 5, no. 2 (1943): 57–88.

———. "Note sur une petite épidémie de typhus tropical survenue dans un groupe de plantations du Cambodge." *Bulletin de la Société de pathologie exotique* 31, no. 6 (1938): 457–60.

———. "Organisation de l'hygiène sur les plantations des Terres-Rouges." *Bulletin de la Société médico-chirurgicale de l'Indochine* 14, no. 6 (1936): 560–71.

———. "Prophylaxie collective par médicaments synthétiques sur les plantations des Terres-Rouges (1934–1936)." *Bulletin de la Société médico-chirurgicale de l'Indochine* 14, no. 6 (1936): 533–59.

———. "Sur le dos du tigre." *Sernamby* 23, 24, 25 (2002–03): 8–69, 31–40, 3–37.

Caresche, L. "Visite de la station agricole d'Ong Yem et de la ferme-école de Ben Cat." *Groupement sud-indochinois des ingénieurs agricoles. Grignon, Montpellier, Rennes, Bulletin de liaison* (Saigon) 1, no. 4 (1933): 12–21.

Carle, Edmond. *L'hévéa brasiliensis en Indochine*. Saigon: Imprimerie nouvelle Albert Portail, 1911.

Carrau, Pierre. "Du commerce et de l'agriculture chez les Moïs." *Excursions et Reconnaissances* 5, no. 14 (1882): 270–93.

Carton, Paul. "Considération sur l'action de la lumière sur les plantes." *Bulletin générale de l'instruction publique* 10 (1934): 1–37.

———. *Cours de climatologie et d'écologie*. Hanoi: Imprimerie Ngo-Tu-Ha, 1938.

———. "L'Ecologie. Importance de son enseignement dans les écoles supérieures d'agricutlure comme 'introduction' à divers cours spéciaux. Importance particulière du point de vue colonial." *l'Agronomie coloniale* 25, no. 227 (1936): 165–75.

———. "Importance des facteurs écologiques 'durée du jour' et 'intensité de la lumière' en agronomie tropicale." *l'Agronomie coloniale*, nos. 183, 184, 185, 186 (1933): 1–23.

———. "Le caoutchouc en Indochine." *Bulletin économique de l'Indochine* 27, no. 167 (1924): 349–449.

———. "Le climat et l'homme." In *Feuillets d'hygiène indochinois*, edited by Henry G. S. Morin, 39–179. Hanoi: IDEO, 1935.

———. "L'œuvre de l'institut des recherches agronomiques et forestières de l'Indochine au cours de la période 1925–1943." *Agronomie Tropicale* 1, nos. 3–4 (1946): 115–24.

Castagnol, E. M. *Problèmes de sol et l'utilisation des terres en Indochine*. 1950 ed. Vol. 7, *Archives de l'Institut des recherches agronomiques de l'Indochine*. Saigon: IDEO, 1951.

Caty, R. "L'amélioration des plantes de culture indigène aux colonies." *l'Agronomie coloniale* 25, nos. 218, 219 (1936): 34–42, 78–89.

———. "Visite de la Plantation de Canque." *Groupement sud-indochinois des ingénieurs agricoles. Grignon, Montpellier, Rennes, Bulletin de liaison* (Saigon) 1, no. 4 (1933): 12–21.

Chambre de commerce de Marseille. *Compte rendu de la situation commerciale et industrielle de la circonscription de Marseille pendant l'année 1900*. Marseille: Imprimerie Marseillaise, 1901.

Chau, Xeng Ua. *L'hévéaculture et le développement du Cambodge*. Phnom-Penh: Faculté de Droit et des Sciences Economiques, 1963.

Chevalier, Auguste. *Catalogue des plantes du jardin botanique de Saigon*. Saigon: Imprimerie nouvelle Albert Portail, 1919.

———. "Economie rurale: les changements de vie et de technique en agriculture et le déséquilibre actuel de la vie rurale." *Revue de botanique appliquée et d'agriculture coloniale* 14 (1934): 49–53.

———. "Le deuxième centenaire de la découverte du caoutchouc par Charles-Marie de la Condamine." *Revue générale du caoutchouc* 13, no. 125 (1936): 26–27.

———. "Les premiers découveurs français du genre hévéa, d'après les collections du Museum de Paris." *Revue générale du caoutchouc* 13, no. 125 (1936): 31.

———. *L'organisation de l'agriculture coloniale en Indochine et dans la métropole*. Saigon: C. Ardin & fils, 1918.

————. "Premier inventaire des bois et autres produits forestiers du Tonkin." *Bulletin économique de l'Indochine* 21, no. 132 (1918): 742–884.

————. "Rapport sur la cinquième Exposition internationale du caoutchouc et des autres produits tropicaux à Londres." *Revue de botanique appliquée et d'agriculture coloniale* 1, no. 4 (1921): 307–74.

————. "Situation de la production du caoutchouc en 1921." *Revue de botanique appliquée et d'agriculture coloniale* 1, no. 2 (1921): 33

Chevalier, Auguste, and Emile Girard. *L'hévéa en Indochine*. Saigon: Ardin, 1918.

Chollet, Raoul. *Planteurs en Indochine française*. Paris: La Pensée universelle, 1981.

————. "Travaux d'assainissement dans la plantation de Snoul." In *Feuillets d'hygiène indochinoise*, edited by Henry G. S. Morin, 445–51. Hanoi: I.D.E.O., 1935.

Cochinchine française. *Rapports au Conseil colonial*. Saigon: Imprimerie nationale, 1880–1943.

Coll. "La conférence internationale du caoutchouc." Bogor, Indonesia: Juillet, 1952.

————. "Nomenclature des principales lianes à caoutchouc, découvertes en Indo-Chine et désignées par leur nom indigène." *Bulletin économique de l'Indochine*, 4, no. 37 (1901): 547–53.

Comité de l' Indochine. *La crise mondiale du caoutchouc et les plantations d'Indochine*. Paris: Comité de l'Indochine, 1931.

Consigny, A. *Considérations sur les feux de brousse, leur méfaits et la possibilité de les enrayer*. s.l., n.d.

Coolidge, Harold J., Jr., and Theodore Roosevelt. *Three Kingdoms of Indo-China*. New York: Thomas Y. Crowell, 1933.

Couffinhal. *La situation actuelle des forêts de la Cochinchine*. Saigon: Imprimerie J. Viet, 1924.

Couzinet, Emile. *Les concessions domaniales et la colonisation européenne en Indochine*. s.l., n.d.

Crémazy, André, and William Bazé. *L'hévéaculture en Indochine*. Paris: É. Larose, 1927.

Cucherousset, Henri. "La question de la main-d'œuvre sur les plantations du sud, Notre réponse à M. Mathieu." *L'Eveil économique de l'Indochine* 11, no. 536 (1927): 1.

————. *Le Tonkin est-il surpeuplé?* Hanoi: Imprimerie Tonkinoise, 1925.

————. "Nos banques agricoles." *L'Eveil économique de l'Indochine* 18, no. 820 (1934): 6–7.

Davis, George N. and Brother. *Illustrated and Descriptive Catalogue and Trade-Price List of India Rubber and Cutta Percha Goods*. Boston: The Company, 1856.

Delamarre, E. *L'émigration et l'immigration ouvrière en Indochine*. Hanoi: IDEO, 1931.

Điệp Liên Anh. *Máu Trắng, Máu Đào: Đời Sống Đọa-Đày của Phu Cao-Su Miền Đất Đỏ*. Sàigòn: Lao-động mới, 1965.

Đoàn, Minh Quan. *Measures for Improving Rubber Production in Vietnam*. Saigon: IRCV, 1970.

Đoàn, Minh Quan, and J. P. Polinière. *Avenir du caoutchouc naturelle*. Saigon: IRCV, 1970.

Doumer, Paul. *Situation de l'Indochine française de 1897 à 1901*. Hanoï: Imprimerie-éd. F.-H. Schneider, 1902.

Du Pasquier, Robert. *Les problèmes d'utilisation des terres et leurs solutions en Indochine.* Saigon: IDEO, 1950.

Durand, Maurice. "Note de technologie vietnamienne." *Bulletin de l'Ecole française d'Extrême-Orient* 46, no. 2 (1954): 651–52.

Duvigneau, Charles André. *Le Thai-Y-Vien, service de santé du palais impérial d'Annam.* Marseille: Ecole d'application du service de santé, 1905.

Eberhardt, Ph., and M. Dubard. *L'arbre à caoutchouc du Tonkin et du nord-Annam, Bleekrodea Tonkinensis (Dub. & Eber.).* Paris: Augustin Challamel, 1910.

Enderlin, L. "Industrie indochinoise: Institut des recherches sur le caoutchouc en Indochine." *Indochine hebdomadaire illustré,* no. 231 (1945): 168–72, 190–91.

Far Eastern Association of Tropical Medicine. *Comptes-rendus des travaux du troisième congrès biennal, Saigon, 8–15 Nov. 1913.* Saigon: A. Portail, 1914.

———. *Comptes-rendus du dixième congrès, Hanoi, 26 Nov.–2 Déc. 1938.* Hanoi: IDEO, 1939.

Farinaud, M. E. "Développement de la lutte contre le paludisme sur les plantations du Sud de l'Indochine." *Bulletin de la Société médico-chirurgicale de l'Indochine* 14 (1936): 488–514.

Farinaud, M. E., and R. Choumara. *Infestation palustre et démographie dans les populations montagnardes du sud-indochinois (P.M.S.I.); Le paludisme dans les P.M.S.I.; Chimioprophylaxie et pulvérisations de D.D.T.* Edited by Services du conseiller à la santé publique Haut-commissariat de France en Indochine, Services du conseiller aux affaires économiques. Saigon: Imprimerie française d'outre-mer, 1950.

Farinaud, M. E., and J. Mesnard. "Recherches sur le réservoir de virus palustre chez les peuplades moïs de l'indochinois méridionale." *Bulletin de la Société de pathologie exotique* 30, no. 8 (1937): 720–27.

Farinaud, M. E., and P. Prost. "Le paludisme chez les Phnongs." *Bulletin économique de l'Indochine* 42, no. 2 (1939): 332–44.

Feyssal, Bernard de. *La réforme foncière en Indochine.* Hanoi: IDEO, 1931.

Fleming, Victor. *Red Dust.* MGM/UA, 1932.

Forces Terrestres en Extrême-Orient. *Le service de santé en Indochine: 1945–1954.* Saigon: IDEO, 1955.

Gaide, Laurent J. *L'assistance médicale et la protection de la sante publique.* Hanoi: IDEO, 1931.

Gaide, Laurent J., and Pierre Dorolle. "Sur des cas de tétanos survenus après injections intramusculaires de sels de quinine." *Bulletin de la Société médico-chirurgicale de l'Indochine* 6, no. 10 (1928): 444–48.

Gautier, A. "Voyage au pays des mois." *Excursions et Reconnaissances* 14 (1882): 219–49.

George, M. P. *Le caoutchouc et les industries du caoutchouc, étude geographique.* Les cours de Sorbonne. Paris: Centre de documentation universitaire, 1950.

Gerber, Théophile. "Coutumier Stieng." *Bulletin de l'Ecole française d'Extrême-Orient* 45, no. 1 (1951): 227–72.

Girard, E. *L'hévéa en Cochinchine.* Edited by Académie des sciences colonials. Paris: Challamel, 1924.

Godart, Justin. *Rapport de mission en Indochine, 1er Jan–14 Mars, 1937.* Edited by Alain Forest. Paris: L'Harmattan, 1994.

Gourou, Pierre. *Land Utilization in French Indochina.* New York: Institute of Pacific Relations, 1945.

———. *Les pays tropicaux, principes d'une géographie humaine et économique.* Paris: Presses universitaires de France, 1946.

———. *L'utilisation du sol en Indochine française.* Paris: P. Hartmann, 1940.

Gouvernement de la Cochinchine. *Notions générales sur le paludisme et les moyens de le combattre dans les centres agricoles et forestiers de la Cochinchine.* Saigon: C. Ardin, 1919.

———. *Recensement générale de la population de la Cochinchine.* Saigon: C. Ardin, 1921.

Gouvernement général de l'Indochine. "Contrat de la colonie avec l'Institut Pasteur, pour l'organisation d'un service de lutte anti-malarienne sur les chantiers des travaux publics de l'Indochine." Hanoi: 1930.

———. *L'Indochine française.* Hanoi: Imprimerie G. Taupin & Cie., 1938.

———. "Réglementation sur la protection de la main d'œuvre indigène et asiatique étrangère employée par contrat sur les exploitations agricoles industrielles et minières en Indochine." *Journal officiel de l'Indochine* 39 (1927): 3033.

Grall, Charles. "Paludisme 'Épidémié.'" *Bulletin de la Société de pathologie exotique* 10, no. 3 (1917): 184–207.

Grima, Roger. *La lutte antipalustre dans les collectivités en Indo-Chine.* Lyon: Intersyndicale Lyonnaise, 1932.

Gueyffier, René. *Essai sur le régime de la terre en Indochine (pays annamites).* Lyon: Imprimerie Bosc et Riou, 1928.

Guillerm, Jean. "L'industrie du nuoc-mam en Indochine." In *Feuillets d'hygiène indochinoise,* edited by Henry G. S. Morin, 583–615. Hanoi: IDEO, 1935.

———. *Rapport à la commission de la ration alimentaire.* Saigon: Gouvernement de la Cochinchine, 1927.

Guillerm, Jean, and Henry G. S. Morin. "Essai de détermination statistique de la ration alimentaire du travailleur annamite." *Bulletin de la Société de pathologie exotique* 9 (1926): 905.

Han, Ji, and Hui-Lin Li. *Nan-Fang Ts'ao-Mu Chuang: A Fourth Century Flora of Southeast Asia: Introduction, Translation, Commentaries.* Hong Kong: Chinese University Press, 1979.

Han, Suyin. *And the Rain My Drink.* Boston: Little, Brown, 1956.

Haumant. "Industrie indochinoise: les planteurs d'hévéas." *Indochine. Hebdomadaire illustré* 5, no. 223 (1944): 4–13.

Henry, Yves. *Économie agricole de l'Indochine.* Hanoi: IDEO, 1932.

———. *Terres rouges et terres noires basaltiques d'Indochine, leur mise en culture.* Hanoi: IDEO, 1931.

Hermant, P. "Fonctionnement du service de vente de la quinine d'Etat dans la province de Nghê-an en 1912." *Bulletin de la Société médico-chirurgicale de l'Indochine* 4 (1913): 231–33.

Ho, Tong Peng. "L'économie de l'hévéaculture en 1965." *Etudes cambodgiennes* 1, no. 1 (1965): 26–28.

Hoenstadt, Gabrielle von. *Three Years in Vietnam (1907–1910): Medicine, Chams and Tribesmen in Nhatrang and Surroundings.* Edited by Walter E. J. Tips. Bangkok, Thailand: White Lotus Press, 1999.

Homberg, Octave. *La France des cinq parties du monde.* Paris: Plon, 1927.

———. *Les grands produits coloniaux: Le caoutchouc.* Paris: Monographies de la Dépêche coloniale, 1925.

Huard, Pierre. "L'Extrême-Asie et le corps humain." *Bulletin de la Société des études indochinoises* 24 (1948): 89–93.

Huard, Pierre, and Alfred Bigot. "Introduction à l'étude des Eurasiens." *Bulletin économique de l'Indochine* 42, no. 4 (1939): 715–58.

Huard, Pierre, and Maurice Durand. "Science au Viet-Nam." *Bulletin de la Société des études indochinoises* 40, nos. 3–4 (1963).

Huard, Pierre, and Đỗ Xuân Hợp. *Morphologie humaine et anatomie artistique.* Hanoï: Imprimerie G. Taupin, 1942.

Huxley, Julian, Alfred C. Haddon, and A. M. Carr-Saunders. *We Europeans; a Survey of "Racial" Problems.* London: J. Cape, 1935.

Indochine. *Adresses, annuaire complet (européen et indigène) de toute l'Indochine, commerce, industrie, plantations, mines, adresses particulière. . . .* Saigon: Imprimerie A. Portail, 1933, 1934–35, 1936–37.

Inspection générale de l'agriculture, de l'élevage et des forêts. *Textes concernant les recherches agronomiques et la police sanitaire végétale en Indochine.* Hanoi: IDEO, 1927.

Inspection générale du travail. *Réglementation du travail en Indochine, supplément annuel, textes parus du 1er juillet 1929 au 31 octobre 1930.* Hanoi: Imprimerie Tonkinoise Le Van Phuc, 1931.

———. *Règlementation du travail en Indochine, textes en vigueur au 1er juillet 1929.* Hanoi: IDEO, 1929.

———. *Réglementation du travail en Indochine, textes en vigueur au 31 octobre 1930.* Hanoi: IDEO, 1931.

Institut des recherches sur le caoutchouc en Indochine (IRCI). *Cahiers de l'Institut des recherches sur le caoutchouc en Indochine.* Saigon: SILI, 1942, 1944.

———. "Procès-verbal de la deuxième réunion du comité technique consultatif du 8 octobre 1941." Lai Khe, Vietnam, 1941.

———. "Procès-verbal de la première réunion de la commission technique du 8 décembre 1948." Lai Khe, Vietnam, 1948.

———. "Procès-verbal de la première réunion de la commission technique du 13 avril 1949." Lai Khe, Vietnam, 1949.

———. *Rapport annuel de l'Institut des recherches sur le caoutchouc en Indochine de 1947.* Saigon: SILI, 1948.

———. *Rapport annuel de l'Institut des recherches sur le caoutchouc en Indochine de 1948.* Saigon: SILI, 1949.

———. *Rapport annuel de l'Institut des recherches sur le caoutchouc en Indochine de 1949*. Saigon: SILI, 1950.

———. *Rapport annuel de l'Institut des recherches sur le caoutchouc en Indochine de 1950*. Saigon: SILI, 1951.

Institut des recherches sur le caoutchouc au Vietnam (IRCV). *Rapport annuel de l'Institut des recherches sur le caoutchouc au Vietnam de 1958*. Saigon: SILI, 1959.

———. *Rapport annuel de l'Institut des recherches sur le caoutchouc au Vietnam de 1959*. Saigon: SILI, 1960.

———. *Rapport annuel de l'Institut des recherches sur le caoutchouc au Vietnam de 1960–61–62*. Saigon: SILI, 1962.

Jumelle, Henri. *Les plantes à caoutchouc et à gutta: Exploitation, culture et commerce dans tous les pays chauds*. Paris: A. Challamel, 1903.

Kérandel, J. "Riziculture et distribution géographique du paludisme en Indochine: Insectes prédateurs de larves de moustiques." *Bulletin de la Société de pathologie exotique* 18 (1925): 815–21.

Lan, M. J. *Notions d'agrologie/Địa Học Yếu Lược. Les sols en agriculture/Những Thủ Đất Trong Vi Cô Canh Nông. Etude Physique/Học Nê Hinh*. s.l., n.d.

———. *Notions de botanique/Thực Vật Học Yếu Lược: les plantes/Cây*. Translated by Công Tiểu Nguyễn, Vân Khuê Vương, and Kim Giám Phạm, Bibliothèque Agricole Indochinoise. Hanoi: IDEO, 1924.

Laprade, C. de. *Simples renseignements sur l'arrondissement de Thudaumot*. Paris: Ecole coloniale, 1898.

Lê Sắc Nghi. *Đất Đỏ Miền Đông: Hồi Ký Cách Mạng*. Biên Hòa, Việt Nam: Công ty Quốc doanh Cao su Đồng Nai, 1979.

Le, Van Tinh. "Le paludisme en Cochinchine et sa prophylaxie." PhD diss., La Faculté de Médecine, 1932.

League of Nations. *Intergovernmental Conference of Far-Eastern Countries on Rural Hygiene. Preparatory Papers: Report of French Indo-China*. Geneva: League of Nations Publications, 1937.

———. *Public Health Services in the French Colonies*. Liége: Publications of the League of Nations, 1926.

———. *Report of the Intergovernmental Conference of Far-Eastern Countries on Rural Hygiene, Held at Bandoeng, Java, August 3rd to 13th, 1937*. Geneva: League of Nations Publications, 1937.

Lecomte, Henri, Henri Humbert, and François Gagnepain, eds. *Flore générale de l'Indo-Chine*. 7 vols. Paris: Masson et Cie, 1907–51.

Lieurade, L. *Essai de démographie des pays "Moïs" du Kontum*. Saïgon: Imprimerie A. Portail, 1937.

Lochard, M. "Science pure et science appliquée: remarques sur le rôle des races blances en Extrême-Orient." *Extrême-Asie* 22, no. 8 (1926): 339–42.

Loubet, Lucien. *Monographie de la province de Kompong-Cham*. Phnom-Penh: Imprimerie Albert Portail, 1939.

Loureiro, João de. *Flora cochinchinensis, sistens plantas in regno Cochinchina nascentes . . . Dispositae secundum sytema sexuale Linnaeanum.* Ulyssipone: Typis et expensis academicis, 1790.

Ly, Binh Huê. *Le régime des concessions domaniales en Indochine.* Paris: F. Loviton et Cie, 1931.

Maître, Henri. *Les jungles moï, exploration et histoire des hinterlands moï du Cambodge, de la Cochinchine, de l'Annam et du bas Laos.* Paris: E. Larose, 1912.

———. *Les régions moï du sud Indo-chinois: Le plateau du Darlac.* Paris: Plon-Nourrit, 1909.

Malye, Maurice. "Les problèmes d'Indochine et le rôle des techniciens d'économie rurale." In *Centre de documentation des ingénieurs agronomes,* 1–14. Paris: Nogent-sur-Marne, 1949.

Marchoux, E. "Afrique Occidentale Française." In *Traité d'hygiène,* edited by A. Chantemesse and E. Mosny. Paris: J. B. Baillière et fils, 1907.

———. "Influence du bien–être sur la régression du paludisme." *Bulletin de la Société de pathologie exotique* 14, no. 8 (1921): 455–59.

Marcille, Guy. "L'extension à l'Indochine de la législation métropolitaine des accidents du travail." PhD diss., Universite de Paris, 1936.

Mesnard, J., and L. A. Bordes. "L'infection paludéenne chez les moïs et son importance épidémiologique dans le sud indochinois." *Bulletin de la Société médico-chirurgicale de l'Indochine* 8 (1930): 919–29.

Mesnard, J., and Constantin Toumanoff. "Note préliminaire sur l'anthropophilie de A. minimus théo avec quelques remarques sur la zoophilie de A. vagus don en l'absence de stabulation parfaire du bétail." *Bulletin de la Société médico-chirurgicale de l'Indochine* 11, no. 8 (1933): 994–1000.

Mingot, René, and Jean Canet. *L'hévéaculture en Indochine. La main d'œuvre contractuelle sur les plantations de caoutchouc en Indochine. Considérations sur l'organisation de la médecine preventive et curative dans les plantations de caoutchouc en Indochine.* Paris: IFC, 1937.

Monet, Paul. *Les Jauniers: histoire vraie.* Paris: Gallimard, 1930.

Montel, M. L. R. "Un essai de quinine préventive à Hatien (Cochinchine)." *Bulletin de la Société de pathologie exotique* 3, no. 9 (1910): 626–28.

Moreau, P. "Notes sur un voyage d'études malariologiques dans l'océan indien (Java–Iles Mascareignes–Afrique du Sud)." *Bulletin de la Société médico-chirurgicale de l'Indochine* (1934).

Morel, E. D. *Red Rubber: The Story of the Rubber Slave Trade Flourishing on the Congo in the Year of Grace 1907.* London: T. F. Unwin, 1907.

Morère, Claude. *Le dialogue interrompu: Auguste Morère, un destin d'exception le journal de marche du gendarme-administrateur au milieu des rebelles Stiengs, Indochine, 1921–1933.* Paris: Connaissances et savoirs, 2008.

Morice, Albert. "Quelques mots sur l'acclimatement des races humaines et des animaux dans la Basse-Cochinchine." *Revue d'Anthropologie* 5 (1876): 479–86.

Morice, Albert, and Walter E. J. Tips. *People and Wildlife in and around Saigon: 1872–1873*. Bangkok: White Lotus Press, 1997.

Morin, Henry G. S. *Enquête malariologique et indications prophylactiques*. Vol. 1. Hanoi: IDEO, 1930.

———. *Entretiens sur le paludisme et sa prevention en Indochine*. Hanoi: IDEO, 1935.

———. *La lutte contre le paludisme dans les collectivités ouvrières*. Hanoi: IDEO, 1931.

———. "La ration alimentaire ouvrière du pont de vue des exploitations agricoles et industrielles." In *Feuillets d'hygiène indochinoise*, edited by Henry G. S. Morin, 552–70. Hanoi: IDEO, 1935.

———. "Le paludisme et sa prophylaxie en Indochine." In *Feuillets d'hygiène indochinoise*, edited by Henry G. S. Morin, 391–412. Hanoi: IDEO, 1935.

———. "Sur l'organisation du service antipaludique des Instituts Pasteur d'Indochine." *Bulletin de la Société médico-chirurgicale de l'Indochine* 9, no. 5 (1931): 271–80.

———. "Vers la prophylaxie sociale du paludisme." *Bulletin de la Société médico-chirurgicale de l'Indochine* 10, no. 1 (1932): 88–128.

Morin, Henry G. S., and J. Bader. "Recherches sur quelques facteurs physico-chimiques de l'oecologie des larves d'anophèles au Tonkin." *Annales de Institut Pasteur* 51, no. 5 (1933): 656–68.

———. "Recherches sur quelques facteurs physico-chimiques de l'oecologie larvaire des anophèles au Tonkin (Deuxième note)." *Annales de Institut Pasteur* 52, no. 3 (1934): 3–51.

Morin, Henry G. S., and P. Carton. "Contribution à l'étude de l'influence des facteurs climatiques sur la répartition de l'endémie palustre en Indochine." *Bulletin économique de l'Indochine* 37 (1934): 459–80.

Morin, Henry G. S., and P. Martin. "Possibilités d'utilisation pratique du Vert de Paris en Indochine." *Archives des Instituts Pasteur d'Indochine* 17 (1933): 103–40.

Morin, Henry G. S., and L. A. Robin. *Essai sur la prévention pratique du paludisme dans les exploitations agricoles en Indochine*. Saigon: Imprimerie Albert Portail, 1933.

———. *Indications pratiques pour le drainage des terres en vue de la prophylaxie antipaludique sur les plantations*. Hanoi: IDEO, 1933.

Motte, Dominique de la. *De l'autre côté de l'eau: Indochine, 1950–1952*. Paris: Tallandier, 2009.

Mus, Paul. *Viêt-Nam, Sociologie d'une guerre*. Collections Esprit: "Frontière Ouverte." Paris: Éditions du Seuil, 1952.

Navarre, P. Just. *Manuel d'hygiène coloniale: guide de l'Européen dans les pays chauds*. Paris: Doin, 1895.

Nghiêm Xuân Yêm. "Điêu-Tra Nhỏ: Những Tá Điên." *Thanh Nghị* 55 (1944): 4–7.

———. "Điêu Tra Nhỏ: Những Tiểu Đồn Điền." *Thanh Nghị* 62 (1944): 7–10.

———. "Nạn Dân Đói (Một Vài Nhận Xét và Thiên Kiến vê Vấn-Đê Thóc Gạo)." *Thanh Nghị* 107 (1945): 18–22.

Nguyễn Công Tiễu. *Những Sự Kỳ-Quan Trong Vũ-Trụ, Loài Thaỏ-Mộc. Les merveilles de la nature, les végétaux*. Hanoi: Imprimerie Kim Duc-Giang, 1924.

———. *Sách Làm Ruộng*. Hà Nội: Direction de l'Instruction Publique en Indochine, 1931.

Nguyễn Hải Trừng. *Địa Ngục Cao-Su*. Hà Nội: NXB Sự thật, 1955.

Nguyễn, Phó Lu. *Rapport stagiaire*. Lai Khe: IRCV, 1961.

Nguyễn, Văn Hoài. "De l'organisation de l'hôpital psychiatrique du sud-Vietnam." PhD diss., University of Saigon, 1954.

Nguyễn, Văn Thịnh. *Etude sur l'étiologie du béribéri*. Paris: Jouve & Cie, 1921.

———. *Note sur les précautions à prendre contre le paludimse*. Saigon: Imprimerie de l'Union, 1929.

Nguyễn, Văn Thịnh, and Hữu Thình Đỗ. *Cahier des vœux et exposés de motifs présentés à la commission d'enquête coloniale, novembre 1937*. Saigon: Imprimerie du Théatre, 1937.

Pasquier, Pierre. *L'Annam d'autrefois, essai sur la constitution de l'Annam avant l'intervention française*. Paris: A. Challamel, 1907.

———. *La colonisation des terres incultes et le problème de la main d'œuvre en Indochine*. Saigon: Ardin et fils, 1918.

Patté, Paul. *Hinterland moï*. Paris: Plon-Nourrit, 1906.

Paucot, and Le Van Chinh. *Notions d'hygiène à l'usage des indigènes*. Hanoi-Haiphong: IDEO, 1908.

Petch, Thomas. *The Diseases and Pests of the Rubber Tree*. London: Macmillan, 1921.

———. *The Physiology & Diseases of Hevea Brasiliensis, the Premier Plantation Rubber Tree*. London: Dulau, 1911.

Pham, Quang Van, Canh Thanh Le, and Tuy Quoc Nguyen. *Vœux présenés par la commission permanente à M. Jules Brevié, Gouverneur général de l'Indochine et à M. Godart, Ministre délégué de la France en mission en Indochine*. Huê: Imprimerie Tieng-Dan, 1937.

Pierre, J. B. Louis. *Flore forestière de la Cochinchine*. Paris: O. Doin, 1882.

———. *Notice sur une espèce d'Isonandra fournissant un produit similaire à la gutta*. Saigon: Imprimerie autographique du collège des stagiaires, 1874.

Pouyanne, Albert Armand. *Historique succinct de l'organisation du service de lutte antimalarienne sur les chantiers des travaux publics de l'Indochine*. Edited by Inspection générale des travaux publics GGI. Vol. 2. Hanoi: IDEO, 1930.

Présentation d'ouvrage. "Les moustiques de la Cochinchine et du Sud-Annam—E. Borel." *Bulletin de la Société médico-chirurgicale de l'Indochine* 9, no. 2 (1931): 135–36.

Robequain, Charles. *The Economic Development of French Indo-China*. Translated by Isabel A. Ward, J. Russell Andrus, and Katrine R. C. Greene. London: Oxford University Press, 1944.

Robin, L. A. "Evolution de l'état sanitaire des collectivitiés ouvrières agricoles importées en région d'hyperendémie palustre: Influence de la 'prémunition acquise.'" *Revue d'hygiène et de médécine preventive* 47, no. 1 (1935): 30–42.

———. "La prophylaxie antipaludique dans les plantations de l'Indochine méridionale: La lutte antilarvaire. Son efficacité." *Bulletin de la Société de pathologie exotique* 27, no. 7 (1934): 691–99.

———. "Sur l'incidence de l'endémie palustre en Indochine méridionale et ses variations de 1926 à 1932." *Archives des Instituts Pasteur d'Indochine* 17 (1933): 11–58.

Rollet, B., Van Hoi Ly, Sam Ol Neang, and Paul Maurand. *Les forêts claires au sud-indochinois: Cambodge, sud-Laos, sud-Viêtnam.* 2 vols. Saigon: Centre de recherches scientifiques et techniques du Cambodge, Laos, Viet-Nam, 1952.

Roubaud, Émile. "À propos des races zoophiles d'anophèles." *Bulletin de la Société de pathologie exotique* 15, no. 1 (1922): 36–38.

———. "La différenciation des races zootropiques d'anophèles et la régression spontanée du paludisme." *Bulletin de la Société de pathologie exotique* 14, no. 9 (1921): 577–94.

Roudeix, H. *Rapport d'enquête sur la situation de l'industrie de transformation du caoutchouc (note préliminaire).* Saigon: IRCV, 1969.

Sarraut, Albert. *La mise en valeur des colonies françaises.* Paris: Imprimerie Charles Colin Payot et Cie, 1923.

Schneider, J., Jean Canet, and R. Dupoux. "Traitement curatif du paludisme par une 2–4 diaminopyrimidine: Premiers résultats." *Bulletin de la Société de pathologie exotique* 45, no. 1 (1952): 33–43.

Schultz, Yvonne. *Dans la griffe des jauniers.* Paris: Plon et Nourrit, 1931.

Service de la statistique générale. *Résumé statistique relatif aux années 1913 à 1940.* Hanoi: IDEO, 1941.

Sharples, Arnold. *Diseases and Pests of the Rubber Tree.* London: Macmillan, 1936.

Shelford, Victor E. *Laboratory and Field Ecology.* Baltimore, MD: Williams & Wilkins Co., 1929.

Simond, P. L. S. "Hygiène de l'Indochine." In *Traité d'hygiène,* edited by A. Chantemesse and E. Mosny, 444. Paris: J. B. Baillière et fils, 1907.

———. "Paludisme." *Annales d'hygiène et de médecine coloniales* 4 (1901): 128–33.

Sở Quân Dân Y Nam Bộ. *Bịnh Sốt Rét.* Hà Nội: Nguyễn Văn Ba, 1954.

Société des études indochinoises. *Essais agricoles et industriels faits en Cochinchine depuis la fondation de cette colonie jusqu'en 1897.* Vols. 1 and 2. Sagion: Imprimerie commerciale Rey, 1897, 1898.

Société des Nations. *Enquête sur les besoins en quinine des pays impaludés et sur l'extension du paludisme dans le monde.* Genève: SDN, 1932.

Spire, A., and C. Spire. *Caoutchouc en Indochine: Etude botanique industrielle et commerciale.* Paris: Challamel, 1906.

Steinmann, Alfred. *De Ziekten En Plagen van Hevea Brasiliensis in Nederlandsch-Indië.* Buitenzorg: Archipel Drukkerij, 1925.

Syndicat des planteurs de caoutchouc de l'Indochine/SPCI. "Annonce du décès de notre regetté [*sic*] collègue M. Georges Vernet." *Bulletin du Syndicat des planteurs de caoutchouc de l'Indochine,* no. 72 (1924): 1141.

———. *Annuaire du Syndicat des planteurs de caoutchouc de L'Indochine.* Saigon: Maison Photo Nadal, 1926.

———. *Annuaire du Syndicat des planteurs de caoutchouc de L'Indochine.* Paris: Braun & Cie, 1931.

Taysum, D. H. "A Possible Basis for Phyto-Immunity." *Nature* 181, no. 4603 (1958): 174–75.

Tirouvanziam, and Phleng. "Le paludisme à Kratié." *Bulletin de la Société médico-chirurgicale de l'Indochine* 11, no. 8 (1933): 930–35.

Trân Hằng Ngôn. "Tieng Annam và *Khoa Học Phổ Thông*, hay là: Đầu Năm Nhìn Con Đường Đi." *Khoa Học Phổ Thông* 73 (1937): 3–5.

Trân Văn Đôn. "Công Trình Khảo Cứu về Bịnh Rét Rừng (Paludisme) tại Núi Bà Đen Tỉnh Tây Ninh." *Khoa Học Phổ Thông* 1, no. 4 (1934): 4–8.

———. "Note sur le paludisme contracté par les pélerins à la grotte de Nui Ba-Den à Tay-Ninh (Cochinchine)." *Bulletin de la Société médico-chirurgicale de l'Indochine* 11, no. 8 (1933): 936–48.

Trịnh, Hoài Đức. *Histoire et description de la basse Cochinchine (Pays de Gia-Dinh)/Gia Định Thành Thông Chí*. Translated by Louis Gabriel Galderic Aubaret. Paris: Imprimerie impériale, 1863.

Union des planteurs de caoutchouc en Indochine. *Annuaire*. Paris: La Librairie technique et économique, 1936.

Vaxelaire, Jean. *Le caoutchouc en Indochine*. Hanoi: IDEO, 1939.

Vernet, Georges. "L'*hévéa brasiliensis*, sa culture et son exploitation dans le sud-Annam." *Bulletin économique de l'Indochine* 8, no. 39 (1905): 687–734.

———. "Observations sur l'hévéa dans le sud-Annam." *Journal d'agriculture tropicale* 51 (1905): 259–62.

Viollis, Andrée. *Indochine S.O.S.* Paris: Gallimard, 1935.

Vũ Thúy. *Công Nhân Cao Su Chiến Đấu, Thành Tích Tranh Đấu của Công Nhân Cao-Su Từ 1920 đến 1948*. Hà Nội: NXB Lao Động, 1950.

Vũ Trọng Phụng. *Kỹ Nghệ Lấy Tây*. Hà Nội: Phương Đông, 1936.

Vũ Văn Quang. "Le problème des Eurasiens en Indochine." PhD diss., Ecole de Médecine et de Pharmacie, 1939.

Wildeman, E. de. "Plantations et maladies de l'hévéa." *Revue de botanique appliquée et d'agriculture coloniale* 6, no. 53 (1926): 18–22.

Wormser, G. *Des quelques facteurs de la situation économique et politique du sud-Indochinois*. Saigon: Imprimerie de l'Union Nguyen-Van-Cua, 1946.

Yersin, Alexandre. "Longue survie d'hévéas brisés par un typhon." *Comptes rendus d'Académie des Sciences* 194 (1932): 1620.

Yersin, Alexandre, and André Lambert. *Essais d'acclimatation de l'arbre à quinquina en Indochine*. Hanoï: IDEO, 1931.

Secondary Sources

Abraham, Itty. "The Contradictory Spaces of Postcolonial Techno-Science." *Economic and Political Weekly*, January 21, 2006.

Ackerman, Evelyn Bernette. "The Intellectual Odyssey of a French Colonial Physician: Jules Regnault and Far Eastern Medicine." *French Historical Studies* 19, no. 4 (1996): 1083–102.

Adas, Michael. *The Burma Delta: Economic Development and Social Change on an Asian Rice Frontier, 1852–1941*. Madison: University of Wisconsin Press, 1974.

————. *Machines as the Measure of Men: Science, Technology, and Ideologies of Western Dominance*. Ithaca, NY: Cornell University Press, 1989.

Affidi, Emmanuelle. "Vulgarisation du savoir et colonisation des esprits par la presse et le livre en Indochine française et dans les Indes Néerlandaises (1908–1936)." In *Moussons: Viêt Nam, Histoire et Perspectives Contemporaines*, edited by Christian Culas and Jean-François Klein, 95–121. Marseille: IRSEA, 2009.

Amicale des anciens planteurs d'hévéas. "L'œuvre d'une Française en Cochinchine, Janie Bertin Rivière de la Souchère (1881–1963)." *Sernamby* 14 (1998): 47–49.

Amicale des planteurs d'hévéas. *Les planteurs d'hévéas en Indochine de 1950 à 1975*. Paris: Charles Lavauzelle, 2006.

————. *Planteurs d'hévéa en Indochine, 1939–1954*. Vichères: Amicale des planteurs d'hévéas, 1996.

Amrith, Sunil S. *Crossing the Bay of Bengal: The Furies of Nature and the Fortunes of Migrants*. Cambridge, MA: Harvard University Press, 2013.

————. *Decolonizing International Health: India and Southeast Asia, 1930–65*. New York: Palgrave Macmillan, 2006.

Anderson, David. *Eroding the Commons: The Politics of Ecology in Baringo, Kenya, 1890s–1963*. Athens: Ohio University Press, 2002.

Anderson, Warwick. *Colonial Pathologies: American Tropical Medicine, Race, and Hygiene in the Philippines*. Durham, NC: Duke University Press, 2006.

————. "Excremental Colonialism: Public Health and the Poetics of Pollution." *Critical Inquiry* 21 (1995): 640–69.

————. "How Far Can East Asian STS Go? A Commentary." *East Asian Science, Technology and Society: An International Journal* 1, no. 2 (2007): 249–50.

————. "Immunities of Empire: Race, Disease, and the New Tropical Medicine, 1900–1920." *Bulletin of the History of Medicine* 70, no. 1 (1996): 94–118.

————. "Introduction: Postcolonial Technoscience." *Social Studies of Science* 32, nos. 5–6 (2002): 643–58.

Andrews, Thomas G. *Killing for Coal: America's Deadliest Labor War*. Cambridge, MA: Harvard University Press, 2008.

Angladette, André. "La vie quotidienne en Indochine de 1939 à 1946." *Comptes rendus trimestriels des séances de l'Académie des sciences d'outre-mer* 39, no. 3 (1979): 482–85.

————. "Les recherches agronomiques en Indochine pendant la première moitie du vingtième siècle, leur impact sur la production rurale, leur évolution ultérieure." *Mondes et cultures* 41, no. 2 (1981): 189–216.

Arnold, David. " 'An Ancient Race Outworn': Malaria and Race in Colonial India, 1860–1930." In *Race, Science and Medicine*, edited by Waltraud Ernst and Bernard Harris, 123–43. London: Routledge, 1999.

————. "Tropical Governance: Managing Health in Monsoon Asia, 1908–1938." ARI Working Paper No. 116, 2009.

Artz, Frederick. *The Development of Technical Education in France, 1500–1850*. Cambridge, MA: MIT Press, 1966.

Aso, Michitake. "De plantations coloniales à la production socialiste: la 'vietnamisation' de l'hévéa (1956–1975)." In *L'Or Blanc: Petits et grands planteurs face au "boom" de l'hévéaculture (Việt Nam-Cambodge)*, edited by Frédéric Fortunel and Christophe Gironde, 65–82. Bangkok: IRASEC, 2014.

———. "How Nature Works: Business, Ecology, and Rubber Plantations in Colonial Southeast Asia, 1919–1939." In *Comparing Apples, Oranges, and Cotton: Environmental Histories of the Global Plantation*, edited by Frank Uekötter, 195–220. New York: Campus Verlag, 2014.

———. "Le caoutchouc au dix-neuvième siècle : un produit industriel et de consommation." In *Histoire du monde au XIXe siècle*, edited by Sylvain Venayre and Pierre Singaravélou, 429–33. Paris: Fayard, 2017.

———. "Les plantations: économies, sociétés et environnements, 1850–1950." In *Les sociétés coloniales à l'âge des empires: des années 1850 aux années 1950*, edited by Dominique Barjot and Jacques Frémeaux, 198–208. Paris: Editions Sedes/Cned, 2012.

———. "Patriotic Hygiene: Tracing New Places of Knowledge Production about Malaria in Vietnam, 1919–75." *Journal of Southeast Asian Studies* 44, no. 3 (2013): 423–443.

———. "Profits or People? Rubber Plantations and Everyday Technology in Rural Indochina." *Modern Asian Studies* 46 (2012): 19–45.

———. "Rubber and Race in Rural Colonial Cambodia (1920s–1954)." *Siksācakr* 12–13 (2010–2011): 127–138.

———. "The Scientist, the Governor, and the Planter: The Political Economy of Agricultural Knowledge in Indochina during the Creation of a 'Science of Rubber,' 1900–1940." *East Asian Science, Technology and Society: An International Journal* 3, nos. 2/3 (2009): 231–56.

Au, Sokhieng. *Mixed Medicines: Health and Culture in French Colonial Cambodia.* Chicago: University of Chicago Press, 2011.

Baird, Ian G. "The Construction of 'Indigenous Peoples' in Cambodia." In *Alterities in Asia: Reflections on Identity and Regionalism*, edited by Leong Yew, 155–76. New York: Routledge, 2011.

———. "Land, Rubber and People: Rapid Agrarian Changes and Responses in Southern Laos." *Journal of Lao Studies* 1, no. 1 (2009): 1–47.

———. "Making Spaces: The Ethnic Brao People and the International Border between Laos and Cambodia." *Geoforum* 41 (2010): 271–81.

Banens, Maks. "Vietnam: A Reconstitution of Its 20th Century Population History." In *Quantitative Economic History of Vietnam, 1900–1990: An International Workshop*, edited by Jean-Pascal Bassino, Jean-Dominique Giacometti, and Kaonosuke Odaka, 1–40. Tokyo: Institute of Economic Research, Hitotsubashi University, 2000.

Bankoff, Greg. "'The Tree as the Enemy of Man': Changing Attitudes to the Forests of the Philippines, 1565–1898." *Philippine Studies* 52, no. 3 (2004): 320–44.

Barham, Bradford, and Oliver Coomes. "Wild Rubber: Industrial Organisation and the Microeconomics of Extraction during the Amazon Rubber Boom (1860–1920)." *Journal of Latin American Studies* 26, no. 1 (1994): 37–72.

Barkan, Elazar. *Retreat of Scientific Racism: Changing Concepts of Race in Britain and the United States between the World Wars*. New York: Cambridge University Press, 1992.

Barlow, Colin. "Changes in the Economic Position of Workers on Rubber Estates and Smallholdings in Peninsular Malaysia, 1910–85." In *The Underside of Malaysian History: Pullers, Prostitutes, Plantation Workers*, edited by Peter J. Rummer and Lisa M. Allen, 25–49. Singapore: Singapore University Press, 1990.

———. "Growth, Structural Change and Plantation Tree Crops: The Case of Rubber." *World Development* 25, no. 10 (1997): 1589–607.

———. *The Natural Rubber Industry: Its Development, Technology, and Economy in Malaysia*. New York: Oxford University Press, 1978.

Barlow, Colin, and J. H. Drabble. "Government and the Emerging Rubber Industries in Indonesia and Malaya, 1900–40." In *Indonesian Economic History in the Dutch Colonial Era*, edited by Anne Booth, William J. O'Malley, and Anna Weidermann, 85–103. New Haven, CT: Yale University Southeast Asia Studies Program, 1990.

Barlow, Colin, and S. K. Jayasuriya. "Stages of Development in Smallholder Tree Crop Agriculture." *Development and Change* 17 (1986): 635–58.

Barlow, Colin, S. K. Jayasuriya, and C. Suan Tan. *The World Rubber Industry*. London: Routledge, 1994.

Barlow, Colin, and Muharminto. "The Rubber Smallholder Economy." *Bulletin of Indonesian Economic Studies* 18, no. 2 (1982): 86–119.

Barlow, Tani E. *Formations of Colonial Modernity in East Asia*. Durham, NC: Duke University Press, 1997.

Basalla, George. "The Spread of Western Science." *Science* 156 (1967): 611–22.

Bassino, Jean-Pascal, and Nakagawa Hironobu. "Exchange Rates and Exchange Rate Policies in Vietnam under French Rule, 1878–1945." In *Quantitative Economic History of Vietnam, 1900–1990: An International Workshop*, edited by Jean-Pascal Bassino, Jean-Dominique Giacometti, and Kåonosuke Odaka, 343–64. Tokyo: Institute of Economic Research, Hitotsubashi University, 2000.

Bayly, Susan. "French Anthropology and the Durkheimians in Colonial Indochina." *Modern Asian Studies* 34, no. 3 (2000): 581–622.

———. "Racial Readings of Empire: Britain, France, and Colonial Modernity in the Mediterranean and Asia." In *Modernity and Culture: From the Mediterranean to the Indian Ocean*, edited by Leila Tarazi Fawaz, C. A. Bayly, and Robert Ilbert, 285–313. New York: Columbia University Press, 2002.

Begon, Michael, John Harper, and Colin Townsend. *Ecology: Individuals, Populations, and Communities*. Cambridge, MA: Blackwell Science, 1996.

Beinart, William, and Lotte Hughes. *Environment and Empire*. New York: Oxford University Press, 2007.

Bergot, Erwan. *Sud lointain*. Paris: Omnibus, 2004.

Bernal, Martin. "The Nghe-Tinh Soviet Movement 1930–1931." *Past & Present*, no. 92 (1981): 148–68.

Bezançon, Pascale. *Une colonisation éducatrice: l'expérience indochinoise, 1860–1945*. Paris: Harmattan, 2002.

Biggs, David. "Breaking from the Colonial Mold: Water Engineering and the Failure of Nation-Building in the Plain of Reeds, Vietnam." *Technology and Culture* 49 (2008): 599–623.

———. "Managing a Rebel Landscape: Conservation, Pioneers, and the Revolutionary Past in the U Minh Forest, Vietnam." *Environmental History* 10, no. 3 (2005): 448–76.

———. *Quagmire: Nation-Building and Nature in the Mekong Delta*. Seattle: University of Washington Press, 2010.

Blaikie, Piers M. *The Political Economy of Soil Erosion in Developing Countries*. London: Longman, 1985.

Blanckaert, Claude. "Of Monstrous Metis? Hybridity, Fear of Miscegenation, and Patriotism from Buffon to Paul Broca." In *The Color of Liberty: Histories of Race in France*, edited by Sue Peabody and Tyler Stovall, 42–70. Durham, NC: Duke University Press, 2003.

Bộ Văn Hóa và Thông Tin. *Tranh Cổ Động Việt Nam, 1945–2000*. Hà Nội: Cục Văn Hóa Thông Tin Cơ Sở, 2002.

Bộ Y Tế. *Bệnh Sốt Rét: Phòng Chống và Tiêu Diệt Sốt Rét ở Việt Nam, 1958–1975*. Hà Nội: Viện Sốt Rét Ký Sinh Trùng và Côn Trùng, 1976.

———. *Sơ Lược Lịch Sử Y Tế Việt Nam*. Hà Nội: NXB Y Học, 1996.

Bone, Jonathan A. "Rice, Rubber, and Development Policies: The Mise en Valeur of French Indochina on the Eve of the Second World War." *French Colonial History* 16 (1990): 154–80.

Bonneuil, Christophe. "Auguste Chevalier, savant colonial, entre science et empire, entre botanique et agronomie." In *Les sciences hors d'occident au XXe siècle*, edited by Roland Waast, 15–35. Paris: ORSTOM, 1995.

———. *Des savants pour l'empire: la structuration des recherches scientifiques coloniales au temps de 'la mise en valeur des colonies françaises, 1917–1945*. Paris: ORSTOM, 1991.

———. "Mettre en ordre et discipliner les tropiques: Les sciences du végétal dans l'empire français, 1870–1940." PhD diss., Université Paris VII-Denis Diderot, 1997.

———. " 'Mise en valeur' de l'empire colonial et naissance de l'agronomie tropicale." In *Du jardin d'essais colonial à la station expérimentale, 1880–1930*, edited by Christophe Bonneuil and Mina Kleiche, 25–28. Paris: CIRAD, 1993.

———. "Science and State Building in Late Colonial and Postcolonial Africa, 1930–1970." In *Nature and Empire: Science and the Colonial Enterprise*, edited by Roy M. MacLeod, 258–81. Chicago: University of Chicago Press, 2000.

Bonneuil, Christophe, and Patrick Petitjean. "Les chemins de la création de l'ORSTOM, du front populaire à la libération en passant par Vichy, 1936–1945." In *Les sciences coloniales: figures et institutions*, edited by Patrick Petitjean, 113–61. Paris: L'Institut français de recherche scientifique pour le développement en coopération, 1996.

Boomgaard, P., *Frontiers of Fear: Tigers and People in the Malay World, 1600–1950.*
New Haven, CT: Yale University Press, 2001.
———, ed. *A World of Water: Rain, Rivers and Seas in Southeast Asian Histories.*
Leiden: KITLV Press, 2007.
Boomgaard, P., Freek Colombijn, and David Henley, eds. *Paper Landscapes:*
Explorations in the Environmental History of Indonesia. Leiden: KITLV Press, 1997.
Boomgaard, P., David Henley, and Manon Osseweijer, eds. *Muddied Waters: Historical*
and Contemporary Perspectives on Management of Forests and Fisheries in Island
Southeast Asia. Leiden: KITLV Press, 2005.
Booth, Anne. *The Indonesian Economy in the Nineteenth and Twentieth Centuries:*
A History of Missed Opportunities. New York: St. Martin's Press in association with
the Australian National University, Canberra, 1998.
Boucheret, Marianne. "Le pouvoir colonial et la question de la main-d'œuvre en
Indochine dans les années vingt." *Cahiers d'histoire. Revue d'histoire critique* 85
(2001): 29–55.
———. "Les organisations de planteurs de caoutchouc indochinois et l'Etat du début
du XXe siècle à la veille de la Seconde Guerre mondiale." In *L'Esprit économique*
impérial (1830–1970): Groupes de pression & réseaux du patronat colonial en France et
dans l'empire, edited by Hubert Bonin, Catherine Hodeir, and Jean-François Klein,
715–34. Paris: Société française d'histoire d'outre-mer, 2008.
———. "Les plantations d'hévéas en Indochine, 1897–1954." PhD diss., Université
Paris 1-Panthéon Sorbonne, 2008.
Bourgois, Philippe I. *Ethnicity at Work: Divided Labor on a Central American Banana*
Plantation. Baltimore: Johns Hopkins University Press, 1989.
Bowd, Gavin, and Daniel Clayton. "French Tropical Geographies: Editors'
Introduction." *Singapore Journal of Tropical Geography* 26, no. 3 (2005): 271–88.
———. "Tropicality, Orientalism, and French Colonialism in Indochina: The Work
of Pierre Gourou, 1927–1982." *French Historical Studies* 28, no. 2 (2005): 297–327.
Bradley, Mark. *Imagining Vietnam and America: The Making of Postcolonial Vietnam,*
1919–1950. Chapel Hill: University of North Carolina, 2000.
Bray, Francesca. *Science and Civilisation in China.* Vol. 6, *Biology and Biological*
Technology, Part 2, Agriculture. Cambridge: Cambridge University Press, 1986.
Bray, Francesca, Peter A. Coclanis, Edda L. Fields-Black, and Dagmar Schaefer, eds.
Rice: Global Networks and New Histories. New York: Cambridge University Press,
2015.
Bréelle, D. "The Regional Discourse of French Geography in the Context of
Indochina: The Theses of Charles Robequain and Pierre Gourou." PhD diss.,
Flinders University, Adelaide, South Australia, 2002.
Brocheux, Pierre. "The Economy of War as a Prelude to a 'Socialist Economy': The
Case of the Vietnamese Resistance against the French, 1945–1954." In *Viêt Nam*
Exposé: French Scholarship on Twentieth-Century Vietnamese Society, edited by
Gisèle L. Bousquet and Pierre Brocheux, 313–30. Ann Arbor: University of
Michigan Press, 2002.

———. "La revue 'Thanh Nghi': Un groupe d'intellectuels Vietnamiens confrontés aux problèmes de leur nation (1941–1945)." *Revue d'histoire moderne et contemporaine* 34, no. 2 (1987): 317–31.

———. "Le prolétariat des plantations d'hévéas au Vietnam méridional: Aspects sociaux et politiques (1927–1937)." *Le mouvement social* 90 (1975): 55–86.

———. *The Mekong Delta: Ecology, Economy, and Revolution, 1860–1960.* Madison, WI: Center for Southeast Asian Studies, 1995.

———. "Moral Economy or Political Economy? The Peasants Are Always Rational." *Journal of Asian Studies* 42 (1983): 791–803.

———. "1908: Le chassé-croisé G. Chiếu-P. Văn Trường." In *Vietnam le moment moderniste,* edited by Gilles de Gantès and Phuong Ngoc Nguyen, 205–9. Aix-en-Provence: Université de Provence, 2009.

———. *Une histoire économique du Viet Nam: 1850–2007: la palanche et le camion.* Paris: les Indes savantes, 2009.

Brocheux, Pierre, and Daniel Hémery. *Indochina: An Ambiguous Colonization, 1858–1954.* Translated by Ly Lan Klein-Dill. Berkeley: University of California Press, 2009.

Brockway, Lucile. *Science and Colonial Expansion: The Role of the British Royal Botanic Gardens in Empire-Building.* New York: Academic Press, 1979.

Brown, Karen, and Daniel Gilfoyle. *Healing the Herds: Disease, Livestock Economies, and the Globalization of Veterinary Medicine.* Athens: Ohio University Press, 2010.

Bruce-Chwatt, Leonard Jan. "Malaria Research and Eradication in the USSR: A Review of Soviet Achievements in the Field of Malariology." *Bulletin of the World Health Organization* 21 (1959): 737–72.

Bruneau, Michel. "From a Centred to a Decentred Tropicality: Francophone Colonial and Postcolonial Geography in Monsoon Asia." *Singapore Journal of Tropical Geography* 26, no. 3 (2005): 304–22.

———. "Pierre Gourou (1900–1999): Géographie et civilisations." *L'Homme,* no. 153 (2000): 7–26.

Buchy, Marlene. "Histoire forestière de l'Indochine (1850–1954): Perspectives de recherche." *Revue française d'histoire d'outre-mer,* no. 299 (1993): 219–49.

Bùi Cát Vũ. "Cọp Ba Móng." In *Lịch Sử Liên Trung Đoàn 301–310 (1945–1950),* edited by Sơn Đài Hổ, 328–54. Hà Nội: NXB Quân Đội Nhân Dân, 2007.

Bunker, Stephen G. "Modes of Extraction, Unequal Exchange, and the Progressive Underdevelopment of an Extreme Periphery: The Brazilian Amazon, 1600–1980." *American Journal of Sociology* 89, no. 5 (1984): 1017–64.

Butcher, John G. *The Closing of the Frontier: A History of the Marine Fisheries of Southeast Asia, c. 1850–2000.* Singapore: Institute of Southeast Asian Studies, 2004.

Bynum, W. F. "Malaria in Inter-War British India." *Parassitologia* 42 (2000): 25–31.

Canguilhem, Georges. "Le Concept et La Vie." *Revue philosophique de Louvain* 64, no. 82 (1966): 193–223.

Carpenter, Kenneth J. *Beriberi, White Rice, and Vitamin B: A Disease, a Cause, and a Cure.* Berkeley: University of California Press, 2000.

Carter, Eric D. " 'God Bless General Perón': DDT and the Endgame of Malaria Eradication in Argentina in the 1940s." *Journal of the History of Medicine and Allied Sciences* 64, no. 1 (2009): 78–122.

―――. "Malaria, Landscape, and Society in Northwest Argentina in the Early Twentieth Century." *Journal of Latin American Geography* 7, no. 1 (2008): 7–38.

Castle, Robert, Jim Hagan, and Andrew Wells. *Quan Hệ Lao Động Trong các Đồn Điền Cao Su Đông Dương, 1910–1940.* TP HCM: Đại học Wollongong, 2000.

Catton, Philip E. *Diem's Final Failure: Prelude to America's War in Vietnam.* Lawrence: University Press of Kansas, 2002.

Chalk, Frank Robert. "The United States and the International Struggle for Rubber, 1914–1941." PhD diss., University of Wisconsin–Madison, 1970.

Chambers, David Wade, and Richard Gillespie. "Locality in the History of Science: Colonial Science, Technoscience, and Indigenous Knowledge." In *Nature and Empire: Science and the Colonial Enterprise,* edited by Roy M. MacLeod, 221–40. Chicago: University of Chicago Press, 2000.

Chandler, David. *A History of Cambodia.* Boulder, CO: Westview Press, 2008.

―――. "Paul Mus (1902–1969): A Biographical Sketch." *Journal of Vietnamese Studies* 4, no. 1 (2009): 149–91.

Chapman, Jessica M. *Cauldron of Resistance: Ngo Dinh Diem, the United States, and 1950s Southern Vietnam.* Ithaca, NY: Cornell University Press, 2013.

Chatelin, Yvon. "Genèse, mutation et éclatement des paradigmes: Le cas de la science des sols tropicaux." In *Nature et environnement,* edited by Christophe Bonneuil and Yvon Chatelin, 141–54. Paris: ORSTOM, 1995.

Choi, Byung Wook. *Southern Vietnam under the Reign of Minh Mạng (1820–1841): Central Policies and Local Response.* Ithaca, NY: Southeast Asia Program Publication, 2004.

Chu, Lai. *Cuộc Đời Dài Lắm.* Hà Nội: NXB Văn Học, 2007.

Cittadino, Eugene. *Nature as the Laboratory: Darwinian Plant Ecology in the German Empire, 1880–1900.* New York: Cambridge University Press, 1990.

Clarence-Smith, William G. "The Rivaud-Hallet Plantation Group in the Economic Crises of the Inter-War Years." In *Private Enterprise during Economic Crises Tactics and Strategies,* edited by Pierre Lanthier and Hubert Watelet, 117–32. New York: LEGAS, 1997.

Claval, Paul. "Colonial Experience and the Development of Tropical Geography in France." *Singapore Journal of Tropical Geography* 26, no. 3 (2005): 289–303.

Cleary, Mark. "Managing the Forest in Colonial Indochina c. 1900–1940." *Modern Asian Studies* 39, no. 2 (2005): 257–83.

―――. " 'Valuing the Tropics': Discourses of Development in the Farm and Forest Sectors of French Indochina, circa 1900–1940." *Singapore Journal of Tropical Geography* 26, no. 3 (2005): 359–74.

Coates, Austin. *The Commerce in Rubber: The First 250 Years.* New York: Oxford University Press, 1987.

Cohen, Paul T. "The Post-Opium Scenario and Rubber in Northern Laos: Alternative Western and Chinese Models of Development." *International Journal of Drug Policy* 20, no. 5 (2009): 424–30.

Cohen, William B. "Malaria and French Imperialism." *Journal of African History* 24, no. 1 (1983): 23–36.

Coll. *Alexandre Yersin (1863–1943): Un demi-siècle au Vietnam.* Hanoi: Le Fonds culturel du Ministère de la culture, de l'information et des sports, 1992.

———. *Hommes et destins: Dictionnaire biographique d'outre-mer.* Paris: Académie des sciences d'outre-mer, 1975, 1977, 1985.

Condominas, Georges. *We Have Eaten the Forest: The Story of a Montagnard Village in the Central Highlands of Vietnam.* New York: Hill and Wang, 1977.

Công Ty Cao Su Đồng Nai, Đảng Bộ, ed. *Lịch Sử Phong Trào Công Nhân Cao Su Đồng Nai 1906–2006.* 2nd ed. Biên Hòa: NXB Đồng Nai, 2008.

Conklin, Alice L. *A Mission to Civilize: The Republican Idea of Empire in France and West Africa, 1895–1930.* Stanford, CA: Stanford University Press, 1997.

Cooper, Frederick. *Colonialism in Question: Theory, Knowledge, History.* Berkeley: University of California Press, 2005.

———. "The Dialectics of Decolonization." In *Tensions of Empire Colonial Cultures in a Bourgeois World,* edited by Frederick Cooper and Ann Laura Stoler, 406–35. Berkeley: University of California Press, 1997.

Cooper, Frederick, and Ann Laura Stoler. "Between Metropole and Colony: Rethinking a Research Agenda." In *Tensions of Empire: Colonial Cultures in a Bourgeois World,* edited by Frederick Cooper and Ann Laura Stoler, 1–56. Berkeley: University of California Press, 1997.

Cronon, William. *Nature's Metropolis: Chicago and the Great West.* New York: W. W. Norton, 1991.

Cullather, Nick. *The Hungry World: America's Cold War Battle against Poverty in Asia.* Cambridge, MA: Harvard University Press, 2010.

———. "Miracles of Modernization: The Green Revolution and the Apotheosis of Technology." *Diplomatic History* 28 (2004): 227–54.

Cường Bùi Thế, ed. *Khoa Học Xã Hội Nam Bộ.* Hà Nội: NXB Khoa Học Xã Hội, 2007.

Curtin, Philip D. *The Rise and Fall of the Plantation Complex.* Cambridge: Cambridge University Press, 1998.

Đặng Phong. *Kinh Tế Miền Nam Thời Kỳ 1955–1975.* Hà Nội: NXB Khoa Học Xã Hội, 2004.

———. *Lịch Sử Kinh Tế Việt Nam, 1945–2000, Tập I: 1945–1954.* Hà Nội: NXB Khoa Học Xã Hội, 2002.

Đặng Văn Đang, and Nguyễn Quế Đăng. *Bệnh Sốt Rét tại Việt Nam.* Sài Gòn: NXB Trung Tâm Học Liệu, 1970.

Đặng Văn Ngữ, A. Y. Lysenko, Van Hyu Ho, and Tumg Toh Dang. "Studies of the Epidemiology of Malaria in Northern Vietnam, Volume 1. Landscape Malariological Studies in Thai-Nguyen Province." *Russian Journal of Geography* 30 (1961): 293–98.

Đặng Văn Ngữ, et al. *Những Kỷ Niệm Sâu Sắc về Giáo Sư Đặng Văn Ngữ và Công Cuộc Phòng Chống Sốt Rét*. Hà Nội: Viện Sốt Rét–Ký Sinh Trùng–Côn Trùng, 1997.

Đặng Văn Vinh. "Etudes sur la contribution du Dr. A. Yersin dans l'agriculture du Viet Nam au debut du XXe siecle." Unpublished manuscript, Ho Chi Minh City, 1991.

———. *100 Năm Cao Su ở Việt Nam*. TP HCM: NXB Nông Nghiệp, 2000.

Daston, Lorraine. "The Moral Economy of Science." *Osiris* 10 (1995): 2–24.

Daughton, J. P. *An Empire Divided: Religion, Republicanism, and the Making of French Colonialism, 1880–1914*. New York: Oxford University Press, 2006.

Davis, Diana K. *Resurrecting the Granary of Rome: Environmental History and French Colonial Expansion in North Africa*. Athens: Ohio University Press, 2007.

Dean, Warren. *Brazil and the Struggle for Rubber: A Study in Environmental History*. New York: Cambridge University Press, 1987.

De Jesus, Ed C., and Alfred W. McCoy, eds. *Philippine Social History: Global Trade and Local Transformations*. Honolulu: University Press of Hawaii, 1982.

De Koninck, Rodolphe. "The Peasantry as the Territorial Spearhead of the State in Southeast Asia: The Case of Vietnam." *Sojourn* 11, no. 2 (1996): 231–58.

Delaye, Karine. "Colonial Co-Operation and Regional Construction: Anglo-French Medical and Sanitary Relations in South East Asia." *Asia Europe Journal* 2 (2004): 461–71.

Del Testa, David Wilson. "Automobiles and Anomie in French Colonial Indochina." In *France and "Indochina": Cultural Representations*, edited by Jennifer Yee, 63–77. Lanham, MD: Lexington Books, 2005.

———. "Paint the Trains Red: Labor, Nationalism, and the Railroads in French Colonial Indochina, 1898–1945." PhD diss., University of California, Davis, 2001.

Deroche, Alexandre. *France coloniale et droit de propriété, les concessions en Indochine*. Paris: l'Harmattan, 2004.

Đỗ Văn Minh. *Vấn Đề Cao Su Việt Nam*. Sài Gòn: Học Viện Quốc Gia Hành Chánh, 1971.

Dove, Michael. "Hybrid Histories and Indigenous Knowledge among Asian Rubber Smallholders." *International Social Science Journal* 54 (2002): 349–59.

———. "Living Rubber, Dead Land, and Persisting Systems in Borneo: Indigenous Representations of Sustainability." *Bijdragen Tot de Taal-, Land- en Volkenkunde* 154, no. 1 (1998): 1–35.

———. "Political vs. Techno-Economic Factors in the Development of Non-Timber Forest Products: Lessons from a Comparison of Natural and Cultivated Rubbers in Southeast Asia." *Society and Natural Resources* 8, no. 3 (1995): 193–208.

———. "Rice-Eating Rubber and People-Eating Governments: Peasant versus State Critiques of Rubber Development in Colonial Borneo." *Ethnohistory* 43, no. 1 (1996): 33–63.

———. "Smallholder Rubber and Swidden Agriculture in Borneo: A Sustainable Adaptation to the Ecology and Economy of the Tropical Forest." *Economic Botany* 47, no. 2 (1993): 136–47.

———. "The Transition from Native Forest Rubbers to Hevea Brasiliensis (Euphorbiaceae) among Tribal Smallholders in Borneo." *Economic Botany* 48 (1994): 382–96.

Drabble, John H. *An Economic History of Malaysia, c. 1800–1990: The Transition to Modern Economic Growth.* Houndmills: Macmillan in association with the Australian National University, Canberra, 2000.

———. *Malayan Rubber: The Interwar Years.* London: Macmillan, 1991.

———. *Rubber in Malaya, 1876–1922: The Genesis of the Industry.* Kuala Lumpur: Oxford University Press, 1973.

Duiker, William J. *The Communist Road to Power in Vietnam.* Boulder, CO: Westview Press, 1996.

Duval, Pierre, and Fabien Chébaut. *L'appel de la rizière.* Paris: Les Indes savantes, 2014.

Edwards, Penny. "The Tyranny of Proximity: Power and Mobility in Colonial Cambodia, 1863–1954." *Journal of Southeast Asian Studies* 37, no. 3 (2006): 421–43.

Elman, Benjamin A. *On Their Own Terms: Science in China, 1550–1900.* Cambridge, MA: Harvard University Press, 2005.

Evans, Grant. "Between the Global and the Local There Are Regions, Culture Areas, and National States." *Journal of Southeast Asian Studies* 33 (2002): 147–51.

Fabre, René. "Les plantations de caoutchouc du Vietnam." *Politique étrangère* 35, no. 4 (1970): 371–403.

Fall, Mamadou. "Investissements publics et politique économique en Indochine 1898–1930." PhD diss., Université Paris 7, 1985.

Fan, Fa-ti. "Redrawing the Map: Science in Twentieth-Century China." *Isis* 98 (2007): 524–38.

Fassin, Didier. "Les économies morales revisitées." *Annales* 64, no. 6 (2009): 1237–66.

Finlay, Mark R. *Growing American Rubber: Strategic Plants and the Politics of National Security.* New Brunswick, NJ: Rutgers University Press, 2009.

Firpo, Christina. *The Uprooted: Race, Children, and Imperialism in French Indochina, 1890–1980.* Honolulu: University of Hawai'i Press, 2016.

Fitzgerald, Deborah Kay. *The Business of Breeding: Hybrid Corn in Illinois, 1890–1940.* Ithaca, NY: Cornell University Press, 1990.

Fogarty, Richard, and Michael A. Osborne. "Constructions and Functions of Race in French Military Medicine, 1830–1920." In *The Color of Liberty: Histories of Race in France,* edited by Sue Peabody and Tyler Stovall, 206–36. Durham, NC: Duke University Press, 2003.

Fortunel, Frédéric. "Les plateaux méridionaux d'Asie du Sud-Est continentale de la marginalité à l'interconnexion." *Espaces, contacts et représentations peninsule, Etudes interdisciplinaires sur l'Asie du Sud-Est Péninsulaire* 57, no. 2 (2008): 19–41.

———. "L'Etat, les paysanneries et les cultures commerciales pérennes dans les plateaux du centre Viêt Nam, L'autochtonie en quête de territoires." PhD diss., Université de Toulouse, Toulouse, 2003.

Foucault, Michel. *Power/Knowledge: Selected Interviews and Other Writings, 1972–1977.* New York: Pantheon Books, 1980.

Fox, R., and G. Weisz, eds. *The Organization of Science and Technology in France (1808–1914).* New York: Cambridge University Press, 1980.

Freed, Libbie. "Networks of (Colonial) Power: Roads in French Central Africa after World War I." *History and Technology* 26, no. 3 (2010): 203–23.

Gant, Jesse J., and Nicholas J. Hoffman. *Wheel Fever: How Wisconsin Became a Great Bicycling State.* Madison: Wisconsin Historical Society Press, 2013.

Gantès, Gilles de, and Phuong Ngoc Nguyen, eds. *Vietnam le moment moderniste.* Aix-en-Provence: Université de Provence, 2009.

Garfield, Seth. *In Search of the Amazon: Brazil, the United States, and the Nature of a Region.* Durham, NC: Duke University Press, 2013.

Garros, Claire, Cam Van Nguyen, Ho Dinh Trung, Wim Van Bortel, Marc Coosemans, and Sylvie Manguin. "Distribution of Anopheles in Vietnam, with Particular Attention to Malaria Vectors of the Anopheles Minimus Complex." *Malaria Journal* 7, no. 11 (2008).

Ghosh, Amitav. *The Glass Palace.* New York: HarperCollins, 2000.

Giebel, Christoph. *Imagined Ancestries of Vietnamese Communism: Ton Duc Thang and the Politics of History and Memory.* Seattle: University of Washington Press, 2004.

Giles-Vernick, Tamara. "Entomology in Translation: Interpreting French Medical Entomological Knowledge in Colonial Mali." *Parassitologia* 50, nos. 3–4 (2008): 281–90.

Gonjo, Yasuo. *Banque coloniale ou banque d'affaires: La banque de l'Indochine sous la IIIe republique.* Paris: Comité pour l'histoire économique et financière de la france, 1993.

Goscha, Christopher E. *Historical Dictionary of the Indochina War (1945–1954): An International and Interdisciplinary Approach.* Singapore: NIAS Press, 2010.

———. *Vietnam: A New History.* New York: Basic Books, 2016.

———. *Vietnam: un Etat né de la guerre, 1945–1954.* Translated by Agathe Larcher. Paris: A. Colin, 2011.

———. *Vietnam or Indochina? Contesting Concepts of Space in Vietnamese Nationalism, 1887–1954.* Copenhagen: Nordic Institute of Asian Studies, 1995.

Goss, Andrew M. "Decent Colonialism? Pure Science and Colonial Ideology in the Netherlands East Indies, 1910–1929." *Journal of Southeast Asian Studies* 40, no. 1 (2009): 187–214.

———. *The Floracrats: State-Sponsored Science and the Failure of the Enlightenment in Indonesia.* Madison: University of Wisconsin Press, 2011.

Gran, Guy. "Vietnam and the Capitalist Route to Modernity: Village Cochinchina 1880–1940." PhD diss., University of Wisconsin–Madison, 1975.

Grandin, Greg. *Fordlandia: The Rise and Fall of Henry Ford's Forgotten Jungle City.* New York: Metropolitan Books/Henry Holt, 2009.

Graveline, François. *Des hévéas et des hommes: l'aventure des plantations Michelin.* Paris: Nicolas Chaudun, 2006.

Grove, Richard. *Green Imperialism: Colonial Expansion, Tropical Island Edens and the Origins of Environmentalism, 1600–1860.* Cambridge: Cambridge University Press, 1995.

Guénel, Annick. "The Conference on Rural Hygiene in Bandung of 1937: Towards a New Vision of Health Care?" Paper presented at the First International Conference on the History of Medicine in Southeast Asia, Siem Reap, Cambodia, January 9–10, 2006.

———. "The Creation of the First Overseas Pasteur Institute, or the Beginning of Albert Calmette's Pastorian Career." *Medical History* 43 (1999): 1–25.

———. "Malaria, Colonial Economics and Migrations in Vietnam." Paper presented at the Fourth Conference of the European Association of Southeast Asian Studies, Paris, France, September 1–4, 2004.

———. "Malaria Control, Land Occupation and Scientific Developments in Vietnam in the 20th Century." Paper presented at the Annual Meeting of the Association for Asian Studies, Boston, MA, March 11–14, 1999.

Guénel, Annick, and Mathieu Guerin. " 'L'ennemi, c'est le moustique': Tirailleurs Cambodgiens et Pastoriens face au paludisme dans le Haut-Chhlong." *Revue historique des armées* 3 trimestre (2004): 113–25.

Guénel, Annick, and Sylvia Klingberg. "Veterinary Science and Cattle Breeding in Colonial Indochina." Paper presented at the Annual Meeting of the History of Science Society, Montreal, Canada, November 4–7, 2010.

Guérin, Mathieu. *Paysans de la forêt à l'époque coloniale: La pacification des aborigènes des hautes terres du Cambodge, 1863–1940.* Caen: Association d'histoire des sociétés rurales, 2008.

Guha, Ramachandra. *The Unquiet Woods: Ecological Change and Peasant Resistance in the Himalaya.* Berkeley: University of California Press, 2000.

Gunn, Geoffrey C. *Rice Wars in Colonial Vietnam: The Great Famine and the Viet Minh Road to Power.* Lanham, MD: Rowman & Littlefield, 2014.

Ha, Marie-Paule. "From 'Nos Ancêtres, Les Gaulois' to 'Leur Culture Ancestrale': Symbolic Violence and the Politics of Colonial Schooling in Indochina." *French Colonial History* 3 (2003): 101–18.

Hahn, Hazel. "Botanical Gardens of French Indochina, 1860s–1930s." Paper presented at the Annual Meeting of the Society for French Historical Studies, Cambridge, MA, April 4, 2013.

Hansen, Peter. "The Virgin Heads South: Northern Catholic Refugees in South Vietnam, 1954–64." PhD thesis, Melbourne College of Divinity, 2008.

Hardy, Andrew. *Red Hills: Migrants and the State in the Highlands of Vietnam.* Copenhagen: NIAS Press, 2003.

Harp, Stephen. *Marketing Michelin: Advertising and Cultural Identity in Twentieth-Century France.* Baltimore: Johns Hopkins University Press, 2001.

Headrick, Daniel R. *The Tentacles of Progress: Technology Transfer in the Age of Imperialism, 1850–1940.* New York: Oxford University Press, 1988.

———. *The Tools of Empire: Technology and European Imperialism in the Nineteenth Century.* New York: Oxford University Press, 1981.

Hecht, Gabrielle. "Rupture-Talk in the Nuclear Age: Conjugating Colonial Power in Africa." *Social Studies of Science* 32, nos. 5–6 (2002): 691–727.

Hecht, Susanna B., and Alexander Cockburn. *The Fate of the Forest: Developers, Destroyers and Defenders of the Amazon.* New York: HarperPerennial, 1990.

Hecht, Susanna B., Kathleen D. Morrison, and Christine Padoch, eds. *The Social Lives of Forests: Past, Present, and Future of Woodland Resurgence.* Chicago: University of Chicago Press, 2014.

Hémery, Daniel. "Tristes tropiques: L'histoire de la nature au miroir colonial." *Revue francaise d'histoire d'outre-mer* 298–299 (1993): 9–24.

Herbelin, Caroline. *Architectures du Vietnam colonial: repenser le métissage.* Paris: INHA/CTHS, 2016.

———. "Des HBM au Việt Nam: La question du logement social en situation coloniale." In *Moussons: Viêt Nam, Histoire et Perspectives Contemporaines*, edited by Christian Culas and Jean-François Klein, 123–46. Marseille: IRSEA, 2009.

Herlihy, David V. *Bicycle: The History.* New Haven, CT: Yale University Press, 2004.

Hickey, Gerald Cannon. *Free in the Forest: Ethnohistory of the Vietnamese Central Highlands, 1954–1976.* New Haven, CT: Yale University Press, 1982.

———. *Sons of the Mountains: Ethnohistory of the Vietnamese Central Highlands to 1954.* New Haven, CT: Yale University Press, 1982.

Ho, Hai Quang. "Le rôle des investissements francais dans la creation du secteur de production capitaliste au Viet-Nam meridional." PhD diss., University de Reims, 1982.

Hồ Sơn Đài. "Căn Cứ Địa Kháng Chiến Chống Thực Dân Pháp ở Miền Đông Nam Bộ (1945–1954)." PhD diss., Viện Khoa Học Xã Hội tại TP HCM, 1995.

———, ed. *Lịch Sử Liên Trung Đoàn 301–310 (1945–1950).* Hà Nội: NXB Quân Đội Nhân Dân, 2007.

Hoàng Phê et al. *Từ Điển Tiếng Việt.* Edited by Trung Tâm Từ Điển Học. Đà Nẵng: NXB Đà Nẵng, 2007.

Học, Viện Sử. *Nông Thôn Việt Nam Trong Lịch Sự.* Hà Nội: NXB Khoa Học Xã Hội, 1977.

Hochschild, Adam. *King Leopold's Ghost.* New York: Houghton Mifflin, 1998.

Hodge, Joseph Morgan. *Triumph of the Expert: Agrarian Doctrines of Development and the Legacies of British Colonialism.* Athens: Ohio University Press, 2007.

Hội Đồng Biên Soạn Lịch Sử Đảng Miền Đông Nam Bộ, ed. "Lịch Sử Đảng Bộ Miền Đông Nam Bộ Lãnh Đạo Kháng Chiến (1945–1975)." Unpublished manuscript, Sài Gòn, 2001.

"Hội thảo về Alexandre Yersin, 1863–1943, Nửa Thế Kỷ ở Nha Trang, Việt Nam, Ngày 1 và 2 Thắng 3 Nam 1991." Biên Soạn Bộ Văn Hóa Thông Tin và Thế Thao. Hà Nội: Quỹ Văn Hóa, 1992.

Huff, W. G. *The Economic Growth of Singapore: Trade and Development in the Twentieth Century.* New York: Cambridge University Press, 1994.

Hunt, Bruce. "Insulation for an Empire: Gutta-Percha and the Development of Electrical Measurement in Victorian Britain." In *Semaphores to Short Waves*, edited by Frank A. J. L. James, 85–104. London: Royal Society for the Encouragement of Arts, Manufactures & Commerce, 1998.

Hunt, Nancy Rose. *A Colonial Lexicon of Birth Ritual, Medicalization, and Mobility in the Congo*. Durham, NC: Duke University Press, 1999.

Huỳnh Lứa, Hồ Sơn Đài, Trần Quang Toại, Hà Xuân Thọ, and Nguyễn Khoa Trung, eds. *Lịch Sử Phong Trào Công Nhân Cao Su Việt Nam, 1906–2001*. 2nd ed. Hà Nội: NXB Lao Động, 2003.

Huỳnh Văn Tòng. *Lịch Sử Báo Chí Việt-Nam: Từ Khởi Thủy đến Năm 1930*. Sài Gòn: Trí Đăng, 1973.

Immerwahr, Daniel. *Thinking Small: The United States and the Lure of Community Development*. Cambridge, MA: Harvard University Press, 2015.

Jackson, Joe. *The Thief at the End of the World: Rubber, Power, and the Seeds of Empire*. New York: Viking, 2008.

Jacobs, Seth. *America's Miracle Man in Vietnam: Ngo Dinh Diem, Religion, Race, and U.S. Intervention in Southeast Asia, 1950–1957*. Durham, NC: Duke University Press, 2004.

Jacoby, Karl. *Crimes against Nature: Squatters, Poachers, Thieves, and the Hidden History of American Conservation*. Berkeley: University of California Press, 2001.

Jain, Ravindra K. *South Indians on the Plantation Frontier in Malaya*. New Haven, CT: Yale University Press, 1970.

Jennings, Eric T. *Imperial Heights: Dalat and the Making and Undoing of French Indochina*. Berkeley: University of California Press, 2011.

Johnson, Walter. "On Agency." *Journal of Social History* 37, no. 1 (2003): 113–24.

Kalikiti, Webby S. "Rubber Plantations and Labour in Colonial Indochina: Interests and Conflicts, 1896–1942." PhD diss., SOAS, 2000.

Kathirithamby-Wells, J. *Nature and Nation: Forests and Development in Peninsular Malaysia*. Honolulu: University of Hawai'i Press, 2005.

Kaur, Amarjit. "Rubber Plantation Workers, Work Hazards, and Health in Colonial Malaya, 1900–1940." In *Dangerous Trade: Histories of Industrial Hazard across a Globalizing World*, edited by Christopher C. Sellers and Joseph Melling, 17–32. Philadelphia: Temple University Press, 2012.

Keith, Charles. *Catholic Vietnam: A Church from Empire to Nation*. Berkeley: University of California Press, 2012.

Kelly, Gail Paradise. *Franco-Vietnamese Schools, 1918–1938: Regional Development and Implications for National Integration*. Madison, WI: Center for Southeast Asian Studies, 1982.

Kiernan, Ben. *Viet Nam: A History from Earliest Times to the Present*. New York: Oxford University Press, 2017.

Kim, Jean. "Empire at the Crossroads of Modernity: Plantations, Medicine, and the Biopolitics of Life in Hawai'i, 1898–1948." PhD thesis, Cornell University, 2005.

Kitron, Uriel. "Malaria, Agriculture, and Development: Lessons from Past Campaigns." *International Journal of Health Services* 17 (1987): 295–326.

Kleiche, Mina. "La professionnalisation des agronomes coloniaux Français: L'ecole de Nogent, 1902–1940." In *Nature et environnement*, edited by Christophe Bonneuil and Yvon Chatelin, 75–92. Paris: ORSTOM, 1995.

Kleinen, John. "Tropicality and Topicality: Pierre Gourou and the Genealogy of French Colonial Scholarship on Rural Vietnam." *Singapore Journal of Tropical Geography* 26, no. 3 (2005): 339–58.

Knight, G. R. "A Precocious Appetite: Industrial Agriculture and the Fertiliser Revolution in Java's Colonial Cane Fields, c. 1880–1914." *Journal of Southeast Asian Studies* 37, no. 1 (2006): 43–63.

Kohler, Robert E. *Landscapes & Labscapes*. Chicago: University of Chicago Press, 2002.

Kramer, Paul A. *The Blood of Government: Race, Empire, the United States, & the Philippines*. Chapel Hill: University of North Carolina Press, 2006.

Krebs, C. J. *Ecology*. New York: Harper and Row, 1972.

Kumar, Prakash. *Indigo Plantations and Science in Colonial India*. New York: Cambridge University Press, 2012.

Lafont, Pierre-Bernard. "The Slash-and-Burn (Ray) Agricultural System of the Mountain Populations of Central Vietnam." In *Pacific Science Congress*, 56–59. Bangkok, Thailand: Secretariat, 9th Pacific Science Congress, 1959.

Lại Nguyên Ân, and Hữu Sơn Nguyễn, eds. *Tạp Chí Tri Tân (1941–1945), Truyện và Ký*. Hà Nội: NXB Hội Nhà Văn, 2000.

Lâm, Văn Thương. *Enquête sur la croissance et le développement de l'hévéa, cultures non en rapport du clone PR107*. Saigon: Sciences agronomiques, Ecole supérieure agronomique, forestière et vétérinaire, 1963.

Lamant, Pierre L. "Les partis politiques et les mouvements de résistance khmers vus par les services de renseignement français (1945–1952)." *Guerres mondiales et conflits contemporains* 148 (1987): 79–96.

Larcher-Goscha, Agathe. "Ambushed by History: Paul Mus and Colonial France's 'Forced Re-Entry' into Vietnam (1945–1954)." *Journal of Vietnamese Studies* 4, no. 1 (2009): 206–39.

Latour, Bruno. *An Inquiry into Modes of Existence: An Anthropology of the Moderns*. Cambridge, MA: Harvard University Press, 2013.

———. *The Pasteurization of France*. Cambridge, MA: Harvard University Press, 1988.

———. *Science in Action: How to Follow Scientists and Engineers through Society*. Cambridge, MA: Harvard University Press, 1987.

———. *We Have Never Been Modern*. Cambridge, MA: Harvard University Press, 1993.

Lawrence, Mark Atwood. *Assuming the Burden: Europe and the American Commitment to War in Vietnam*. Berkeley: University of California Press, 2005.

———. "Recasting Vietnam: The Bao Dai Solution and the Outbreak of the Cold War in Southeast Asia." In *Connecting Histories: Decolonization and the Cold War in*

Southeast Asia, edited by Christopher E. Goscha and Christian F. Ostermann, 15–38. Washington, DC: Woodrow Wilson Center Press, 2009.

Lê Quang Hiền. *Khảo Luận Kinh Tế Nông Nghiệp Viễn-Tượng Ngành Cao-Su Việt-Nam*. Sài Gòn: Viện Quốc Gia Hành Chánh, 1970.

Lê Văn Khải, Toản Trân, Văn Thịnh Trân, Văn Nhà Ngô, and Tấn Tự Nguyễn, eds. *Những Chặng Đường Đấu Tranh Cách Mạng của Công Nhân Cao Su Đồng Nai*. TP HCM: Công Ty Cao Su Đồng Nai, 1985.

Lê Văn Khoa, and Ban Giám Đốc Công Ty Cao su Dầu Tiếng Đảng ủy, eds. *Lịch Sử Phong Trào Công Nhân Cao Su Dầu Tiếng (1917–1997)*. 2nd ed. TP HCM: NXB Lao Động, 2000.

Li, Tana. *Nguyen Cochinchina: Southern Vietnam in the Seventeenth and Eighteenth Centuries*. Ithaca, NY: Southeast Asia Program Publications, 1998.

Li, Tania Murray. "Beyond 'the State' and Failed Schemes." *American Anthropologist* 107, no. 3 (2005): 383–94.

———. *The Will to Improve: Governmentality, Development, and the Practice of Politics*. Durham, NC: Duke University Press, 2007.

Lieberei, Reinhard. "South American Leaf Blight of the Rubber Tree (Hevea Spp.): New Steps in Plant Domestication Using Physiological Features and Molecular Markers." *Annals of Botany* 100 (2007): 1125–42.

Liew, Kai Khiun. "Making Health Public: English Language Newspapers and the Medical Sciences in Colonial Malaya." *East Asian Science, Technology and Society: An International Journal* 3, nos. 2–3 (2009): 209–29.

Lo, Ming-cheng Miriam. *Doctors within Borders: Profession, Ethnicity, and Modernity in Colonial Taiwan*. Berkeley: University of California Press, 2002.

Lottman, Herbert R. *The Michelin Men: Driving an Empire*. New York: I. B. Tauris, 2003.

Lowe, Celia. *Wild Profusion: Biodiversity Conservation in an Indonesian Archipelago*. Princeton, NJ: Princeton University Press, 2006.

Lương Văn Lựu. *Biên Hòa Sử Lược Toàn Biên*. 4 vols. Biên Hòa: Kim-Anh, 1971.

Mạc Đường, Bạch Trần Đăng, Công Kiệt Lê, and Nguyễn Đình Đầu, eds. *Địa Chí Tỉnh Sông Bé*. Thủ Dầu Một: NXB Tổng Hợp Sông Bé, 1991.

Mạc Đường, et al., eds. *Vấn Đề Dân Tộc ở Sông Bé*. Thủ Dầu Một: NXB Tổng Hợp Sông Bé, 1985.

Malarney, Shaun Kingsley. "Germ Theory, Hygiene, and the Transcendence of 'Backwardness' in Revolutionary Vietnam (1954–60)." In *Southern Medicine for Southern People: Vietnamese Medicine in the Making*, edited by Laurence Monnais, Claudia Michele Thompson, and Ayo Wahlberg, 107–32. Newcastle upon Tyne: Cambridge Scholars Pub., 2012.

Manderson, Lenore. *Sickness and the State: Health and Illness in Colonial Malaya, 1870–1940*. Hong Kong: Cambridge University Press, 1996.

Manguin, Sylvie, P. Kengne, L. Sonnier, Ralph E. Harbach, Visut Baimai, Ho Dinh Trung, and Marc Coosemans. "Scar Markers and Multiplex PCR-Based Identification of Isomorphic Species in the *Anopheles Dirus* Complex in Southeast Asia." *Medical and Veterinary Entomology* 16 (2002): 46–54.

Manivong, V., and R. A. Cramb. "The Adoption of Smallholder Rubber Production by Shifting Cultivators in Northern Laos." In *Smallholder Tree Growing for Rural Development and Environmental Services*, edited by D. Snelder and R. Lasco, 117–38. The Netherlands: Springer, 2008.

Mann, Charles C. *1493: Uncovering the New World Columbus Created.* New York: Knopf, 2011.

Mann, Charles C., and Josh D'Aluisio-Guerrieri. "Addicted to Rubber." *Science* 325 (2009): 564–66.

Marr, David. *Vietnam: State, War, and Revolution.* Berkeley: University of California Press, 2013.

———. *Vietnamese Anticolonialism: 1885–1925.* Berkeley: University of California Press, 1971.

———. "Vietnamese Attitudes Regarding Illness and Healing." In *Death and Disease in Southeast Asia: Explorations in Social, Medical, and Demographic History*, edited by Norman G. Owen, 162–86. Singapore: Oxford University Press, 1987.

———. *Vietnamese Tradition on Trial, 1920–1945.* Berkeley: University of California Press, 1981.

———. *Vietnam 1945: The Quest for Power.* Berkeley: University of California Press, 1995.

McCook, Stuart. "Ephemeral Plantations: The Rise and Fall of Liberian Coffee, 1870–1900." In *Comparing Apples, Oranges, and Cotton: Environmental Histories of the Global Plantation*, edited by Frank Uekötter, 85–112. Frankfurt am Main: Campus Verlag, 2014.

———. "Global Rust Belt: Hemileia Vastatrix and the Ecological Integration of World Coffee Production since 1850." *Journal of Global History* 1 (2006): 177–95.

———. "Managing Monocultures: Coffee, the Coffee Rust, and the Science of Working Landscapes." In *Knowing Global Environments: New Historical Perspectives on the Field Sciences*, edited by Jeremy Vetter, 87–107. New Brunswick, NJ: Rutgers University Press, 2011.

McElwee, Pamela D. *Forests Are Gold: Trees, People, and Environmental Rule in Vietnam.* Seattle: University of Washington Press, 2016.

McHale, Shawn Frederick. *Print and Power: Confucianism, Communism, and Buddhism in the Making of Modern Vietnam.* Honolulu: University of Hawai'i Press, 2004.

McIndoe, K. G. *A Preliminary Survey of Rubber Plantations in South Vietnam.* New York: Development and Resources Corporation, 1969.

McIntyre, Kevin T. "Eating the Nation: Fish Sauce in the Crafting of Vietnamese Community." PhD diss., University of Wisconsin–Madison, 2002.

McNeill, J. R. *Something New under the Sun: An Environmental History of the Twentieth-Century World.* New York: Norton, 2000.

Méry, Jacques. *History of the Mountain People of Southern Indochina up to 1945.* Washington, DC: Agency for International Development, [1950s].

Michon, Michel M. *Indochina Memoir, Rubber, Politics, and War in Vietnam and Cambodia 1955–1972.* Tempe, AZ: Program for Southeast Asian Studies, 2001.

Miller, Edward G. *Misalliance: Ngo Dinh Diem, the United States, and the Fate of South Vietnam.* Cambridge, MA: Harvard University Press, 2013.

Miller, Ian J. *The Nature of the Beasts: Empire and Exhibition at the Tokyo Imperial Zoo.* Berkeley: University of California Press, 2013.

Mintz, Sidney Wilfred. *Sweetness and Power: The Place of Sugar in Modern History.* New York: Viking, 1985.

Mitchell, Timothy. *Rule of Experts: Egypt, Techno-Politics, Modernity.* Berkeley: University of California Press, 2002.

Mitman, Gregg. *Breathing Space: How Allergies Shape Our Lives and Landscapes.* New Haven, CT: Yale University Press, 2007.

———. "In Search of Health: Landscape and Disease in American Environmental History." *Environmental History* 10, no. 3 (2005): 184–210.

———. *Reel Nature: America's Romance with Wildlife on Film.* Cambridge, MA: Harvard University Press, 1999.

Mitman, Gregg, and Paul Erickson. "Latex and Blood: Science, Markets, and American Empire." *Radical History Review,* no. 107 (2010): 45–73.

Mitman, Gregg, Michelle Murphy, and Christopher Sellers, eds. *Landscapes of Exposure: Knowledge and Illness in Modern Environments.* Vol. 19, *Osiris.* Chicago: University of Chicago Press, 2004.

Mollaret, Henri Hubert, and Jacqueline Brossollet. *Alexandre Yersin, le vainqueur de la peste.* Les inconnus de l'histoire. Paris: Fayard, 1985.

Monnais, Laurence. "In the Shadow of the Colonial Hospital: Developing Health Care in Indochina, 1860–1939." In *Viet-Nam Expose: French Scholarship on Twentieth-Century Vietnamese Society,* edited by Gisèle L. Bousquet and Pierre Brocheux, 140–84. Ann Arbor: University of Michigan Press, 2002.

———. *Médecine et colonisation: l'aventure indochinoise 1860–1939.* Paris: CNRS Editions, 1999.

Montesano, Michael J. "War Comes to Long An, the Classic We Hardly Know?" *Journal of Vietnamese Studies* 6 (2011): 87–122.

Moon, Suzanne. "Takeoff or Self-Sufficiency: Ideologies of Development in Indonesia 1957–1961." *Technology and Culture* 39, no. 2 (1998): 187–212.

———. *Technology and Ethical Idealism: A History of Development in the Netherlands East Indies.* Leiden: CNWS Publications, 2007.

Morlat, Patrice. *Indochine années vingt: Le rendez-vous manqué (1918–1928): la politique indigène des grands commis au service de la mise en valeur.* Paris: Les Indes savantes, 2005.

———. *Les affaires politiques de l'Indochine, 1895–1923.* Paris: Editions l'Harmattan, 1995.

Moulin, Anne Marie. "The Pasteur Institutes between the Two World Wars: The Transformation of the International Sanitary Order." In *International Health Organisations and Movements, 1918–1939,* edited by Paul Weindling, 244–65. Cambridge: Cambridge University Press, 1995.

———. "Patriarchal Science: The Network of the Overseas Pasteur Institutes." In *Science and Empires: Historical Studies about Scientific Development and European Expansion*, edited by Patrick Petitjean, Catherine Jami, and Anne Marie Moulin, 307–22. Boston: Kluwer Academic Publishers, 1992.

Murray, Martin J. *The Development of Capitalism in Colonial Indochina (1870–1940)*. Berkeley: University of California Press, 1980.

———. "'White Gold' or 'White Blood'? The Rubber Plantations of Colonial Indochina, 1910–40." In *Plantations, Proletarians, and Peasants in Colonial Asia*, edited by E. Valentine Daniel, Henry Bernstein and Tom Brass, 41–67. London: Frank Cass, 1992.

Nam, Cao. *Chi Pheo and Other Stories*. Hanoi: Red River, Foreign Languages Publishing House, 1983.

Nash, Linda. *Inescapable Ecologies: A History of Environment, Disease, and Knowledge*. Berkeley: University of California Press, 2006.

National Academy of Sciences. *The Effects of Herbicides in South Vietnam*. Washington, DC: National Academy of Sciences, 1974.

Neill, Deborah. *Networks in Tropical Medicine: Internationalism, Colonialism, and the Rise of a Medical Specialty, 1890–1930*. Stanford, CA: Stanford University Press, 2012.

Ngo, Vinh Long. *Before the Revolution: The Vietnamese Peasants under the French*. Cambridge, MA: MIT Press, 1973.

Nguyễn Đình Thống, Hồng Tuấn Nguyễn, Huy Xuân Nguyễn, Hữu Thuận Nguyễn, Xuân Thụ Nguyễn, and Thị Thanh Huyền Đỗ, eds. *Lịch Sử Phong Trào Công Nhân Viên Chức Lao Động và Hoạt Động Công Đoàn Bà Rịa - Vũng Tàu (1930-2006)*. Bà Rịa–Vũng Tàu: Sở Văn hóa–Thông tin tỉnh Bà Rịa–Vũng Tàu, 2008.

Nguyễn Đình Tư. *Từ Điển Địa Danh Hành Chính Nam Bộ*. TP HCM: NXB Chính Trị Quốc Gia, 2008.

Nguyễn Khắc Đạm. *Những Thủ Đoạn Bóc Lột của Tư Bản Pháp ở Việt-Nam*. Hà-Nội: NXB Văn Sử Địa, 1958.

Nguyễn, Khắc Viện. *Tradition and Revolution in Vietnam*. Berkeley, CA: Indochina Resource Center, 1975.

———. *Vietnam: A Long History*. Hanoi: Gioi Publishers, 1993.

Nguyễn Phong. *Tư Bản Pháp và Vấn Đề Cao Su ở Miền Đông Việt Nam*. Hà Nội: NXB Khoa Học, 1963.

Nguyễn Phương Đỗ. *Tôn Thất Tùng, Cuộc Đời và Sự Nghiệp*. Hà Nội: NXB Y Học, 1997.

Nguyễn Q. Thắng, and Bá Thế Nguyễn. *Từ Điển Nhân Vật Lịch Sử Việt Nam*. TP HCM: Viện Khoa Học Xã Hội, 2006.

Nguyễn, Thế Anh. "La campagne nord-vietnamienne, de la dépression économique de 1930 à la famine de 1945." *Revue francaise d'histoire d'outre-mer* 274 (1987): 43–54.

———. "L'élite intellectuelle vietnamienne et le fait colonial dans les premières années du XXe siècle." *Revue francaise d'histoire d'outre-mer* 268 (1985): 291–307.

———. "Traditional Vietnam's Incorporation of External Cultural and Technical Contributions: Ambivalence and Ambiguity." *Southeast Asian Studies* 40, no. 4 (2003): 444–58.

———. "The Vietnamese Confucian Literati and the Problem of Nation-Building in the Early Twentieth Century." In *Religion, Ethnicity and Modernity in Southeast Asia.* Edited by Myung-Seok Oh and Hyung-Jun Kim, 231–50. Seoul: Seoul National University Press, 1998.

Nguyễn Thế Nghĩa, Văn Giàu Trần, and Andrew Wells. *Tư Tưởng Hồ Chí Minh–Lịch Sử Giai Cấp Công Nhân và Công Đoàn Việt Nam.* TP HCM: Viện Khoa Học Xã Hội, 2000.

Nguyễn, Thị Định. *No Other Road to Take: Memoir of Mrs Nguyen Thi Dinh.* Translated by Mai Elliot. Ithaca, NY: Southeast Asia Program Publications, 1976.

Nguyễn Thị Mộng Tuyên. *Phong Trào Đấu Tranh của Công Nhân Cao Su Thủ Dầu Một Trong Kháng Chiến Chống Thực Dân Pháp.* 2nd ed. TP HCM: NXB Lao Động, 2003.

Nguyen, Thi Tuyet Mai, and Monique Senderowicz. *The Rubber Tree: Memoir of a Vietnamese Woman Who Was an Anti-French Guerrilla, a Publisher, and a Peace Activist.* Jefferson, NC: McFarland, 1994.

Nguyễn, Tung. "Les Vietnamiens et les oiseaux." In *Les messagers divins: aspects esthétiques et symboliques des oiseaux en Asie du Sud-Est,* edited by Pierre Le Roux and Bernard Sellato, 237–69. Paris: Connaissances et savoirs, IRASEC, 2006.

Nguyen, Van Ky. *La société vietnamienne face à la modernité: le Tonkin de la fin du XIXe siècle à la Seconde Guerre mondiale.* Paris: Harmattan, 1995.

Nguyễn Viết Đức. *Thực Tế Khai-Thác Cao Su của Người Pháp tại Bình Long.* Sài Gòn: Học-Viện Quốc-Gia Hành-Chánh, 1973.

Nguyen-Marshall, Van. *In Search of Moral Authority: The Discourse on Poverty, Poor Relief, and Charity in French Colonial Vietnam.* New York: Peter Lang, 2008.

Ninh, Kim Ngoc Bao. *A World Transformed: The Politics of Culture in Revolutionary Vietnam, 1945–1965.* Ann Arbor: University of Michigan Press, 2002.

Norindr, Panivong. *Phantasmatic Indochina: French Colonial Ideology in Architecture, Film, and Literature.* Durham, NC: Duke University Press, 1996.

Nyhart, Lynn K. *Modern Nature: The Rise of the Biological Perspective in Germany.* Chicago: University of Chicago Press, 2009.

Oberländer, Christian. "The Rise of Western 'Scientific Medicine' in Japan: Bacteriology and Beriberi." In *Building a Modern Japan: Science, Technology, and Medicine in the Meiji Era and Beyond,* edited by Morris Low, 13–36. New York: Palgrave Macmillan, 2005.

Ong, Thi Đan Thanh. *Một Số Vấn Đề về Tổ Chức Lãnh Thổ Sản Xuất Cao Su ở Việt Nam.* TP HCM: Trường Đại Học Sư Phạm, 1985.

Opinel, A., and G. Gachelin. "Theories of Genetics and Evolution and the Development of Medical Entomology in France (1900–1939)." *Parassitologia* 50 (2008): 267–78.

Orsenna, Erik. *L'Exposition coloniale.* Paris: Editions du Seuil, 1988.

Osborne, Michael A. *The Emergence of Tropical Medicine in France*. Chicago: University of Chicago Press, 2014.

———. *Nature, the Exotic, and the Science of French Colonialism*. Bloomington: Indiana University Press, 1994.

Packard, Randall M. "Agricultural Development, Migrant Labor and the Resurgence of Malaria in Swaziland." *Social Science and Medicine* 22, no. 8 (1986): 861–67.

———. "Maize, Cattle and Mosquitoes: The Political Economy of Malaria Epidemics in Colonial Swaziland." *Journal of African History* 25, no. 2 (1984): 189–212.

———. *The Making of a Tropical Disease: A Short History of Malaria*. Baltimore: Johns Hopkins University Press, 2007.

———. "Malaria Dreams: Visions of Health and Development in the Third World." *Medical Anthropology* 17 (1997): 279–96.

———. *White Plague, Black Labor: Tuberculosis and the Political Economy of Health and Disease in South Africa*. Berkeley: University of California Press, 1989.

Padirac, Raymond de. *L'Institut de recherches sur le caoutchouc, 1936–84*. Montpellier: CIRAD, 1993.

Panthou, Eric, and Tran Tu Binh. *Les plantations Michelin au Viêt Nam*. Vertaizon: La Galipote Editeur, 2013.

Parmer, J. Norman. *Colonial Labor Policy and Administration: A History of Labor in the Rubber Plantation Industry in Malaya, c. 1910–1941*. Locust Valley, NY: J. J. Augustin, 1960.

Pelley, Patricia. " 'Barbarians' and 'Younger Brothers': The Remaking of Race in Postcolonial Vietnam." *Journal of Southeast Asian Studies* 29, no. 2 (1998): 374–91.

Perrow, Charles. *Normal Accidents: Living with High-Risk Technologies*. Princeton, NJ: Princeton University Press, 1999.

Pesenti, Sophie. "De l'administration à la mise en valeur. Les tentatives de modernisation d'une agriculture coloniale. Annam, Tonkin, 1886–1919." Master's thesis, Université d'Aix-Marseille, 1994.

Peycam, Philippe. *The Birth of Vietnamese Political Journalism: Saigon, 1916–1930*. New York: Columbia University Press, 2012.

Phạm, Cao Dương. "L'accaparement des terres au Viêt Nam pendant la periode coloniale." *Revue francaise d'histoire d'outre-mer* 268 (1985): 257–90.

———. *Vietnamese Peasants under French Domination, 1861–1945*. Berkeley, CA: Center for South and Southeast Asia Studies, 1985.

Phạm Phi Thăng. "Hiện Trạng Đời Sống Sinh Hoạt của Công Nhân Nhập Cư ở Nông Trường Cao Su Xa Trạch, Huyện Bình Long, Tỉnh Bình Phước." PhD diss., Trường Đại học Khoa Học Xã Hội, 2007.

Phan, Quang Đán. *The Republic of Vietnam's Environment and People*. Sài Gòn: Self published, 1975.

Phan, Văn Trường. *La Fraternité, association d'Indochinois. Notes pour nos compatriotes*. Paris: Vigot frères, 1913.

Pholsena, Vatthana, and Oliver Tappe, eds. *Interactions with a Violent Past: Reading Post-Conflict Landscapes in Cambodia, Laos, and Vietnam*. Singapore: NUS Press in association with IRASEC, 2013.

Pick, Daniel. *Faces of Degeneration: A European Disorder, c. 1848–c.1918*. Cambridge: Cambridge University Press, 1989.

Popkin, Samuel L. *The Rational Peasant: The Political Economy of Rural Society in Vietnam*. Berkeley: University of California Press, 1979.

Prakash, Gyan. *Another Reason: Science and the Imagination of Modern India*. Princeton, NJ: Princeton University Press, 1999.

Pyenson, Lewis. *Civilizing Mission: Exact Sciences and French Overseas Expansion, 1830–1940*. Baltimore: Johns Hopkins University Press, 1993.

Raj, Kapil. *Relocating Modern Science: Circulation and the Construction of Knowledge in South Asia and Europe, 1650–1900*. Basingstoke: Palgrave Macmillan, 2007.

Rajan, S. Ravi. *Modernizing Nature: Forestry and Imperial Eco-Development 1800–1950*. Oxford: Oxford University Press, 2006.

Ramasamy, P. *Plantation Labour, Unions, Capital, and the State in Peninsular Malaysia*. New York: Oxford University Press, 1994.

Reid, Anthony. "Humans and Forests in Pre-Colonial Southeast Asia." In *Nature and the Orient: The Environmental History of South and Southeast Asia*, edited by Richard Grove, Vinita Damodaran and Satpal Sangwan, 106–26. New York: Oxford University Press, 1998.

Richards, John F. *The Unending Frontier: An Environmental History of the Early Modern World*. Berkeley: University of California Press, 2003.

Rogaski, Ruth. *Hygienic Modernity: Meanings of Health and Disease in Treaty-Port China*. Berkeley: University of California Press, 2004.

Russell, Edmund. "The Garden in the Machine: Toward an Evolutionary History of Technology." In *Industrializing Organisms: Introducing Evolutionary History*, edited by Philip Scranton and Susan R. Schrepfer, 1–16. New York: Routledge, 2004.

Saada, Emmanuelle. "Race and Sociological Reason in the Republic: Inquiries on the Métis in the French Empire (1908–37)." *International Sociology* 17, no. 3 (2002): 361–91.

Salemink, Oscar. *The Ethnography of Vietnam's Central Highlanders: A Historical Contextualization, 1850–1990*. Honolulu: University of Hawai'i Press, 2003.

Santiago, Myrna I. *The Ecology of Oil: Environment, Labor, and the Mexican Revolution, 1900–1938*. New York: Cambridge University Press, 2006.

Sasges, Gerard Henry. "Contraband, Capital, and the Colonial State: The Alcohol Monopoly in Northern Việt Nam, 1897–1933." PhD diss., University of California, 2006.

Schiebinger, Londa L. *Plants and Empire: Colonial Bioprospecting in the Atlantic World*. Cambridge, MA: Harvard University Press, 2004.

Schiebinger, Londa L., and Claudia Swan, eds. *Colonial Botany: Science, Commerce, and Politics in the Early Modern World*. Philadelphia: University of Pennsylvania Press, 2005.

Schmalzer, Sigrid. *Red Revolution, Green Revolution: Scientific Farming in Socialist China*. Chicago: University of Chicago Press, 2016.

Schneider, William H. "The Eugenics Movement in France, 1890–1940." In *The Wellborn Science: Eugenics in Germany, France, Brazil, and Russia*, edited by Mark B. Adams, 69–109. New York: Oxford University Press, 1990.

———. *Quality and Quantity: The Quest for Biological Regeneration in Twentieth-Century France*. Cambridge: Cambridge University Press, 1990.

Scott, James C. *The Art of Not Being Governed: An Anarchist History of Upland Southeast Asia*. New Haven, CT: Yale University Press, 2009.

———. *The Moral Economy of the Peasant: Rebellion and Subsistence in Southeast Asia*. New Haven, CT: Yale University Press, 1976.

———. *Seeing Like a State: How Certain Schemes to Improve the Human Condition Have Failed*. New Haven, CT: Yale University Press, 1998.

———. *Weapons of the Weak: Everyday Forms of Peasant Resistance*. New Haven, CT: Yale University Press, 1985.

Sears, Paul B. "Ecology—a Subversive Subject." *BioScience* 14, no. 7 (1964): 11–13.

Serier, Jean-Baptiste. *Les barons du caoutchouc*. Paris: CIRAD, Karthala, 2000.

Shaw, David Gary. "A Way with Animals." *History and Theory* 52, no. 4 (2013): 1–12.

Shepherd, Chris John. "Participation, Authority, and Distributive Equity in East Timorese Development." *East Asian Science, Technology and Society: An International Journal* 3, nos. 2–3 (2009): 315–42.

Showers, Kate Barger. *Imperial Gullies: Soil Erosion and Conservation in Lesotho*. Athens: Ohio University Press, 2005.

Slocomb, Margaret. *Colons and Coolies: The Development of Cambodia's Rubber Plantations*. Bangkok: White Lotus Press, 2007.

Smith-Howard, Kendra. *Pure and Modern Milk: An Environmental History since 1900*. New York: Oxford University Press, 2014.

Stanfield, Michael Edward. *Red Rubber, Bleeding Trees: Violence, Slavery, and Empire in Northwest Amazonia, 1850–1933*. Albuquerque: University of New Mexico Press, 1998.

Stapleton, Darwin H. "Internationalism and Nationalism: The Rockefeller Foundation, Public Health, and Malaria in Italy, 1923–51." *Parassitologia* 42 (2000): 127–34.

Stepan, Nancy. *"The Hour of Eugenics": Race, Gender, and Nation in Latin America*. Ithaca, NY: Cornell University Press, 1991.

———. *The Idea of Race in Science: Great Britain, 1800–1960*. Hamden, CT: Archon Books, 1982.

———. "Race, Gender, Science and Citizenship." *Gender & History* 10 (1998): 26–52.

Sterling, Eleanor J., Martha Maud Hurley, and Minh Duc Le. *Vietnam: A Natural History*. New Haven, CT: Yale University Press, 2006.

Stern, Alexandra M. *Eugenic Nation: Faults and Frontiers of Better Breeding in Modern America*. Berkeley: University of California Press, 2005.

———. "From Mestizophilia to Biotypology: Racialization and Science in Mexico, 1920–1960." In *Race & Nation in Modern Latin America*, edited by Nancy P. Appelbaum, Anne S. Macpherson, and Karin Alejandra Rosemblatt, 188–96. Chapel Hill: University of North Carolina Press, 2003.

Stewart, Mart A. *"What Nature Suffers to Groe": Life, Labor, and Landscape on the Georgia Coast, 1680–1920*. Athens: University of Georgia Press, 1996.

Stoler, Ann Laura. *Capitalism and Confrontation in Sumatra's Plantation Belt, 1870–1979*. Ann Arbor: University of Michigan Press, 1995.

Tạ Thị Thúy. *Đồn Điền của Người Pháp ở Bắc Kỳ, 1884–1918*. Hà Nội: Trung tâm Khoa Học Xã Hội và Nhân Văn Quốc Gia, Viện Sử Học, 1996.

———. "Les concessions agricoles françaises au Tonkin de 1884 à 1918." PhD diss., Ecole de Hautes Etudes en Sciences Sociales, Paris, 1993.

———. *Việc Nhượng Đất, Khẩn Hoang ở Bắc Kỳ từ 1919 đến 1945*. Hà Nội: Thế Giới, 2001.

Tai, Hue-Tam Ho. *Millenarianism and Peasant Politics in Vietnam*. Cambridge, MA: Harvard University Press, 1983.

———. *Radicalism and the Origins of the Vietnamese Revolution*. Cambridge, MA: Harvard University Press, 1992.

Takahata, Isao. *Pom Poko*. Walt Disney Home Entertainment, 2005.

Tan, Stan B. H. "'Swiddens, Resettlements, Sedentarizations, and Villages': State Formation among the Central Highlanders of Vietnam under the First Republic, 1955–1961." *Journal of Vietnamese Studies* 1, nos. 1–2 (2006): 210–52.

Tate, D. J. M. *The RGA History of the Plantation: Industry in the Malay Peninsula*. New York: Oxford University Press, 1996.

Taussig, Michael. *The Devil and Commodity Fetishism in South America*. Chapel Hill: North Carolina University Press, 1980.

Taylor, Keith W. *A History of the Vietnamese*. Cambridge: Cambridge University Press, 2013.

———. "Surface Orientations in Vietnam: Beyond Histories of Nation and Region." *Journal of Asian Studies* 57, no. 4 (1998): 949–78.

Taylor, Philip. *The Khmer Lands of Vietnam: Environment, Cosmology, and Sovereignty*. Honolulu: Asian Studies Association of Australia in association with University of Hawai'i Press, 2014.

Tertrais, Hugues. *La piastre et le fusil: Le coût de la guerre d'Indochine, 1945–1954*. Paris: Ministère de l'économie, des finances et de l'industrie, comité pour l'histoire économique et financière de la France, 2002.

Thành Nam. *Phong Trào Đấu Tranh Cách Mạng của Công Nhân Cao Su Miền Đông Nam Bộ*. Hà Nội: NXB Lao Động, 1982.

Thomas, Frédéric. "Ecologie et gestion forestiere dans l'Indochine francaise." *Revue francaise d'histoire d'outre-mer* 319 (1998): 59–86.

———. *Histoire du regime et des services forestiers français en Indochine de 1862 à 1945*. Hanoi: The Gioi, 1999.

———. "Protection des forêts et environnementalisme colonial: Indochine, 1860–1945." *Revue d'histoire moderne et contemporaine* 56, no. 4 (2009): 104–36.

Thomas, Julia Adeney. *Reconfiguring Modernity: Concepts of Nature in Japanese Political Ideology*. Berkeley: University of California Press, 2001.

Thomas, Lynn M. "Modernity's Failings, Political Claims, and Intermediate Concepts." *American Historical Review* 116, no. 3 (2011): 727–40.

Thompson, C. Michele. "Medicine, Nationalism, and Revolution in Vietnam: The Roots of a Medical Collaboration to 1945." *East Asian Science, Technology, and Medicine* 21 (2003): 114–48.

———. "Mission to Macau: Smallpox, Vaccinia, and the Nguyen Dynasty." *Portuguese Studies Review* 9 (2001): 194–231.

———. *Vietnamese Traditional Medicine: A Social History*. Singapore: NUS Press, 2015.

Thompson, Edgar T. *The Plantation*. Columbia: University of South Carolina Press, 2010.

Thompson, Robert. *No Exit from Vietnam*. London: Chatto & Windus, 1969.

Thompson, Virginia. *French Indo-China*. New York: Macmillan, 1937.

———. *Labor Problems in Southeast Asia*. New Haven, CT: Yale University Press, 1947.

Thongmanivong, Sithong, Yayoi Fujita, Khamla Phanvilay, and Thoumthone Vongvisouk. "Agrarian Land Use Transformation in Northern Laos: From Swidden to Rubber." *Southeast Asian Studies* 47, no. 3 (2009): 330–47.

Tilley, Helen. *Africa as a Living Laboratory: Empire, Development, and the Problem of Scientific Knowledge, 1870–1950*. Chicago: University of Chicago Press, 2011.

Tran, Angie Ngoc. *Ties That Bind: Cultural Identity, Class, and Law in Vietnam's Labor Resistance*. Ithaca, NY: Southeast Asia Program Publications, 2013.

Trần Kháng Chiến, Trung Quốc Dương, Đình Trọng Phạm, Kiến Quốc Trần, and Việt Trung Trần, eds. *Trần Tử Bình: Từ Phú Riềng Đỏ đến Mùa Thu Hà Nội* . . . Hà Nội: NXB Lao Động, 2007.

Trần Lục. *Nghiệp-Đoàn Công-Nhân Đồn-Điền tại Việt-Nam*. Sài Gòn: Học-Viện Quốc-Gia Hành-Chánh, 1970.

Trần Quang Toại, Quang Hữu Nguyễn, and Toàn Trần, eds. *Phong Trào Công Nhân Cao Su Bình Sơn (1923–1993)*. 2nd ed. TP HCM: Công Ty Cao su Đồng Nai, 1993.

Trần Quang Toại, Văn Thịnh Trần, and Tấn Quốc Nguyễn, eds. *Phong Trào Công Nhân Cao Su Ông Quế*. 2nd ed. Biên Hòa: NXB Tổng Hợp Đồng Nai, 2001.

Trần Thị Minh Hoàng, ed. *Địa Chí Đồng Nai*. Biên Hòa: NXB tổng hợp Đồng Nai, 2001.

Trần Thị Thu Lương, ed. *Chế Độ Sở Hữu và Canh Tác Ruộng Đất ở Nam Bộ Nửa Đầu Thế Kỷ XIX*. TP HCM: NXB TP HCM, 1994.

Trần Thúc Kỳ. "Tiến Độ Khoa-Học và Kỹ Nghệ ở Xứ Ta." *Khoa Học Phổ Thông*, no. 27, 1 octobre 1935: 77–78.

Tran, Tinh Hien, Nicholas P. J. Day, Nguyen Hoan Phu, Nguyen Thi Hoang Mai, Tran Thi Hong Chau, Pham Phu Loc, Dinh Xuan Sinh, Ly Van Chuong, Ha Vinh,

Deborah B. M. Waller, Timothy E. A. Peto, and Nicholas J. White. "A Controlled Trial of Artemether or Quinine in Vietnamese Adults with Severe Falciparum Malaria." *New England Journal of Medicine* 335, no. 2 (1996): 76–83.

Trần Toản. *Sự Hình Thành và Phát Triển của Đội Ngũ Công Nhân Cao Su Đồng Nai Qua các Thời Kỳ Lịch Sử, 1906–1991.* TP HCM: Viện Khoa Học Xã Hội, 1994.

Trần, Tử Bình. *The Red Earth: A Vietnamese Memoir of Life on a Colonial Rubber Plantation.* Translated by John Spragens Jr. Athens, OH: Center for Southeast Asian Studies, 1985.

Trịnh, Ngọc Chuyển. "Contribution à la mise en pratique du 'Programme de la santé rurale.'" PhD diss., Université de Saigon, 1960.

Trịnh Quang Quỹ. *Phong Trào Lao-Động Việt Nam.* Sài Gòn: s.n., 1971.

Trinh, Van Thao. *L'école française en Indochine.* Paris: Karthala, 1995.

Trung, Ho Dinh, Wim Van Bortel, Tho Sochantha, Kalouna Keokenchanh, Olivier J. T. Briet, and Marc Coosemans. "Behavioural Heterogeneity of *Anopheles* Species in Ecologically Different Localities in Southeast Asia: A Challenge for Vector Control." *Tropical Medicine and International Health* 10, no. 3 (2005): 251–62.

Trung, Ho Dinh, W. Van Bortel, T. Sochantha, K. Keokenchanh, N. T. Quang, L. D. Cong, and M. Coosemans. "Malaria Transmission and Major Malaria Vectors in Different Geographical Areas of Southeast Asia." *Tropical Medicine and International Health* 9, no. 2 (2004): 230–37.

Trương, Như Tảng, David Chanoff, and Van Toai Doan. *A Vietcong Memoir.* San Diego: Harcourt Brace Jovanovich, 1985.

Tschirley, Fred H. "Defoliation in Vietnam." *Science* 163 (1969): 779–86.

Tsing, Anna Lowenhaupt. *Friction: An Ethnography of Global Connection.* Princeton, NJ: Princeton University Press, 2005.

Tully, John A. *The Devil's Milk: A Social History of Rubber.* New York: Monthly Review Press, 2011.

Uekötter, Frank. "Introduction." In *Comparing Apples, Oranges, and Cotton: Environmental Histories of the Global Plantation,* edited by Frank Uekötter, 7–25. Frankfurt am Main: Campus Verlag, 2014.

Van Bortel, Wim, Ho Dinh Trung, P. Roelants, T. Backeljau, and Marc Coosemans. "Population Genetic Structure of the Malaria Vector *Anopheles Minimus* A in Vietnam." *Heredity* 91 (2003): 487–93.

Van Bortel, Wim, Ho Dinh Trung, Tho Sochantha, Kalouna Keokenchanh, Patricia Roelants, Thierry Backeljau, and Marc Coosemans. "Eco-Ethological Heterogeneity of the Members of the *Anopheles Minimus* Complex (Diptera: Culicidae) in Southeast Asia and Its Consequences for Vector Control." *Journal of Medical Entomology* 41, no. 3 (2004): 366–74.

Vandergeest, Peter, and Nancy Lee Peluso. "Empires of Forestry: Professional Forestry and State Power in Southeast Asia, Part 1." *Environment and History* 12 (2006): 31–64.

Vann, Michael. "White Blood on Rue Hue: The Murder Of 'le négrier' Bazin." *Proceedings of the Western Society for French History* 34 (2006): 247–62.

Verney, Sébastien. "Le nécessaire compromis colonial: le cas de la plantation Michelin de Dầu Tiếng de 1932 à 1937." In *Les administrations coloniales XIXe–XXe Siècles, Esquisse d'une histoire comparée*, edited by Samia El Mechat, 163–74. Rennes: Presses universitaires de Rennes, 2009.

Viollis, Andrée. *Indochine SOS*. Pantin: Les bons caractères, 2008.

Võ Nguyên. *Phong Trào Công Nhân Miền Nam*. Hà Nội: NXB Sự Thật, 1961.

Vogüé, Arnaud de. *Ainsi vint au monde . . . la S.I.P.H. (1905–1939)*. Vichères: Amicale des anciens planteurs d'hévéa, 1993.

Voon, Phin Keong. *American Rubber Planting Enterprise in the Philippines, 1900–1930*. London: University of London, SOAS, 1977.

———. *Western Rubber Planting Enterprise in Southeast Asia, 1876–1921*. Kuala Lumpur: Penerbit Universiti Malaya, 1976.

Vorapheth, Kham. *Commerce et colonisation en Indochine, 1860–1945*. Paris: Les Indes savantes, 2004.

Walker, Brett L. "Meiji Modernization, Scientific Agriculture, and the Destruction of Japan's Hokkaido Wolf." *Environmental History* 9, no. 2 (2004): 248–74.

———. *Toxic Archipelago: A History of Industrial Disease in Japan*. Seattle: University of Washington Press, 2010.

Walton, Catherine, Jane M. Handley, Frank H. Collins, Visut Baimai, Ralph E. Harbach, Vanida Deesin, and Roger K. Butlin. "Genetic Population Structure and Introgression in *Anopheles Dirus* Mosquitoes in South-East Asia." *Molecular Ecology* 10 (2001): 569–80.

Walton, Catherine, Jane M. Handley, Willoughby Tun-Lin, Frank H. Collins, Ralph E. Harbach, Visut Baimai, and Roger K. Butlin. "Population Structure and Population History of Anopheles Dirus Mosquitoes in Southeast Asia." *Molecular Biology and Evolution* 17, no. 6 (2000): 962–74.

Wargnier, Regis. *Indochine*. Columbia TriStar Home Video, 1993.

Wehrle, Edmund F. *Between a River and a Mountain: The AFL-CIO and the Vietnam War*. Ann Arbor: University of Michigan Press, 2005.

Weinstein, Barbara. *The Amazon Rubber Boom, 1850–1920*. Stanford, CA: Stanford University Press, 1983.

White, Luise. *Speaking with Vampires: Rumor and History in Colonial Africa*. Berkeley: University of California Press, 2000.

Wilder, Gary. "Colonial Ethnology and Political Rationality in French West Africa." In *Ordering Africa: Anthropology, European Imperialism and the Politics of Knowledge*, edited by Helen Tilley and Robert J. Gordon, 336–77. Manchester: Manchester University Press, 2007.

Woodside, Alexander. *Community and Revolution in Modern Vietnam*. Boston: Houghton Mifflin, 1976.

———. "The Development of Social Organizations in Vietnamese Cities in the Late Colonial Period." *Pacific Affairs* 44, no. 1 (1971): 39–64.

Worster, Donald. *Nature's Economy: A History of Ecological Ideas*. New York: Cambridge University Press, 1994.

Yacob, Shakila. "Model of Welfare Capitalism? The United States Rubber Company in Southeast Asia, 1910–1942." *Enterprise and Society* 8, no. 1 (2007): 136–74.

Young, Marilyn Blatt. *The Vietnam Wars, 1945–1990.* New York: HarperCollins, 1991.

Ziegler, Alan D., Jefferson M. Fox, and Jianchu Xu. "The Rubber Juggernaut." *Science* 324, no. 5930 (2009): 1024–25.

Zinoman, Peter. *The Colonial Bastille: A History of Imprisonment in Vietnam, 1862–1940.* Berkeley: University of California Press, 2001.

Index